GREEN ACCOUNTING IN EUROPE – FOUR CASE STUDIES

Dedication

To our parents

FONDAZIONE ENI ENRICO MATTEI (FEEM) SERIES ON ECONOMICS, ENERGY AND ENVIRONMENT

This series serves as an outlet for the main results of FEEM's research programmes in the areas of economics, energy and environment.

The titles published in this series are listed at the end of this volume.

Green Accounting in Europe – Four case studies

Edited by

Anil Markandya

University of Bath, Bath, UK

and

Marcella Pavan

Fondazione Eni Enrico Mattei, Milan, Italy

KLUWER ACADEMIC PUBLISHERS
DORDRECHT / BOSTON / LONDON

A C.I.P. Catalogue record for this book is available from the Library of Congress

ISBN 0-7923-5470-2

Published by Kluwer Academic Publishers,
P.O. Box 17, 3300 AH Dordrecht, The Netherlands

Sold and distributed in North, Central and South America
by Kluwer Academic Publishers,
101 Philip Drive, Norwell, MA 02061, U.S.A.

In all other countries, sold and distributed
by Kluwer Academic Publishers Group,
P.O. Box 322, 3300 AH Dordrecht, The Netherlands.

Printed on acid-free paper

Printed and bound in Great Britain by Antony Rowe Limited.

Table of Contents

PREFACE

The GARPI project was initiated in 1993 and completed in 1995. It was a pioneering exercise at the time it was carried out. Since then, thinking on some dimensions of the exercise, in particular aspects of valuation, has advanced. We have not incorporated these advances in the present book. We believe that, broadly, the estimates provided for damages here are correct but, more importantly they provide some important insights into the issues of damage estimation and its role in an overall accounting framework for environmental accounting.

Readers will also notice that damage estimates are missing in key categories. In particular health damage estimates are incomplete. Doubts about the accuracy of data had a limiting effect on the extent of damages covered. This has been rectified in follow-up work.

The team that worked on GARPI has been engaged, since late 1996, in a follow up study – GARPII – which takes the issues further and brings the estimation of effects up to date. Key areas where research has been focused are: (a) the attribution of damages to sources of emissions, (b) treatment of damages from chronic health effects of air pollution, (c) effects of additional sources of damages, such as heavy metals and carbon monoxide and (d) developments in the valuation of damages to water and forests.

The philosophy behind both GARPI and GARPII is to provide estimates of damages that are useful to policy makers in the EU. In this respect, the concern is to ensure the maximum degree of consistency and comparability in the estimates for different countries, and to review how damage estimates relate to other estimates of environmental impact, such as defensive expenditures and mitigation costs. Unfortunately, as will be clear from this report, there is still a lot of work to be done before we can have estimates that are fully comparable across countries and over time. But this study represents an important step in that direction.

Many people have contributed to the ideas developed in this book and we cannot mention all of them by name. All members of the core ExternE team should, however, be singled out for mention, as should the US team with whom much of the early work on impact pathways and damage estimation was done. In particular researchers at Resources for the Future and Oakridge National Laboratory provided continued support and advice. Mention should also be made of reviewer comments received from Alan Krupnick and Stale Navrud. Finally we thank Ian Milborrow who was responsible for the careful editing and preparing of this final version.

A. Markandya *Bath, UK,*
M. Pavan *Milan, Italy*

May, 1998

SYNTHESIS REPORT

Prepared by Metroeconomica* in collaboration with the other country teams.

This study has been undertaken by the following persons:

Metroeconomica:	A. Markandya, I. Milborrow, R. Nurick
Ifo Institute:	U. Triebswetter, R. Sprenger
Fondazione Eni Enrico Mattei (FEEM):	M. Pavan, L. Del Furia, D. Guidi
Institute for Environmental Studies (IVM):	O. Kuik, F. Spaninks, C. Dorland
AEA Technology (ETSU):	M. Holland, C. Johnson, P. Watkiss
Eyre Energy and Environment:	N. Eyre
Institute of Occupational Medicine (IOM):	F. Hurley
Institute of Terrestrial Ecology (ITE):	M. Hornung, D. Howard

The overall direction of the study has been the responsibility of Professor Anil Markandya.

In addition the following persons have contributed to the report:

NTCEN, Culham, Oxon.	G. Campbell, D. Lee, J. Stedman
IER, University of Stuttgart	W. Krewitt, R. Friedrich

The work was carried out as part of the JOULE II Programme of DGXII of the European Commission, under contract JOULE-CT-0316.

Legal Notice

This study was funded by Directorate General XII – Science, Research and Development – of the European Commission. It does not, however, express the official views of the European Commission.

* Metroeconomica Ltd, Economic and Environmental Consultants, 108, Bloomfield Road, Bath, BA2 2AR, UK. Tel: +(44) 1225 336978/826954; Fax: +(44) 1225 461678/323423; E-mail: hssam@bath.ac.uk

CHAPTER 1

SCOPE OF STUDY AND BACKGROUND

1.1. Scope of study

This study (referred to under the acronym GARP I[1]) was undertaken with the objectives of:

(a) Assessing the feasibility of preparing, at the EU level, monetary valuations of environmental damages caused by economic activities. The assessment should focus on the *scope* for making the valuations in a consistent manner for all the countries, on the *accuracy* of the estimates obtained, and on the *extent* of the damages covered by the monetary valuation. Issues to be discussed include the transferability of studies of damage across countries and over time, and the extent to which damages are already internalized through the costs of pollution control.

(b) Reporting some initial estimates of national level damages for the countries studied for detailed analyses: Germany, Italy, Netherlands and the United Kingdom. Where monetary valuation is not feasible, reporting the results in terms of physical impacts.

(c) Comparing the results of monetary valuation methods with other methods of assessing environmental damage or 'pressure', such as the expert based approaches.

The present report makes extensive use of a methodology developed by the EXTERNE project of the EC Directorate General XII JOULE II Programme. The EXTERNE project is a major multi-disciplinary study which has developed a detailed 'bottom-up' approach for the assessment of the social costs of fuel cycles (see European Commission, 1995a–f). By following this methodology the present study has made use of state-of-the-art models and data which have been the subject of international peer review.

As can be seen from the results, the study has covered a very wide range of impacts. In several cases, it was not possible to exploit the full potential of the data, or indeed to collect all the relevant information. Fortunately the same team of institutes has been awarded a follow-on contract (GARP II) under which

[1] Green Accounting Research Project.

several of the gaps in the present work will be filled. Where work is ongoing under the new contract, this is reported in the text.

One of the objectives that has not been fulfilled is (c) above – that of comparing the monetary valuation approach with expert based approaches. This is because the study that was to undertake the expert approach was delayed, and in fact the results of that exercise are still not available. The study on the use of expert systems and pressure indicators is being undertaken by EUROSTAT and was initiated in January 1995. It is expected that some comparison with the pressure indicators project will be possible under GARP II.

1.2. The case for green accounting

There is a great deal of interest, both internationally and within the EU, in the preparation of a coherent set of environmental accounts that establish the impacts of economic and social activities within the Union on the environment within Europe. At the international level, a set of guidelines for the preparation of a set of 'satellite accounts' that complement the national income accounts has been prepared (United Nations, 1993). This sets out the methodology to be followed for the valuation of natural resources and environmental degradation caused by anthropogenic activity. These guidelines (and others which are similar) are being used in a number of countries in preparing environmental accounts, both at the monetary and non-monetary level. These include: Brazil, Canada, Costa Rica, France, Germany, the Netherlands and Norway. Some attempts at adjusting national income accounts for environmental effects have also been undertaken in Australia, Japan, India, Indonesia, Mexico, New Zealand, Papua New Guinea, Sweden, the United Kingdom, the United States and Zimbabwe. References to the related studies are given at the back of this chapter.

Within the EU, interest in obtaining coherent environmental accounts is evidenced by the support for this project on environmental accounting, the recently initiated project on environmental pressure indicators (under EUROSTAT), as well as a number of research projects supported by DG12 on sustainable indicators and green accounting. The Commission has made its position clear in a few recent publications.[2] The publication on "Directions for the EU on Environmental Indicators and Green National Accounting" (COM 94, 670) states that that there is a need for '*a harmonised European system of integrated economic and environmental indicators and accounts which addresses the problems of the various economic sectors and policy fields at various scales and which will allow for comparison between Member States*'. The same publication also states

[2] For a review of related EC research on sustainability indicators and 'green' accounting see Proops, (1995).

that *'the Commission has developed a set of complementary actions establishing a framework for 'green accounting' which will provide ... Integrated Economic and Environmental Indices of economic performance and environmental pressures of economic sectors within 2–3 years [and] more fundamental work on the 'greening' of National Accounts in a satellite format'*."

Thus it is apparent that there is a strong demand for consistent environmental accounts to be prepared. At the same time there is some scepticism about the feasibility of producing consistent, credible environmental damage estimates in monetary terms, that can be integrated into the environmental accounts (Hueting, 1989, van Tongeren *et al.*, 1992). Even the UN Statistical Office is lukewarm about the idea, and the guidelines provided by the United Nations say very little about how environmental damages should be measured.[3] By undertaking a comprehensive, consistent valuation of damages, this study can provide direct evidence on the feasibility of damage estimation in an accounting framework and thereby contribute significantly to the development of a methodology for the establishment of such accounts.

1.3. Measuring environmental damage

1.3.1. *National income as a measure of welfare*

This study aims to use a welfare-theoretic approach in the valuation of damages. This means that we are seeking to measure human welfare, or more precisely changes in human welfare, *in the countries that we are dealing with* (Italy, Germany, Netherlands and the UK). The damage estimates made are presented as a percentage of GDP. In our view the purpose of this is for illustration and comparison. We are not advocating integrating the estimates to produce a measure of *green* GDP. The techniques have not advanced sufficiently to undertake such work and even if this were the case questions would arise as to whether this was methodologically defensible.

It is generally recognized that GDP itself is not an ideal measure of welfare, often being considered more as a measure of national throughput or service flows. However, it is widely used as such and is an important guide to policy. GDP is an important dimension of welfare; and whether we decide to modify it in the light of estimates of national damage, or whether we simply present these data alongside the national income accounts, it is desirable that our calculations be based on a clearly defined economic principle. The appropriate method offered by welfare economics for the measurement of environmental damages is willingness-to-pay (WTP). This is discussed in more detail in

[3] The guidelines recommend the use of 'contingent value methods' but such advice is very general and not appropriate to the valuation of some impacts. In fact the experience of the EXTERNE project provides a much more solid basis for deciding on how such valuation should be carried out.

Chapter 2 of the synthesis report. Some commentators would suggest that taking a micro-level concept such as WTP and using it to conduct a macro assessment is not credible. The GARP team recognise there are problems here, but our purpose has been to see whether plausible damage estimates at the national level can be made. This report aims to show that they can.

1.3.2. *Limits to valuation and other measures of welfare*

This study aimed to apply WTP valuation techniques to as many types of environmental impact as possible. The argument is that with monetary values it is easier to integrate environmental costs with other costs and benefits, a stated goal of the European Commission, as well as that of many other bodies. However, we are aware that not all impacts can be valued in monetary terms. This study has helped in identifying those areas where such valuation is not possible or credible. In such cases a careful, consistent reporting of the physical impacts is seen as important and has been carried out.

As a result of the problems described in the previous section, there are attempts at producing better measures of welfare that include environmental, social and other considerations. Examples include the UNDP index of welfare (UNDP, 1990 et seq.), or the Daly–Cobb index on sustainable welfare (Daly and Cobb, 1990). The GARP team recognize the importance of these measures. Indeed, we would argue that a careful assessment of the environmental damages will be an important input into the construction of such an index, particularly those concerned with environmental pressure indicators and sustainability. Hence the work carried out here has applicability to a wider range of indices of welfare.

1.4. Methodological issues

In preparing environmental accounts many important issues arise including:

- Assessing flows of environmental services
- Assessing and presenting estimates of environmental damage
- Defining background levels of pollution
- Treatment of transboundary pollution effects
- Treatment of defensive expenditures
- Treatment of intertemporal damages
- Natural resource depletion and sustainability questions

Some of these are considered in more detail in this section and there many problems associated with all. For further discussion of the issues see Ahmad, El Serafy and Lutz (1989), United Nations (1993), Markandya (1994). This study did not seek to resolve the controversies surrounding the questions of whether or not any of the above should be added or subtracted from the national income accounts. Instead it focuses on the question, *if the item under consideration is to be treated as part of the national income accounting framework, what is the most practicable way of measuring it, how should be valuation be carried out, and what issues need to be flagged in the conduct of the valuation.*

Early on in this study it was decided that the approach taken would be based on a spatially disaggregated analysis of damages in each country. Such a disaggregation was essential, in our opinion, to achieve the kind of detail that is required for an adequate valuation, as well as providing regional information likely to be of greater value to policy makers (just as regional income accounts are widely used in economic analysis). Thus the land area in each country was zoned according to factors that varied from pollutant to pollutant. Given data on the levels of these pollutants, damage estimates were made using dose–response functions and unit valuations based on a WTP approach. At the first stage it was decided that we would not seek to estimate the *sources* of the pollution. The approach is shown in Figure 1.1 below.

The use of spatial data and dose–response functions reflected a methodology that was originally employed for the assessment of the impacts of the major fuel cycles for electricity generation, including fossil, nuclear and renewable technologies, and has previously been applied almost solely to quantify the effects associated with individual power plants. This is generally referred to as the EXTERNE Project (see European Commission 1995a–f). In applying this methodology to quantify national level damages we thus seek to find out whether there are major barriers to its application for a purpose different to that for which it was developed.

The use of the above methodology has meant that the focus of much of the work has been the impacts of air pollution. The team recognise that there are other impacts associated with other pollution problems, such as eutrophication of freshwater bodies and wastes, and these have been addressed in some studies. However, the constraints of time and resources would not permit such a comprehensive estimation across four countries. Furthermore, the focus on air pollution can be defended by the fact that most of the previous literature has suggested that it is in this area where the largest damages would arise. The coverage has not, however, been restricted solely to air quality impacts, since impacts of noise have also been investigated. Table 1.1 below indicates the main impacts that have been analysed in this study.

From the Table it can be seen that not all the relevant damage related

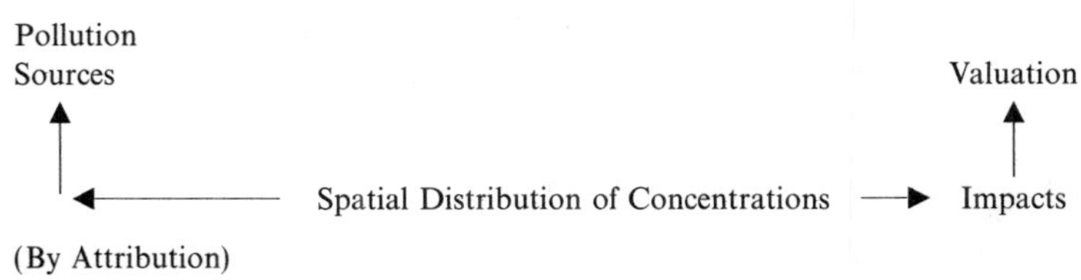

Figure 1.1. Impact pathway approach

Table 1.1. Impacts assessed in this study

	Health	Materials	Crops	Forests	Amenity (2)	Ecosystems (3)
PM_{10}	A	A	NE	NE	NA	NE
SO_2 (1)	A	A	A	A	NA	A
NO_x (1)	A	A	A	A	NA	A
Ozone	A	A	A	NA	NE	NE
GHG (3)	A	A	A	A	A	A
Noise	NA	NE	NE	NE	A	NE

Notes:
A Assessed in at least some studies.
NA Not assessed, although we believe they may be important.
NE No effect of significance is anticipated.
(1) SO_2 and NO_x include acid deposition impacts.
(2) Effects of PM_{10}, NO_x and SO_2 on amenity arise with respect to visibility. In previous studies these have not been found to be significance in Europe, although they are important in the US.
(3) Greenhouse gases (i.e. those that affect climate globally).

impacts have been covered, *although we believe that we have covered the major ones.* In addition, it has not been possible to value all impacts marked 'A' for all countries.

1.4.1. *Defining background levels of pollution*

In order to commence an assessment of environmental damages, background levels of pollutants for the various media need to be established. This is an important and often contentious issue. Background levels, defined as levels that would exist in the absence of human activity, will vary from region to region, let alone from country to country. The calculated damages are very sensitive to the choice of these values. If we are trying to measure changes in service flows resulting from environmental damage, a baseline definition is irrelevant. The GARP team adopted a pristine environment as a baseline for defining background levels but acknowledge that this may be questionable. The main problem is that there is great uncertainty on the nature of dose–response functions when the particular emission is very near to the background level.

It is clear from the work carried out here that further research will be required on this issue. For the purposes of this report, however, we decided to report the results for more than one baseline value, basing the range on considerations of the plausibility of different values.

1.4.2. *How should transboundary pollution be treated?*

National accounts attempt to measure the flow of welfare of the residents of a given country. Damages caused by country 1's pollution to country 2 are not directly relevant to the welfare of country 1's residents, although the received pollution from other countries is relevant. Hence estimates should be *receptor based* rather than *source based*. This is particularly important for the valuation of damages related to climate change. Most damage estimates for CO_2 are global damages which cannot be subtracted from a national level calculations of benefits. For this reason we undertook to make an estimate of the likely damages caused within each of the countries selected for study by climate change. This work was undertaken by the Dutch team and is reported in Chapter 9.

1.4.3. *How should defensive expenditures be treated?*

Given different levels of expenditure on pollution abatement across the EU, and given the increasing levels of such abatement expenditures over time, it would be very difficult to make cross-country or intertemporal comparisons of the costs of environmental damage if one did not allow for changes in the levels of defensive expenditure. Hence, it is our view that such information is important for the purposes of environmental accounting, irrespective of whether or not national income accounts are adjusted for it. This study has made a limited start on collecting the relevant information, which is reported here. The issue of whether such data can be collected at reasonable cost and within a common framework is addressed in the concluding section.

1.4.4. *How should intertemporal damage impacts of pollution be treated?*

In some cases, particularly forest impacts and climate change, current damages are a function of both present and past pollution concentrations, and present emissions cause damages both now and in the future. Conceptually, national income measures the flow of welfare, both now and in the future, emanating from the current flow of activities (Weizmann, 1976). Hence in so far as future damages are being generated, they should be measured and allowed for. In practice, this is difficult to do, and the case of climate change is the only one where serious attention is paid to this issue. Where it arises, the issue of what discount rate to apply to future damages also arises. This question has been discussed at length elsewhere (see European Commission, 1995b) and is not covered here again. The basic argument presented there is that a 'social' rate of around 3% in real terms is justified for most costs, but that costs which occur over very long periods, or which involve significant changes in the sustainability of eco-systems, should be treated differently. One possible method of treating them is to look at the costs of restoring the system or limiting the damage at a level consistent with overall 'sustainability'. In the case of climate change the choice of the discount rate is discussed in Chapter 9.

1.4.5. *Natural resource depletion and sustainable income*

In the current discussion of environmental accounting there is much interest in the concept of *sustainable income*. The idea is that we should seek to measure the flow of goods and services that the present asset base would permit, in perpetuity. This is the value of all that *is produced*, less the amount that has to be set aside to allow for depletion in the asset base through depreciation, degradation etc. The asset base includes physical capital, human capital and natural capital. For examples of this approach see Costanza and Daly, 1992, da Motta *et al.* (for Brazil), 1992; Bryant and Cook (for the UK), 1992; Young (for Australia), 1992; Repetto *et al.*, (for Indonesia), 1989; Soloranzo *et al.* (for Costa Rica), 1991; Uno (for Japan), 1989; van Tongeren *et al.* (for Mexico), 1991.

This study contributes to the measurement of sustainable income by correcting the measure of the goods and services that flow from the present capital stock. Thus an economy that is generating increases in emissions should account for those emissions as part of the flow of goods and services (negative ones in this case).

An additional dimension of the environmental damages is to view the emissions as causing damage to the stock of natural capital, and thereby depleting the stock of that capital. Consequently, there will be a fall in the flow of services in the future, other things being equal. In some cases, the emissions generated do cause damages to an identified natural asset. Examples would be damages caused by agricultural runoff to water reservoirs, or damages caused by acidification to forest growth. One can then calculate the damages by estimating the reduced value of the natural capital stock, and the reduced flow of services that the stock will generate.

Two points need to be made here. First, not all environmental 'damages' impact on the natural (or other) capital stock. Noise and crop damages are examples. For these it is necessary to measure the damages in terms of the reduced flow of goods and services. Second, where there is an impact on the 'capital stock', the present calculations intend, *in principle,* to pick up effects of that impact on the flow of goods and services over time. Hence, if the calculations are done correctly, the estimation of damages as carried out here does pick up the effects over time, and hence the impact of the pollution on the value of the natural (or other) asset. This can be seen as follows:

Suppose an asset has current value A, and will generate a stream of services of value $H_1, H_2, \ldots, H_n$ over n time periods. By the definition of the value of an asset:

$$A = \sum_i H_i(1+r)^{-i}$$

where r is the discount rate.

If an increase in pollution reduces the flow of services to $L_1, L_2, \ldots, L_n$, then our methodology seeks to measure environmental damage D, where

$$D = \sum_i (H_i - L_i) \cdot (1+r)^{-i}$$

The change in the value of the asset, however, is also *D*. A sustainable income approach would then deduct from present national income an amount equal to *D*, to allow for the depreciation in the asset base (and therefore what has to be put aside to maintain that base).

Thus the two approaches are equivalent in terms of the correction to national income. It does imply, however, that, in correcting measures of GDP for environmental damages, we should work with *net national income* – income after depreciation. In practice, of course, both a sustainable income approach from a capital perspective or from the flow of services perspective needs careful measurement of changes in environmental service flows, choices of discount rates etc. Often approximations are taken which will result in a capital approach and a flow of services approach giving different answers.

1.5. Coverage of the study

A broad description of the coverage of this study was given in Table 1.1. Table 1.2 overleaf provides a more detailed account of the scope of the assessment. The main points that emerge from that table are:

1. The most consistent coverage is for health impacts, although even there, it was not possible to get all four teams to agree enough on background data. In Germany an evaluation of PM_{10} damages was not possible because data on an EMEP basis were not available. By and large, however, agreement was reached on the relevant dose–response functions and the same ones were used in all studies.
2. Coverage of ozone impacts of health is not complete but this was due to lack of time in collecting all the data. With time it should be possible to get consistent coverage of these impacts.
3. Damage to materials was not based on a common set of dose–response functions. In Italy there were problems in assessing the stock as risk which will take some time to resolve. Damage to cultural monuments, which is significant, was only carried out on a partial basis for Italy and Germany.
4. Coverage of crop damage was carried out in a consistent manner for SO_2 damages but damages from ozone were only treated in the Netherlands and the UK. Again it is a matter of time for the necessary data to be collected for a fuller coverage to be obtained.
5. Forest damage coverage is partial in that damages to recreation and non-timber use have only be assessed in Germany. As can be seen from that study, these values could be substantial.
6. Amenity valuation focussed on noise damages which could not be valued in Italy. More data need to be collected for such a valuation to be undertaken there.
7. All eco-system assessments were in physical terms.

Table 1.2. Details of coverage under this study

		Health				Forests			
		Morb.	Mort.	Material	Crops	Timber	Other	Amenity	Eco-systems
PM_{10}	I	A (1)	A (1)	NE	NE	NE	NE	NE	NE
	DE	NA (1)	NA (1)	NE	NE	NE	NE	NE	NE
	NL	A (1)	A (1)	NE	NE	NE	NE	NE	NE
	UK	A (1)	A (1)	A	NE	NE	NE	NA	NE
SO_2	I	A (2)	A (2)	NA (4)	A	A (6)	NA	NE	PA (7)
	DE	A (2)	A (2)	PA (4)	A	A (6)	A	NE	NA
	NL	NA	NA	PA (4)	A	A (6)	NA	NE	NA
	UK	A (2)	A (2)	PA (4)	A	A	NA	NA	PA (7)
Ozone	I	A (3)	A (3)	NA	NA	A (6)	NA	NE	NA
	DE	PA (3)	PA (3)	NA	NA	A (6)	A	NE	NA
	NL	A (3)	A (3)	NA	A	A (6)	NA	NE	NA
	UK	A (3)	A (3)	PA	A	A (6)	NA	NE	PA (7)
GHG	ALL	A	A	A	A	A	A	NA	PA
Noise	I	NA	NE	NE	NE	NE	NE	NA	NE
	DE	NA	NE	NE	NE	NE	NE	A	NE
	NL	NA	NE	NE	NE	NE	NE	A	NE
	UK	NA	NE	NE	NE	NE	NE	A	NE
Def. Exp.	I	NA	NA	PA	NA	NA	NA	NA	NA
	DE	NA	NA	PA	NA	A	A	A	NA
	NL	NA	NA	NA	NA	NA	NA	A	NA
	UK	NA	NA	NA	A (5)	NA	NA	NA	NA

Notes:

(1) For PM_{10} both chronic and acute impacts have been estimated in I, NL and UK. In Germany an evaluation of PM_{10} on an EMEP basis was not available.
(2) Note that SO_2 levels are not additive to the levels of PM_{10}. Coverage of morbidity effects is not for the same end-points in all countries.
(3) For ozone both chronic and acute impacts have been estimated in I, NL and UK. Germany reports only a single estimate of ozone damages taken from other studies.
(4) Materials coverage is not based on the same dose–response functions in all countries. Cultural assets are not covered on a WTP approach in any country. In Germany, data are for West Germany only. Coverage of defensive expenditures is for cultural buildings only and is acknowledged to be incomplete.
(5) Defensive expenditures for the UK include value of nitrogen from NO_x.
(6) All valuations were undertaken by the UK team using a single model.
(7) There is no monetary valuation undertaken here. Impacts include those from NO_x.

Abbreviations:

A	Assessed
PA	Partially assessed
NA	Not assessed, although we believe they may be important
NE	No effect of significance is anticipated
Def. Exp	Defensive expenditures
Morb.	Morbidity impacts
Mort.	Mortality impacts
I	Italy
DE	Germany
NL	Netherlands
UK	United Kingdom

8. Impacts of greenhouse gases are based on a review of existing studies of global damage, supplemented by an assessment in each of the four countries. No primary data collection was involved.
9. Defensive expenditure coverage is very patchy. The main coverage was for noise damages although some estimates were made for materials damages and crop protection

The coverage as it exists at present would not permit a full comparison across countries. It is expected that better coverage will be obtained as a result of GARP II but even then, one cannot expect all impacts to be equally covered in all cases. This should not be seen, however, as a reason for rejecting environmental accounts. Coverage of many aspects of national income accounts is not equally good in all EU countries. As long as the gaps are recognized, some cross-country comparisons and changes in damages over time can be evaluated.

1.6. Air quality data requirements

1.6.1. *Air pollution concentration data*

This study makes use of spatially disaggregated data on pollution concentrations to estimate damages. The level of disaggregation selected varied from country to country. Although pan-European data sets exist with resolutions of 50–100 km^2 finer resolutions are available in some countries, and for some problems. Where they were available and accessible, they were chosen.

For Italy data were taken for PM_{10}, SO_2, NO_x, and ozone from the network of monitoring stations. Coverage is variable, and considerable amount of interpolation was required.[4] For ozone data are really only available for Lombardy. For acid depositions, the EMEP data with resolution of 50 × 50 km was used.

In the case of Germany the EUROGRID coordinate system was used for presenting SO_2 and NO_x concentration levels. In moving from point sources to grid values models of atmospheric transport were used (as opposed to the interpolation method used in Italy). Due to limitations such as no availability of geographically resolved data for PM_{10} and a lack of dose–response-functions for NO_x, only SO_2 impacts were calculated.

For Netherlands air quality is measured at 34 stations, from which the distribution of pollutants is made using an interpolation method. For particulate matter the stations measure black smoke, so a conversion from black smoke to PM_{10} was developed. The resulting pollution data are reported for a grid that is 5 km × 5 km.

[4] For details of the interpolation methods used see the individual reports.

In the UK data on NO_x was obtained at a resolution of 5 km × 5 km. This was then used to estimate PM_{10} concentrations on the same grid. For ozone and SO_2 the same grid was employed, with data collected from different pollution monitoring stations. For acid depositions a map was generated on a grid of 20 km × 20 km, derived from interpolation of measured data, combined with information on rainfall. Acid depositions for all of Europe were modelled using the Harwell Trajectory model, which is a Lagrangian plume model that estimates acid depositions at each receptor point, based on the arrival of 24 trajectories weighted by the frequency of the wind in each 15° sector. The resolution achieved for the final data set was 100 km × 100 km.

1.6.2. *Background concentrations*

The objective of the current study, to relate environmental impacts to human activity, dictates that the damages resulting from 1990 emissions of pollution are compared to the level that would exist in the absence of human activity. For some pollutants an estimate of these levels can be gauged from the concentrations measured in remote areas.

The most appropriate data for SO_2 and NO_x levels in Northern Europe seems likely to come from North West Scotland. This region is remote from major industrial activity and the prevailing south westerly winds are unlikely to carry a high pollution load from other countries. The Review Group on Acid Rain (RGAR, 1990) provide values of about 1 ppb SO_2 and 2 ppb NO_2 for this region. Data modelled for the UK using the Harwell Trajectory Model agree well with these figures. Background levels of acid deposition (H^+) were estimated from a 20 × 20 km rainfall map for the UK and the assumption of pH 5 rain in pristine areas (RGAR, 1990). These values were taken by the German, Dutch and UK teams. The Italian team took the view that background levels of SO_2 may well be higher due to volcanic activity in the South of the country, and hence it took a value of 1.3 ppb for that pollutant.

Background PM_{10} concentrations are more difficult to predict, because of diverse sources of particulates of natural origin. Each team was asked to make its own assessment of background levels. The UK and Dutch teams concluded that no single figure could be reliably estimated and a range of values should be taken – from 5–15 $\mu g/M^3$. The Italian team selected a figure of 10 $\mu g/M^3$, based on consultations with national experts, but it also recognized difficulties in selecting the relevant value, and carried out sensitivity analysis based on a figure of 7 $\mu g/M^3$. The German team could not agree on background levels and decided not to estimate damages from this pollutant.

Finally background levels had to be agreed for ozone. These are also difficult to estimate as they will vary considerably with altitude, sunlight and other climatic factors. The UK team took annual mean concentrations to be between 10 and 20 ppb, which translate to a range of between 20–30 ppb for the annual mean hourly maximum level for each day. The Italian team took a similar but single value – of 25 ppb. The Dutch team took background levels of 20 ppb, and the German team did not need to consider this issue as it did not make direct estimates of ozone damages.

It is clear that damages will be highly dependent on background levels and that, as of now, we do not have a firm grip on what values should be taken. This area will be researched further in GARP II.

1.7. Conclusions

This study has undertaken an ambitious research programme: to estimate environmental damages at a detailed level of each of four countries, and to do it within a single coherent framework. It is not surprising that the first attempt at such an endeavour should turn out to face a number of problems; indeed, identifying these problems was a primary objective if the study. This introduction has presented the difficulties, and indicated what measures, if any, were taken to resolve them. In a number of areas further work is required, and we have indicated where this is the case.

The estimation of monetary values of environmental damages is an important component of an overall exercise in environmental accounting. It will be useful whether the accounts are presented separately or as part of a single national accounting framework. Within the EC, there is a clear demand for the integration of environmental and economic data, a demand that this research effort explicitly addresses.

The environmental estimation of damages also has wider applications. It is useful, for example, to those developing alternative measures of national well being, based on combinations of economic and social indicators, or on the concept of sustainable development.

As noted earlier, At the same time there is some scepticism about the feasibility of producing consistent, credible environmental damage estimates in monetary terms, that can be integrated into the environmental accounts – even the UN Statistical Office is lukewarm about the idea, and the guidelines provided by the United Nations say very little about how environmental damages should be measured.[5] By undertaking a comprehensive, consistent valuation of damages, this study can provide direct evidence on the feasibility of damage estimation in an accounting framework and thereby contribute significantly to the development of a methodology for the establishment of such accounts.

[5] The guidelines recommend the use of 'contingent valuation methods' but such advice is very general and not appropriate to the valuation of some impacts. In fact the experience of the EXTERNE project provides a much more solid basis for deciding on how such valuation should be carried out.

References

Adger, W.N. (1993). Sustainable National Income and Natural Resource Degradation: Initial Results for Zimbabwe, in Turner, R.K. (ed.) Sustainable Environmental Economics and Management: Principles and Practice (London: Belhaven Press).

Adger, W.N. and Whitby, M. (1993). National Resource Accounting in the Land Use Sector: Theory and Practice. European Review of Agricultural Economics, 20: 77–97.

Ahmad, Y.J., El Serafgy, S. and Lutz, E. (eds.) (1989). Environmental Accounting for Sustainable Development (Washington DC: The World Bank).

Anielski, M. (1992). Resource Accounting: Indicators of the Sustainability of Alberta's Forest Resources. Paper presented to the International Society of Ecological Economics Conference, Stockholm.

Bartelmus, P. et al. (1992). Integrated Environmental and Economic Accounting: A Case Study for Papua New Guinea. Environment Department Working Paper No. 54 (Washington DC: The World Bank).

Bryant, C. and Cook, P. (1992). Environmental Issues and National Accounts. Economic Trends, 469: 99–122.

Clough, P. (1991). Natural Resource Accounting for New Zealand's Indigenous Forests. Report to the Ministry of Environment, New Zealand Institute of Economic Research.

Commission Interministerielle Des Comptes Du Patrimonie Naturel (1986). Les Collection du Patrimonie Naturel, Collection de L'Insee serie C, n. 137–138, Paris.

CBS Netherlands (1993). Environmental Statistics in the Netherlands. Central Bureau of Statistics, Voorburg.

CBS Norway (1992). Natural Resources and the Environment (Central Bureau of Statistics, Oslo).

Daly, H.E. and Cobb, J.B. Jr. (1990). For the Common Good (Beacon Press).

Da Motta Seroa, R. and Young, P.H. (1992). Natural Resources and National Accounts: Sustainable Income from Mineral Extraction (Rio de Janeiro: IPEA).

European Commission DGXII (1995a–f). Science, Research and Development Programme (JOULE). Externalities of Energy *EXTERNE* PROJECT. Report numbers: EUR 16520 (Summary), EUR 16521 (Methodology), EUR 16522 (Coal & Lignite), EUR 16523 (Oil & Gas), EUR 16524 (Nuclear), EUR 16525 (Wind & Hydro).

German Department of Environment (1992). The Federal Statistical Office Presents the Idea of an Economic Valuation of the Environment. Press Release 34/92, Bonn.

Hueting, R. (1989). Correcting National Income for Environmental Losses: Toward a Practical Solution, in Ahmad et al. (op cit.).

Hultkrantz, L. (1992). National Account of Timber and Forest Environmental Resources in Sweden. Environmental and Resource Economics, 2: 283–305.

Markandya, A. and Costanza, C. (eds.) (1993). Environmental Accounting: A Review of the Current Debate. Environmental Economics Series, No 8 (Nairobi: UNEP).

Nordhaus, W. (1992). Is Growth Sustainable? Reflections on the Concept of Sustainable Economic Growth. Paper presented to the International Economics Society, Varenna, Italy.

Parikh K. et al. (1992). Natural Resource Accounting: a Framework for India, Indira Ghandi Institute of Development Research, Bombay.

Prince, R. and Gordon, P.L. (1994). Greening the National Accounts. Congressional Budget Office, Washington DC.

Proops, J. (1995). Research Workshop on Indicators of Sustainable Development and Green Accounting. Rapporteur's Report, Brussels.

Repetto, R. et al. (1989). Wasting Assets: Natural Resources in the National Accounts. World Resources Institute, Washington DC.

Solorzano, R. et al. (1991). Accounts Overdue: Natural Resource Depreciation in Costa Rica. World Resources Institute, Washington DC.

United Nations Development Programme (1990). Human Development Report. United Nations Development Programme, New York.

United Nations (1993). Integrated Environmental and Economic Accounting. Department for Economic and Social Information and Policy Analysis – Statistical Division. UN, New York.

Uno, K. (1989). Economic Growth and Environmental Change in Japan: Net National Welfare and Beyond, in Archibucji, F. and Nijkamp, P. (eds.), Economy and Ecology: Towards Sustainable Development. Kluwer Academic Publishers, London.

van Tongeren, J. et al. (1991). Integrated Environmental and Economic Accounting: A Case Study for Mexico. Environment Department Working Paper No. 50. The World Bank, Washington DC.

Weizmann, M.L. (1976). On the Welfare Significance of National Product in a Dynamic Economy. Quarterly Journal of Economics, 90: 156–162.

Young, M.D. (1992). Natural Resource Accounting: Some Australian Experiences and Observations, Working Paper 92/1, CSIRO, Canberra.

CHAPTER 2

VALUATION OF DAMAGES

2.1. Methods of valuation

The valuation approach adopted in this study is demonstrated in Figure 2.1 below. Ideally one would like to start with the basic relationship between the pollution sources and the environmental impacts (going from emissions to concentrations to impacts). Then one would go on to value the impacts associated with specific sources. In fact, at the national level, such an approach would be extremely time consuming. Instead, what has been done is to start with the spatial distribution of concentrations and value their impacts. At a later stage (in GARP II) the concentrations will be attributed to different sources.

Having estimated the physical impacts, the next step is to value the damages. Over the last 25 years or so, a number of techniques have been developed for estimating environmental effects. A survey of these may be found in Markandya

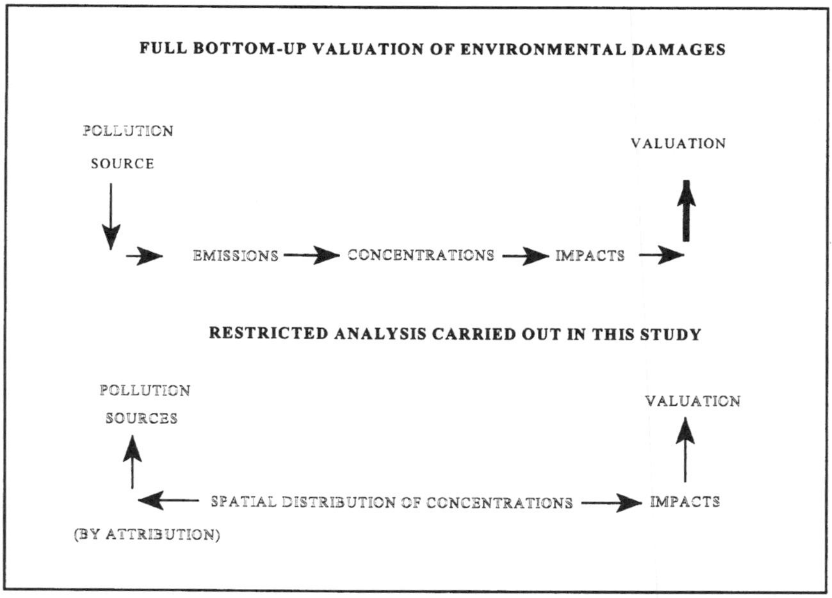

Fig. 2.1. Pathway for valuing damage.

and Richardson (1993). This study does not attempt to summarize them in any comprehensive way. For each of the impacts that are valued in this study, however, it does discuss the method or methods of valuation that are appropriate, and the issues that arise in their application.

As discussed briefly in the previous chapter, the underlying principle in monetary valuation is to obtain the *willingness to pay* (WTP) of the affected individual to avoid the negative impact, or the *willingness to accept* (WTA) payment as compensation if a negative impact takes place. The rationale is that values should be based on individual preferences, which are translated into money terms through individual WTP and WTA.

Once the impacts have been identified in physical terms, they can be valued using market prices, where the things impacted (crops, materials etc.) have a market price, although even in this simple case there are problems and issues that arise, which are discussed further later in this report. For a wide range of impacts, however, such as increased risk of death or loss of recreational values, there are no direct market prices that can be used. Three techniques are widely used in this context. One is to elicit the WTP or WTA by direct questionnaire. This is termed the *contingent valuation method* (CVM) and has been developed into a sophisticated procedure for valuing a number of environmental impacts. Another is to look at the WTP as expressed in related markets. Frequently environmental effects are reflected in property values. Thus an increase in noise or a reduction in visibility will 'show up' in reductions in the value of properties affected by the changes. This approach is called the *hedonic price method* (HPM) and is widely used for noise and aesthetic effects.

Where the individuals undertake expenditures to benefit from a facility such as a park or a fishing area one can determine their WTP through the expenditures on the recreational activity, including costs of travel to the park, any fees paid etc. Economists have developed quite sophisticated procedures for estimating the values of changes in environmental facilities using such data. This is the third approach and is known as the *travel cost method.* It is particularly useful for valuing recreational impacts.

2.2. Categories of value

The WTP/WTA numbers can be expressed for a number of categories of value. The most important distinction is between values arising from the use of the environment by the individual and values that arise even when there is no identifiable use made of that environment. These are called *use values* and *non-use values* respectively. Non-use values are also sometimes referred to as *existence values.*

Within the category of use values there are many different categories. *Direct use values* arise when an individual makes use of the environment (e.g. s/he breathes the air) and derives a loss of welfare if that environment is polluted.

Indirect use values arise when an individual's welfare is affected by what happens to another individual. For example, if I feel a loss of welfare as a result of the death or illness of a friend or relation, resulting from increased levels of air pollution, then this loss of welfare translates into a cost through my WTP. It can and has been measured in limited cases and is referred to as an *altruistic value* (see later in this chapter for details). Both direct and indirect use values have a time dimension; an environmental change today can result in such values now and in the future.

Another category of use value that is potentially important is that of *option value*. This arises when a action taken now can result in a change in the supply or availability of some environmental good in the future. For example, flooding a region to impound water for a hydro project would result in that area not being available for hiking. A person might have a WTP for the option to use that hiking area, even if s/he was not sure that it would be used. This WTP is the sum of the expected gain in welfare from the use of the area, plus a certain gain in welfare from the knowledge that s/he *could* use it even if it is not actually used. The latter is referred to as the option value. The literature on environmental valuation shows that, in certain cases the option value will be positive but in general it is not an important category of value (see Freeman, 1991). There are very few estimates of such values, and in the context of the fuel cycle study it was felt that estimating future use values was difficult enough; estimating option values was not considered an important category to address.

Non-use value is particularly controversial, although values deriving from the existence of a pristine environment are real enough, even for those who never make any use of it. In some respects what constitutes 'use' and what constitutes 'non-use' is not clear. If someone sees a programme about a wilderness area but never visits it, that represents a use value, however indirect. Pure non-use value must not involve any welfare from any sensory experience related to the item being valued. In fact some environmentalists argue that such non-use or existence values are unrelated to human appreciation or otherwise of the environment, but are embedded in, or intrinsic to, the things being valued. For a sympathetic review of this position, see the discussion on the 'new naturalist ethics' by Turner (1988). However, that is not the position taken in this study. The basis of valuation remains an anthropocentric one which, however, does not imply an anti-environmentalist stance.

The difficulty in defining non-use values extends, not unnaturally, to measuring them. The only method available for this category is that of the questionnaire approach, or contingent valuation method. This method has been tested and improved extensively in the past 20 years, and the general consensus is that the technique works effectively where 'market conditions' of exchange can be simulated effectively and where the respondent has considerable familiarity with the item being valued. For most categories of non-use value this is simply not the case. Hence, for the present, non-use values are extremely difficult to value with any accuracy.

2.3. Issues arising in the use of monetary values

2.3.1. *Income constraints are ignored*

As noted above, the the basic philosophy underlying the valuation is based on individual preferences, which are expressed through the willingness to pay (WTP) for something that improves individual welfare, and willingness to accept payment (WTA) for something that reduces individual welfare. The total value of environmental impacts is taken as the sum of the WTP or WTA of the individuals comprising it. Thus no special weight is given to any particular group. This approach contrasts, for example, with that of values based on expert opinion, or values based on the costs of making good any damage done to the environment. Such mitigation costs will only provide a valid measure of cost if society is collectively willing to pay for the mitigation, rather than suffer the damage. In such cases, mitigation based estimates can provide important values, and have in fact been used in the study in selected areas. However, the validity of that use is dependent on the assumption that society is willing to pay for the mitigation.

Although the valuation of environmental impacts using money values is widespread and growing, there are still many people who find the idea strange at best and distasteful and unacceptable at worst. Given the central role being played by monetary valuation in this exercise, a justification of the method is warranted.

One objection often voiced in the use of WTP is that it is 'income constrained'. Since you cannot pay what you do not have, a poorer person's WTP is less than that of a richer person, other things being equal. This occurs most forcefully in connection with the valuation of a statistical life (VOSL) where the WTP to avoid an increase in the risk of death is measured in terms of a VOSL. In general one would expect the VOSL for a poor person to be less than that of a rich person. But this is no more or less objectionable than saying that a rich person can and does spend more on health protection than a poor person; or that individuals of higher social status and wealth live longer on average than person of lower status; or that better neighbourhoods will spend more on environmental protection than poorer neighbourhoods.

The basic inequalities in society result in different values being put on the environment by different people. One may object to these inequalities, and make a strong case to change them but, as long as they are there, one has to accept the consequences. Society takes the view that most decisions about allocation of resources are predicated on the existing inequality of income and wealth, both between and within societies. This is discussed in greater detail in Chapter 3.

2.3.2. *Environmental damages are already accounted for in other prices*

A related issue that arises in making environmental damage estimates is that of 'internalization'. Damages may occur to certain groups who have already fully taken account of such damages in deciding on what actions to take. A central case is when workers in certain industries suffer from higher incidence of diseases, or when their occupational mortality rates are higher. If they are aware of these factors, and if the contract of employment reflects proper compensations for these added risks, then there is no relevant environmental effect in the sense defined above. In technical terms it is said that the environmental cost has been internalized through a contract that has been entered into freely. It may still be of interest to measure the 'environmental costs' for such impacts, but care has to be taken in what interpretation is placed on the resulting damage figures.

Two matters of contention in deciding whether or not an environmental cost has been internalized or not, are those of information and market imperfections. It is argued that if a worker takes a job not knowing the risks involved, and then suffers from work-related health effects, there is a genuine environmental effect that needs to be accounted for. It is unarguable, for example, that the presence of an unexpected impact on a working population which was not foreseen is an important environmental effect.[1] What is less clear is the significance of such effects for policy. Clearly, if neither party is aware of a possible effect, and the same applies to the regulating authority, then no measurable external damages can be included in the analysis. If the regulating authority is aware of certain health risks, and believes the working population is not, the appropriate measure is to disseminate the information. The resources that are devoted to that, however, could be influenced by the size of the environmental damages caused to this group. In this sense then, data on work related environmental effects may be relevant to the analysis.

A related issue is that of market imperfections. Staying with the example of occupational health effects, there may be macroeconomic difficulties in the region, in which case those seeking work would not have the luxury of choosing between a riskier occupation with a higher wage, and a less risky one with a lower wage. In the presence of structural or long term unemployment, as is the case in some parts of Europe, the occupational health damages will form part of the external costs of the related fuel cycles. In other situations, however, where the labor force is more mobile, and where unemployment is less long term, the environmental damages may be said to be internalized.

As far as this study is concerned, the issue of internalization has not been addressed. It was decided that *all* significant environmental damages be calculated in the first instance. This will include those arising from occupation-related environmental impacts, which are the main area of contention. A separate exercise should then determine how relevant they are to *policy*. This

[1] In economics, environmental effects that have not been accounted for through markets or special deals and regulations are called *external effects*. The issue is whether the environmental effect is an external effect, or is something that has been accounted for.

will depend on how poor the information base on which workers make employment decisions is, and on how strong the structural labor-market problems are.

2.4. Shortcuts in valuing damages

2.4.1. *Benefit transfer*

The environmental damages associated with a particular pollutant will depend on a great deal of spatially disaggregated information as well as estimates of dose response functions and damages per unit of pollutant. Clearly, it would be infeasible to estimate all environmental damages for each location and pollutant *ab initio*. Much of the work required is extremely time consuming and expensive, making the transfer of estimates from one study to another an important part of the exercise. The difficult issue is to know when a damage estimate is transferable and what modifications, if any, need to be made before it can be used in its new context.

Benefit transfer is "an application of monetary values from a particular valuation study to an alternative or secondary policy decision setting, often in another geographic area than the one where the original study was performed" (Navrud, 1994). There are three main biases inherent in transferring benefits to other areas:

2.4.2. *Meta analysis*

Where several studies, reporting a similar final estimate of environmental damage, exist, and where there are significant differences between them in terms of the background variables, a procedure known as *meta-analysis* has been developed to transfer the results from one study across to other applications. What such an analysis does is to take the estimated damages from a range of studies of, for example, coal fired plants and see how they vary systematically, according to affected population, building areas, crops, level of income of the population, etc. The analysis is carried out using econometric techniques, which yield estimates of the responsiveness of damages to the various factors that render them more transferable across situations. This can then be used to derive a simple formula relating environmental costs to per capita income, which could then be employed to calculate damages in countries where no relevant studies were available.

Estimates of damages based on meta-analysis have been provided in a formal sense in two studies carried out in the US on recreation demand (Smith and Kaoru, 1990), and on air pollution (Smith and Huang, 1993). The results in the recreation studies indicate that, as one would expect, the nature of the site is significant for the WTP attached to a visit, as are the costs of substitutes and the opportunity cost of time. Choice of functional form in the estimating equations also appears to play a part. In the air pollution study referred to above, it was found that damages per unit of concentration vary inversely with the average price of property in the study (the higher the price the lower the unit value of damage). If correct, it would enable an adjustment to the estimated value to be made on the basis of the average prices of properties in the area being investigated. However, the authors are cautious about the validity of the estimates obtained.

A formal meta-analysis is difficult to carry out, and has not proved possible in the context of the this study. However, some of the 'expert' adjustments do make an informal meta analysis. For example, adjusting estimates of damages for size of population to obtain a per capita estimate and transferring that to the new study implicitly assumes that damages are proportional to population. Such adjustments are frequently made.

2.4.3. *Conclusions on benefit transfer*

Transferability depends on being able to use a large body of data from different studies and estimating the systematic factors that would result in variations in the estimates. In most cases the range of studies available are few. More can be done to carry out meta-analysis of the type indicated, but it will take time. The best practice in the meantime is to use estimates from sources as close to the one in which they are being applied and adjust them for differences in underlying variables where that is possible. Often the most important obstacle to systematic benefit transfer, however, is a lack of documentation in the existing valuation studies.

From the environmental damage-energy source linkages identified above, one can identify an increasing order of difficulty (in terms of modifications that have to be made) with which estimates can be transferred from the original study to the situation in which they are to be used;

(a) for most receptor categories, the most easily transferred data is the dose–response function itself, relating environmental impacts adjusted for population. Thus numbers in the form 0.8×10^{-6} excess deaths per $\mu g/m^3$ would be transferable across studies as long as adjustments to the other variables in the dose–response function were made (e.g., relative humidity). The *additional* local information that is required to use such data is simply local market conditions, costs and prices. The exception is ecosystems where such functions are usually very site specific;

(b) the next ones in order of difficulty are monetary estimates of damages per unit of pollutant *by concentration.* Results are reported, e.g., in ECU/$\mu g/m^{-3}$, or in ECU/km/person of lost visibility. Estimates may vary according to population affected, in which case an analysis of such variations would be desirable. Other socioeconomic variables that would be of relevance are income levels of the affected population, age, background environmental variables such as rainfall etc., and socioeconomic variables such as medical services and how they are paid for. If enough studies are available, a meta-analysis can be performed (see below), in which the mean estimated value is regressed against these variables. Then the relevant adjustment to the estimates are made, given the local values of the explanatory variables. No additional local variables should be required;

(c) similar to (b) above are estimates of monetary damages in terms of emissions or units of energy produced. In such cases one needs all the information listed above, plus details of how the emissions or energy units relate to the concentrations or whatever impacts are responsible for the damages. For example, damages may be quoted as $x/kWh for coal. The relevance of this estimate to a different situation will depend on how the kWh is related to emissions and how the emissions are converted into concentrations in the area where the impacts were measured, *plus* the variables with which the relationship between concentrations and damages vary. Thus most work will have to be done in these cases and, in the context of this fuel cycle study it is unlikely that such estimates could be used at all.

It is important to note that national boundaries themselves are not of any relevance in transferring estimates, except that there may be cultural differences that will influence factors such as frequency with which a person visits a doctor, or how he perceives a loss of visibility. In this sense there is no reason why the EXTERNE Project should not draw on the US and other studies, or transfer estimates from one country to another within Europe, as long as the above consideration is taken into account.

2.4.4. *Use of avoidance costs*

Different to the approach advocated in this study for valuing damages is the 'Avoidance Cost' approach. This has been used by Bernow *et al.* (1990) of the Tellus Institute and recommended by, among others, Hueting (1989), for the construction of environmental accounts. Their reasoning is that it is difficult to estimate social costs based on damages, and in any event the numbers are too uncertain and not credible. Instead they suggest that abatement costs may be a reasonable surrogate for damages. In this approach, existing and proposed environmental regulations are analyzed to estimate the value that society implicitly places on different environmental impacts. According to Tellus, the marginal cost of abating emissions, when they are at the limit imposed by regulation, reflects the preference of regulators to require that particular level of abatement and the corresponding incremental cost, rather than allow emissions to exceed that limit and subsequently to have adverse impact on the public. The reasoning used by Tellus is that since these regulators represent the public, their views represent the costs placed on those emissions by the public.

We take the view that such reasoning cannot be applied generally. The premise that marginal control costs represent the costs of air emissions to society implies that regulators know what individual environmental damages are and always decide on the optimal policy where the marginal costs of control equal the marginal damages. In fact it is quite clear that they do not know these costs, and the political processes by which policy decisions are made do not generally have the property that they equate social damages to costs of abatement. In specific cases, however, the use of avoidance costs is justified. This is when the damages are clearly in excess of the avoidance costs, and when the avoidance measures are actually implemented.

The question of whether credible and consistent damage estimates can be constructed based on a WTP approach is, we argue, an open question. This study will provide some important evidence as to whether this can be done or not.

2.5. Valuation and discounting

Discounting is the practice of placing lower numerical values on future benefits and costs as compared to present benefits and costs. In the context of this study it is an important issue because many of the environmental damages of present actions will occur many years from now and the higher the discount rate, the lower the value that will be attached to these damages. This can have major implications for policy. This section reviews the arguments for and against different discount rates, including the case for discounting environmental damages at a different rate from other projects.

2.5.1. *The rationale for discounting and choice of discount rate*

The practice of *discounting* arises because individuals attach less weight to a benefit or cost in the future than they do to a benefit or cost now. Impatience, or 'time preference', is one reason why the present is preferred to the future. The second reason is that, since capital is productive, an ECU's worth of resources now will generate more than an ECU's worth of goods and services in the future. Hence an entrepreneur would be willing to pay more than one ECU in the future to acquire an ECU's worth of these resources now. This argument for discounting is referred to as the 'marginal productivity of capital' argument; the use of the word marginal indicates that it is the productivity of additional units of capital that is relevant.

If a form of damage, valued at ECU X today, but which will occur in T

years time is to be discounted at a rate of r percent, the value of X is reduced to:

$$X/(1+r)^T.$$

Clearly the higher is r and the greater is T, the lower will be the value of the discounted damages. Typically discount rates for low risk borrowers in EC countries run at around 4–7% *in real terms*. The latter means that when future values are being computed, no allowance is made for general inflation of money values, and all damages are calculated in present prices. The real rate can be computed by deducting the rate of inflation from the nominal or actual rate. At the same time the typical real discount rates for lenders investing in low risk assets range from a negative value to around 5%. Thus, even looking at market data, and excluding questions of risk which are discussed later, there is a wide range of discount rates to choose from.

The discounting debate from an environmental perspective

Although much of the environmental literature argues against discounting *in general* and high discount rates in particular, (see Parfit, 1983 or Goodin, 1986) there is in fact no unique relationship between high discount rates and environmental deterioration. High rates may well shift the cost burden to future generations but, as the discount rate rises, so falls the overall level of investment, thus slowing the pace of economic development in general. Since natural resources are required for investment, the demand for such resources is lower at higher discount rates. High discount rates may also discourage development projects that compete with existing environmentally benign uses, e.g., watershed development as opposed to existing wilderness use. Exactly how the choice of discount rate impacts on the overall profile of natural resource and environment use is thus ambiguous. This point is important because it indicates the invalidity of the more simplistic generalizations that discount rates should be lowered to accommodate environmental considerations.[2]

What are the arguments for modifying the discount rates when valuing future damages in the context of a study such as this? They can be classified into the following groups:

(a) individuals discount the future at too high a rate compared to society's needs;
(b) the rates of return on capital investments (which tend to be quite high) are not relevant to the discounting of environmental damages.

(a) *Individual discount rates.* A number of economists and philosophers have commented that individual discounting is irrational from a social point of view and that what individuals want carries no necessary implications for public

[2] This prescription has been challenged at an intuitive level by Krutilla (1967). For further discussions see Markandya and Pearce (1988), Krautkraemer (1988).

policy. Many countries, for instance, compulsorily force savings behaviour on individuals, e.g., through state pensions, indicating that the state overrides private preferences concerning savings behaviour.

In many respects the state acts as a guardian of future generations' interests. If these interests are not adequately protected by using discount rates based on individual preferences it will consider it within its powers to change those rates. The extent to which the interests of future generations are safeguarded when using positive discount rates in the ranges indicated is a matter of debate within the literature. Theoretical models with overlapping generations have shown that the discount rate that emerges as a result of private discounting is not necessarily efficient, i.e., it is not the one that takes the economy on a long run welfare maximizing path.

Some economists such as Becker (1988) and Sen (1982) have argued that individuals make decisions in two contexts, 'private' decisions reflecting their own interests and 'public' decisions in which they act with responsibility for fellow beings and for future generations. Market discount rates, it is argued, reflect the private context, whereas social discount rates should reflect the public context. Hence, for a variety of reasons relating to future generations' interests, the social discount rate may be below the market rate based on individual preferences. The implications for the choice of the discount rate are that there is a need to look at an individual's public role behavior, or to leave the choice of the discount rate to the state, or to try and select a rate based on a collective savings contract. However, none of these options appears to offer a practical procedure for determining the discount rate in quantitative terms. What they do suggest is that market rates will not be proper guides to social discount rates once future generations' interests are incorporated into the social decision rule.

(b) *Opportunity cost of capital.* The opportunity cost of capital is obtained by looking at the rate of return on the best investment of similar risk that is displaced as a result of the particular project being undertaken. It is only reasonable to require the investment undertaken to yield a return at least as high as that on the alternative use of funds. This is the basic justification for an opportunity cost discount rate. In developing countries where there is a shortage of capital, such rates tend to be very high and their use is often justified on the grounds of the allocation of scarce capital.

The environmental literature has also made some attempts to discredit discounting on opportunity cost grounds. (Parfit, 1983 and Goodin, 1986). The first criticism is that opportunity cost discounting implies a reinvestment of benefits at the opportunity cost rate, and this is often invalid. For example, at a 10% discount rate ECU 100 today is comparable to ECU 121 in 2 years time if the ECU 100 is invested for 1 year to yield ECU 10 of return and then both the original capital and the return are invested for another year to obtain a total of ECU 121. Now, if the return is consumed but not reinvested then, the critics argue, the consumption flows have no opportunity cost. What, they ask, is the relevance of a discount rate based on assumed reinvested profits if in fact the profits are consumed?

The second environmental critique of opportunity cost discounting relates to compensation across generations. Suppose an investment today would cause environmental damages of ECU X, T years from now. The argument for representing this damage in discounted terms by the amount ECU $X/(1+r)^T$ is the following. If this latter amount were invested at the opportunity cost of capital discount rate r, it would amount to ECU X in T years time. This could then be used to compensate those who suffer the damages in that year. Parfit argues, however, that using the discounted value is only legitimate if the compensation is *actually* paid. Otherwise, he argues, we cannot represent those damages by a discounted cost. (Parfit, 1983).

The problem in both cases is that actual and 'potential' compensation are being confused. The fact that there is a sum that could be generated by the project and that could be used for the *potential* compensation of the victim is enough to ensure its efficiency. Whether the investment is made and whether compensation is *actually* carried out is a separate question and one which is not relevant to the issue of how to choose a discount rate.

Related to these criticisms, however, is a very real issue. It is that in some cases the investment fund needs to be held for very long periods before the potential compensation becomes due. With some forms of nuclear waste the period could run into thousands of years. Human experience with capital investments and their rates of return does not run into more than a couple of hundred years at the most. To give an example, if 1 ECU were invested today a real rate of return of 7% per annum, it would amount to nearly ECU 491 trillion in 500 years time. Such amounts are inconceivable and it is not reasonable to base decisions to protect future generations on them. In the course of the EXTERNE project this issue had to be grappled with. It became evident quite quickly that where future damages were occurring very far into the future, the use of discount rates did not offer a reasonable way of dealing with them. At any positive rate the costs became insignificant and a zero rate they were so large as to dominate all the calculations. One possible solution is to define the rights of future generations and use those to circumscribe the options in terms of what impacts can be imposed on them.

Discount rates and future environmental benefits

Both the above arguments suggest that the discount rate that is observed in the market may be too high from a social point of view. However, one factor that reduces the impact of discounting in future years is that of increasing values associated with future benefits and costs. If an economy is growing at, say, 3% in real terms per capita, and if population is constant, one can expect that the value attached to a loss of environmental facilities will increase at 3% or more per annum. This is because the 'income elasticity' for environmental goods takes values that are generally in excess of one. Hence, if future benefits are being discounted at, say, 5%, the numerical value that will be entered for the services of an environmental amenity will only fall at 2% per annum compared to the value today. And if the environmental values grow faster than 3% future values will be effectively discounted at an even lower rate. Finally, as the environmental amenities become more scarce, as is happening, or as the population grows, the value attached to the amenity will rise even faster than the income elasticity would suggest. These points, initially observed by Krutilla and Fisher (1975), can, to a large extent reduce the force of the argument that discounting is damaging to the environment. If future values are properly measured and future scarcity allowed for, this should be less valid.

A sustainability approach

Although the environmental debate has contributed to discussion on the rationale for discounting, it has not been successful in demonstrating a case for rejecting discounting as such. The arguments by economists and philosophers that private discount rates are too high in social terms have considerable validity and have to be taken seriously. At the same time, these discussions do not provide practical advice on how the social rate should be determined. It is also important to note here that these arguments do not apply only to environmental costs and benefits but to all costs and benefits.

The environmentalists' arguments against the use of opportunity cost of capital discount rates were largely shown to be invalid. Particularly if future benefits and costs of environmental amenities are properly measured, the discount rate need not be the enemy of the environment that it is sometimes considered to be. However, the point that there are no historical data to apply productivity of capital discount rates for impacts that have very long gestation periods is important and has to be acknowledged. It was suggested that in situations where impacts over hundreds of years need to be valued another route might be followed.

One such approach is through a 'sustainability constraint'. This could be used to place physical constraints on the amount of risk and type of damage that future generations would have to endure. Then future costs can be included in the analysis with an acceptable discount rate applied but that discount rate cannot be said to have influenced key decisions about the environmental liabilities handed on to future generations.

To some extent, a 'sustainability' approach is already followed, when environmentalists have argued that, in some key cases, protection for key resources and environments has to be guaranteed, *irrespective* of whether it can be justified on cost-benefit grounds at conventional discount rates. Although there are merits in favor of such an argument, what is being called for here is more than that. What is needed is a *systematic procedure* by which a sustainability criterion can be invoked in support of certain actions. Such a procedure does not exist, but it would be desirable to develop one.

Conclusions on discounting

This section has reviewed the arguments for different discount rates and concluded that the arguments against any discounting at all are not valid. At the same time the arguments that individual discount rates are too high have some validity, as does the argument that discounting is not the appropriate way to deal with very long term impacts (e.g. over one hundred years). For the latter a sustainability approach is recommended. For the former, it is recommended that the environmental costs be calculated for *rates of around 3%*. The global damage estimates reported in Chapter 9 are based on discount rates of 1–3%. In computing the health costs on the basis of life years lost, a discount rate of 3% has been used. Likewise for other damages where discounting is relevant, a rate of 3% has been adopted.

2.6. Uncertainty

Uncertainty in its various manifestations is arguably the most important and least satisfactorily-addressed aspect of the whole assessment of the environmental impacts (not just the economic component). There are two distinct aspects of the uncertainty. One is the uncertainty about the different parameters of the impact pathway, including the monetary valuation; and the other is the impact of uncertainty on the valuation of impacts, such as accidents, risks of disease etc. For the first kind we recognize that overall uncertainty is the product of uncertainty in each of the four stages of the impact pathway: estimation of emissions, calculation of the spatial dispersion of the emissions, estimation of the physical impacts of the dispersed pollutants, and valuation of these impacts.[3] If one could quantify the uncertainty at each of these stages, one could quantify the overall uncertainty in terms of some probability distribution. Some work along these lines is being undertaken as part of the development of the EXTERNE project (based on lognormal distributions of the different parameters and some subjective estimates of the geometric standard deviation of those distributions). At this stage, however, it is not generally possible to quantify the uncertainty in terms of probability distributions.

The alternative that has been adopted is to take a scenario approach. A 'base case' value for each key parameter is combined with low and high case scenarios. A range of values has been taken for key parameters and the results reported not only for the central estimate but for the range of values. This does not provide a statistically valid estimate of the range of values for the external costs, with a given confidence interval, or resolve the uncertainty issue, but it will be a useful indication of where the uncertainties lie.

A separate but equally important aspect of the uncertainty dimension in the

[3] In this report only the last two stages are analysed: estimation of the physical impacts of the dispersed pollutants, and valuation of these impacts.

valuation of environmental impacts arises from the fact that, for health related damages and accidents, one is valuing changes in risk of damage. Attitudes to risk aversion should therefore play an important part in determining such values. In the valuation literature reviewed in later sections of this synthesis report there is, unfortunately, little evidence that risk aversion has been taken explicitly into account. In the EXTERNE study this was most serious in the valuation of accidents, especially nuclear accidents. Hence an effort was made to develop a methodology for the valuation of nuclear accidents. (European Commission, 1995b). The results indicate that taking explicit account of risk aversion can result in considerably higher values for, for example, nuclear accidents, but that it depends on the precise formulations of the risk aversion function. A more important factor seems to be the *assessment* of risk, as given by the estimated probability of the accident.
This study has not addressed this issue. The risk related costs of environmental damages have not been explicitly valued. Such costs could be significant and this is an area where further research is required under GARP II.

2.7. Conclusions

This chapter has provided the background on the valuation issues that have arisen in this study. There are several categories of value that should be estimated, and a range of techniques for estimating them. The greatest problems arise in the valuation of non-use damages, and this is reflected in this study by the much more limited coverage for such damages, compared to 'use' damages, such as health effects, materials corrosion etc.

The basis of the valuation is willingness to pay for improvements in the environment or willingness to accept payment for damages suffered. There are questions that arise with such an approach but, in our opinion, it is the most appropriate method for valuation of damages. Issues such as income distribution which are ignored by the approach can be handled in other ways. There is a question of whether damages are 'double counted' if we estimate them using the methods proposed. This is because they may be reflected in higher wages and prices for the products that cause the damage. This is an issue but not one that we address, leaving it to the policy analysts to decide on whether specific damages are internalized or not.

In conducting damage estimation we have to rely on existing studies and estimates and the process of transferring estimates from one type of study to another is called benefits transfer. In the chapter we discuss the issues that arise in making such transfer and recommend best practices that should be followed.

Some researchers have recommended using avoidance costs to estimate damages. We take the view that this is not the correct procedure in general, although there may be circumstances where avoidance costs can be employed.

They are appropriate when the damages are clearly in excess of mitigation costs, and when the mitigation is actually carried out.

This chapter has also addressed the question of discounting. Discounting is necessary in comparing costs at different points in time. It is recommended that the discount rate for future damages be in the region of 3%. Such a rate will not address the problems of very long term damage, or damage that could threaten the sustainability of an eco-system. For such cases a sustainability approach, where a full repairing cost is estimated and debited against the source of the damage is warranted.

Finally, we have addressed the question of uncertainty, which in its various manifestations, is arguably the most important, and least satisfactorily dealt with aspect of the whole assessment of the environmental impacts (not just in the context of this study). It has to be recognized at the outset that all the estimates of damage in money terms will have large elements of uncertainty about them. This is unavoidable and there is evidence to suggest that further scientific work on the physical relationships will not reduce that estimate by much. To capture the range of results that are possible it is recommended that ranges of values of key parameters be used where possible.

Another issue that is of great importance is the cost associated with the uncertain environmental impacts that arise from economic activities. This applies particularly to accidents. We have not dealt with such environmental costs in this study although we know, from other work undertaken as part of the EXTERNE study that such costs could be significant.

References

Becker, G.S. (1988). Family Economics and Macro Behavior. American Economic Review.

Bernow, S. et al. (1990). Valuation of Environmental Externalities for Energy Planning and Operations. Tellus Institute, Boston Mass.

European Commission DGXII (1995b). Science, Research and Development Programme (JOULE). Externalities of Energy 'EXTERNE' Project. Methodology Report EUR 16521 EN.

Freeman, M. (1991). The Measurement of Environmental Resource Values. Johns Hopkins University Press, Baltimore, MD.

Goodin, R.E. (1986). Protecting the Vulnerable (Chicago: University of Chicago Press).

Hueting, R. (1989). Correcting National Income for Environmental Losses: Towards a Practical Solution, in Ahmad, J.J., Lutz, E. and El Sarafy (1989), Environmental Accounting for Sustainable Development, World Bank, Washington D.C.

Krautkaemer, J.A. (1988). The Rate of Discount and the Preservation of Natural Environments. Natural Resources Modelling, 2(3): 421–38.

Krutilla, J.V. (1967). Conservation Reconsidered. American Economic Review, 57: 777–786.

Krutilla, J.V. and Fisher, A.C. (1975). The Economics of Natural Environments. Resources for the Future, Washington DC.

Markandya, A. and Pearce, D.W. (1988). Environmental Considerations and the Choice of the Discount Rate in Developing Countries. Environment Department Working Paper No. 3. The World Bank, Washington DC.

Markandya, A. and Richardson, J. (1993). The Earthscan Reader in Environmental Economics. Earthscan Publications, London.

Navrud, S. (1994). Economic Valuation of External Costs of Fuel Cycles: Testing the Benefit Transfer Approach. In A.T. de Almeida (ed.), Models for Integrated Electricity Resource Planning (Amsterdam: Kluwer Academic Publishers).

Parfitt, D. (1983) Energy Policy and Further Future: The Social Discount Rate. In Maclean, D. and Brown, P. (eds.), Energy and the Future. Rowman and Littlefield, Totowa, NJ.

Sen, A.K. (1982). Approaches to the Choice of Discount Rates for Social Benefit Cost Analysis. In Lind R.C. (ed.), Discounting for Time and Risk in Energy Policy. Johns Hopkins Press, Baltimore.

Smith, V.K. and Kaoru, Y. (1990). Signals or Noise? Explaining the Variation in Recreation Benefit Estimates. American Journal of Agricultural Economics, 68: 280–290.

Smith, V.K. and Huang, Y (1993). Hedonic Models and Air Pollution: Twenty Five Years and Counting. Environmental and Resource Economics, 3, 4: 381–394.

Turner, R.K. (1988). Wetland Conservation: Economics and Ethics. in: D. Collard, D.W. Pearce and D. Ulph (eds.), Economics, Growth and Sustainable Environments. Macmillian, London.

CHAPTER 3

HEALTH IMPACTS

3.1. Introduction

Health impacts are probably the most important of all environmental impacts but also among the most difficult to measure. This chapter provides a discussion of the different health impacts, which are divided into mortality impacts, morbidity impacts, and accidents.

The literature on the dose response function is strongly dependent on studies carried out in the US, which have been used extensively in this study. This is an acceptable procedure in an area where the literature is world-wide, and where the research carried out in the US has resulted in careful, sound studies. There is a need, however, for more European epidemiological studies in this area. New studies are being undertaken in the EU under the 4th Framework Programme (in particular the APHEA project) and we intend to access these in the second phase of our work. For the present report, the growing epidemiological literature was re-reviewed fully, leading to some differences both in model assumptions and in detailed exposure–response functions relative to earlier work (EC DG XII, 1995b). The final set of exposure–response functions is still, however, strongly based on North American studies.

A similar dependence on US research applies to economic valuation. Although there are a number of studies of the value of mortality impacts in Europe, there are virtually no serious attempts to value morbidity impacts. Hence reliance on the US work is inevitable, but it is important to look at the 'cultural dependence' of the estimates and methodologies more closely.

3.2. Exposure–response relationships

The damage-function approach is the logical way to proceed with an evaluation of health impacts provided that the state-of-the-art is sufficiently well developed that suitable functions and background data exist. It is useful to consider to what extent the implementation of the available exposure–response functions really does express the full health impacts of the pollutants that we are considering. While the damage-function approach is critically discussed in a recent report for the EXTERNE Project (European Commission, 1995b), the main issues are briefly addressed in order to point the main limitations of the approach.

- Lung function health impacts from epidemiological studies have been ignored, because at present there are no usable economic valuations of lung function changes. These effects are relatively minor in health terms however; and the inclusion and valuation of symptoms means that some related, relatively minor, effects have been assessed.
- Chronic mortality and morbidity has been included in a preliminary way only; but sufficiently to indicate that these effects may be important.
- For ease of implementation, several of the key relationships are linearised and annualised, assuming an independence of background levels. The real situation obviously is more complex. It is assumed that the linearised functions are a good approximation of the reported non-linear functions, although some scepticism is warranted about whether the linearised functions are appropriate to other contexts.
- Following the impact pathway approach, it is difficult to quantify an impact of an air pollution mixture where little is known about the independence or not of the effects of the component parts. The epidemiological evidence suggests however that the acute effects of particulates and of ozone may reasonably be considered as additive in many situations, the confounding between these two pollutants being generally less than that between particulates, SO_2 and NO_x.
- The transferability of an exposure–response function depends partly on the health endpoint being considered. Transferability is likely to be better to the extent that the endpoint represents a biological rather than a social event. Health indicators like Restricted Activity Days, Hospital Admissions, Emergency Room Visits etc. partly depend on socio-cultural factors. There are also general problems with the transferability of exposure–response functions from one air pollution context to another.

However, it should be emphasised that there is a strong body of evidence associating daily variation in pollution with a wide range of acute health effects. Exposure–response relationships reflecting that evidence have been identified and implemented here for particulates, SO_2 and ozone. Relationships for particulates take account not only of primary emissions, but also of secondary sulphates and nitrates. All relationships have been implemented over wide geographical areas. Additivity of effects across the health endpoints and pollutants studied is a reasonable approximation to the full acute effects of the incremental air pollution, within limits of existing knowledge

For each of the four pollutants that this assessment of public health effects is concerned with, (see Table 1.1, Chapter 1) we have identified, where appropriate, exposure–response relationships describing changes in health endpoints associated with unit changes in pollutant. The assessment, while conducted for the pollutants and for endpoints individually, has however been an integrated exercise, with decisions and conclusions in any one area impacting on another. All of the exposure–response relationships proposed are based on epidemiological studies of general air pollution; experimental studies (whether human or animal) are used to provide contextual information only. Within this framework, we have primarily selected studies which show a clear (statistically significant) relationship between pollutant and effect, through the use of appropriate statistical methods, adjusting for the effects of possible confounding factors, including other pollutants and climate.

These generally high standards have been modified in some instances where, all things considered, it seems likely that a real relationship exists between the pollutant and endpoint of interest; but no particularly strong study of that relationship is available currently. In these circumstances, use of a somewhat weaker study may be better than no attempt at quantifying that relationship.

Epidemiological study of the health effects of ambient air pollution has been a very active area of research in recent years, and is likely to continue to be so in the coming period. Thus, identifying and implementing a suitable set of exposure–response functions within the present project has meant taking definite positions on issues which are still under active research and are matters of legitimate debate. The purpose of this introductory section is to note what are the principal issues of judgement to be addressed and what positions have been adopted as working assumptions at this time. The reasons for the choices made are described more fully in the methodology report for the EXTERNE project, upon which this study has drawn extensively (see European Commission, 1995b). It is important to note that research on some of these functions is very much ongoing and new results are coming in all the time. We should expect, therefore, to make changes to the estimates as the science advances. In particular, new European studies will soon be made available.

3.2.1. *Acute and chronic effects*

It is very important to distinguish between *acute* effects (which occur on the same day as increases in pollution, or very soon thereafter), and *chronic* effects (which are the delayed effects of long-term exposure). With acute mortality, higher air pollution days contribute to a higher number of deaths on the same day or on immediately following days. In this case, the 'at-risk' population consists mainly of elderly people (>65) with existing (serious) cardio-respiratory problems. The expectation is that persons affected are already quite ill and have only a short life expectancy.

With chronic mortality, long term exposure to air pollution leads to disease, which contributes to premature death. In this case, it is formally irrelevant whether that death follows a higher pollution day. The cohort studies show increased mortality from cardio-respiratory disease, and possibly from lung cancer.

As described below, acute effects of several pollutants across a range of health endpoints are well established. There is some degree of informed speculation but no established understanding of the mechanisms by which these effects occur.

It is more difficult to establish reliable relationships in chronic effects studies, and so there are fewer potentially usable exposure–response functions. It may well be however that in terms of public health impacts and their economic valuation, chronic effects are more important. We have therefore attempted to include chronic mortality and morbidity relationships within the overall implementation.

3.2.1.1. *Particulates: acute effects*

There are now numerous studies which associate particulate air pollution with a wide range of acute effects on health. These associations have been found at normal background levels in a wide range of locations. In the present study we have assumed that there is no threshold level of pollution for these effects. There is some evidence, though not compelling, that the health effects per unit incremental exposure are higher when background pollution levels are lower. Because this is not well-established, and for ease of implementation, effects are estimated in this project as independent of background.

Within this project these associations are treated as causal, though there is no established understanding of the mechanism of action. It is not known for sure what size range or what components of respirable particulate air pollution are relevant. It seems however that particulates emitted from combustion sources, or formed subsequently (e.g. sulphates, nitrates) are more dangerous to health than wind-blown natural (crustal) particulates.

Particulate air pollution may arise due to direct emissions of particles from industrial sources, vehicles, homes, etc., or due to the formation of aerosols such as ammonium sulphate and ammonium nitrate following the release of NO_x or SO_2. We have followed a widespread current convention in using PM_{10}, i.e. particulates of less that 10 μm aerodynamic diameter, as the most relevant index of ambient particulate concentrations. A case can however be made for relationships based on sulphates or on fine particulates ($PM_{2.5}$). This is one of many areas of debate in this area of research at the present time.

3.2.1.2. *Oxides of nitrogen*

Some studies link NO_x or NO_2 with acute effects. These studies have not been used, because the apparent NO_x effect is arguably in reality not an effect of NO_x as such. Rather, NO_x may be a surrogate or marker for a mixture of pollutants not otherwise well measured, including particulates from combustion sources, especially traffic. Some role for NO_x/NO_2 is not discounted, but the direct effect if any is likely to be small and not quantifiable reliably on the basis of available studies. NO_x is however implicated indirectly via nitrates and ozone.

3.2.1.3. *Sulphur dioxide*

To some extent this also appears to be true for SO_2. However, the range and diversity of positive studies linking SO_2 with acute health effects is quite substantially greater than for NO_x, and human experimental studies are more suggestive of a real link especially for asthmatics. A limited number of exposure–response functions have thus been adopted for SO_2.

3.2.1.4. *Ozone: acute effects*
There are positive studies linking ozone with a wide range of acute health effects endpoints. Most of these are summertime studies, in high-ozone areas. It seems that a real relationship exists in these circumstances, and exposure–response relationships are proposed accordingly. Daily 1-hr maximum ozone concentration is used, though many recent studies use longer time-periods which possibly are more biologically relevant. Most of these studies do not suggest a threshold for ozone effects. It is nevertheless unclear whether these results generalise to other locations and seasons, where direct investigation often fails to show an ozone effect. We have treated the results as generalisable, without threshold, on the assumption that ozone is the active agent.

3.2.1.5. *Particulates: chronic effects*
Recent studies, adjusting for confounders at the individual level, have shown associations between chronic health effects and ambient particulate levels, though the adequacy of the adjustment remains a matter of debate. Quantification of the chronic effects is complicated, and possibly compromised, by use of recent pollution levels only, rather than the higher historical levels which may be biologically relevant to chronic mortality and morbidity. (Use of recent levels only will over-estimate the risk per unit exposure to an unknown extent).

Evidence of a chronic effect on mortality is given by Dockery *et al.* (1993) and Pope *et al.* (1995). These cohort studies found clear relationships with PM_{10}, fine particulates ($PM_{2.5}$) and with sulphates. Risk estimates from Pope *et al.* (1995), translating from $PM_{2.5}$ to PM_{10}, are used as the best available estimate of chronic mortality. They show that chronic effects on mortality may have a significant impact on the overall evaluation.

It must be said that this view is not universal at the moment. For example these studies have been cited as evidence that short term fluctuations in particulate levels reduce the life expectancy of those included under the 'acute' exposure–response functions by possibly several years, rather than as evidence for an effect of chronic exposure. Clearly this issue needs to be openly debated, and for this reason results from the function derived by Pope *et al.* are given in this chapter and in the summary tables to the report. However, significantly greater confidence is associated with our estimates of acute effects of particulates on mortality, because of the greater body of literature from which the functions used to calculate them are drawn.

Some aspects of chronic morbidity are also evaluated. These have been included, on the assumption that morbidity is less affected by the 'historical-to-recent' exposure problem. This is certainly true for childhood disease. It is

Table 3.1. Exposure–response functions used for the assessment of mortality effects

Pollutant: PM_{10}						
Schwartz (1993a)	Percentage change in acute mortality	=	Low: Mid: High:	0.064 0.104 0.145	*	Change in annual PM_{10} concentration (in $\mu g/m^3$)
Pope *et al.* (1995)[1]	Percentage change in chronic mortality	=	Low: Mid: High:	0.295 0.386 0.477	*	Change in annual PM_{10} concentration (in $\mu g/m^3$)
Pollutant: SO_2						
Touloumi *et al.* (1994)	Percentage change in acute mortality	=	Low: Mid: High:	0.065 0.096 0.127	*	Change in annual SO_2 concentration (in $\mu g/m^3$)
Pollutant: O_3						
Kinney *et al.* (1994)	Percentage change in acute mortality	=	Low: Mid: High:	0.010 0.015 0.020	*	Change in annual O_3 concentration (in ppb)

[1] The relationship for chronic effects of PM_{10} on mortality is thought to implicitly include acute effects. Chronic effects have therefore been calculated by subtracting the Schwartz (1993a) function from the Pope *et al.* (1995) function. PM_{10} and SO_2 levels are expressed as $\mu g/m^{-3}$, whilst O_3 levels are expressed in ppb.

arguable whether the childhood conditions estimated here should be called chronic or acute.

3.2.1.6. *Other pollutants: chronic effects*

The situation regarding chronic effects of ozone is unclear. There may be some effects. However, relationships are not proposed. Nor are chronic effects estimated for NO_x. Two chronic functions are given for SO_2.

3.2.2. *Summary of exposure–response relationships proposed*

Effects quantified now include mortality, hospital admissions, (hospital) emergency room visits (ERVs), restricted activity days (RADs), asthma attacks and (mostly respiratory) symptoms. Lung function effects are not described because there is currently no good means of economic valuation. The relationships proposed are summarised in the following Tables 3.1 to 3.5.

3.2.3. *Unquantifiable uncertainties with exposure–response functions*

There are unquantifiable uncertainties in transferring the results of mostly North American studies to a European context. Some aspects have been considered earlier. Others include differences in pollution sources and mixtures, and in the relationships between pollutants and confounding factors. It is

Table 3.2. Exposure–response functions used for the assessment of acute effects on morbidity of PM_{10} levels (μg/m^{-3})

Schwartz (1994) and Burnett *et al.* (1994)	Change in hospital admissions for respiratory infections per 100,000 persons (all ages) per year	=	Low: Mid: High:	0.124 0.187 0.251	*	Change in annual PM_{10} concentration (in μg/m^3)
Schwartz (1994) and Burnett *et al.* (1994)	Change in hospital admissions for COPD per 100,000 persons (all ages) per year	=	Low: Mid: High:	0.161 0.227 0.293	*	Change in annual PM_{10} concentration (in μg/m^3)
Sunyer *et al.* (1993)	Change in ERVs for COPD per 100,000 persons (all ages) per year	=	Low: Mid: High:	0.58 0.72 0.86	*	Change in annual PM_{10} concentration (in μg/m^3)
Schwartz (1993) and Bates *et al.* (1990)	Change in ERVs for asthma per 100,000 persons (all ages) per year	=	Low: Mid: High:	0.4 0.64 0.86	*	Change in annual PM_{10} concentration (in μg/m^3)
Schwartz *et al.* (1991)	Change in hospital visits for childhood croup per 100,000 persons (all ages) per year	=	Low: Mid: High:	2.18 2.91 3.82	*	Change in annual PM_{10} concentration (in μg/m^3)
Ostro (1987)	Change in RADs per 1000 adults per year	=	Low: Mid: High:	31.8 49.9 78.3	*	Change in annual PM_{10} concentration (in μg/m^3)
Ostro *et al.* (1991)	Change in 'shortness of breath' days per asthmatic per year	=	Low: Mid: High:	0.07 0.14 0.21	*	Change in annual PM_{10} concentration (in μg/m^3)
Krupnick *et al.* (1990)	Change in symptom days per 100,000 persons (all ages) per year	=	Low: Mid: High:	221.9 465.0 686.9	*	Change in annual PM_{10} concentration (in μg/m^3)

Notes: COPD – Chronic Obstructive Pulmonary Disease.
ERV – Emergency Room Visit.
RAD – Restricted Activity Day.

however a strength of the studies that relationships, expressed as percentage change in health effect per unit exposure, seem remarkably invariant to changes in population, location and pollution mixtures.

If the exposure–response functions were to be implemented in the form of percentage change in health effects per unit exposure, it would be necessary to know (or to estimate reliably) the background levels of mortality and morbidity over the full population at risk. Death rates are quite readily available; and mortality has therefore been estimated using percentage change functions. It was however impractical to estimate background morbidity rates across the range of endpoints, in a way which took account of possible variations across

Table 3.3. Exposure–response functions used for the assessment of acute effects on morbidity of incremental ozone pollution, expressed as the annual mean of the daily 1 hour maximum level (ppb)

Schwartz (1994) and Burnett *et al.* (1994)	Change in hospital admissions for respiratory infections per 100,000 persons (all ages) per year	=	Low: Mid: High:	0.636 0.848 1.011	*	Change in annual O_3 concentration (in ppb)
Schwartz (1994) and Burnett *et al.* (1994)	Change in hospital admissions for COPD per 100,000 persons (all ages) per year	=	Low: Mid: High:	0.387 0.617 0.848	*	Change in annual O_3 concentration (in ppb)
Thurston *et al.* (1994) and Burnett *et al.* (1994)	Change in hospital admissions for asthma per 100,000 persons (all ages) per year	=	Low: Mid: High:	0.283 0.571 0.858	*	Change in annual O_3 concentration (in ppb)
Cody *et al.* (1992) and Bates *et al.* (1990)	Change in ERVs for asthma per 100,000 persons (all ages) per year	=	Low: Mid: High:	1.72 2.63 3.50	*	Change in annual O_3 concentration (in ppb)
Ostro and Rothschild (1989)	Change in minor RADs per 1000 adults per year	=	Low: Mid: High:	0 15.6 52.3	*	Change in annual O_3 concentration (in ppb)
Holguin *et al.* (1984)	Change in asthmatic attacks per asthmatic per year	=	Low: Mid: High:	0.365 0.582 0.798	*	Change in annual O_3 concentration (in ppb)
Krupnick *et al.* (1990)	Change in symptom days per 100,000 persons (all ages) per year	=	Low: Mid: High:	26.9 52.8 78.8	*	Change in annual O_3 concentration (in ppb)

Notes: COPD – Chronic Obstructive Pulmonary Disease.
ERV – Emergency Room Visit.
RAD – Restricted Activity Day.

grid-cells. Instead, a single background rate was estimated from an available epidemiological study (often, but not always, the same study as was used to provide the relationship of percentage change), and combined with the estimated percentage change to give estimated numbers of impacts per unit incremental exposure and unit population.

This approach, while easing implementation, is inexact in that it assumes both that background rates do not vary importantly within the population at risk, and that the average background rates can be estimated adequately using studies in other locations. These assumptions ignore population-related variations in true morbidity levels, as well as socio-economic and cultural differences affecting health service usage, restricted activity, and symptoms reporting. Better estimation of background levels of morbidity may be one useful area of improving the present estimates.

Despite these limitations, we consider that an informative quantified assessment can be made with current knowledge; though the uncertainties involved

Table 3.4. Exposure–response functions used for the assessment of chronic effects on morbidity of PM_{10} levels ($\mu g/m^{-3}$)

Schwartz (1993b)	Change in prevalence of adults with chronic bronchitis per 100,000 adults per year	=	Low: Mid: High:	45 70 94	*	Change in annual PM_{10} concentration (in $\mu g/m^3$)
Schwartz (1993b)	Change in prevalence of adults with respiratory illness per 100,000 adults per year	=	Low: Mid: High:	60 95 129	*	Change in annual PM_{10} concentration (in $\mu g/m^3$)
Dockery *et al.* (1989)	Change in prevalence of children with bronchitis per 100,000 children per year	=	Low: Mid: High:	85 161 238	*	Change in annual PM_{10} concentration (in $\mu g/m^3$)
Dockery *et al.* (1989)	Change in prevalence of children with chronic cough per 100,000 children per year	=	Low: Mid: High:	103 207 313	*	Change in annual PM_{10} concentration (in $\mu g/m^3$)

Table 3.5. Exposure–response functions used for the assessment of acute effects on morbidity of SO_2 pollution levels ($\mu g/m^{-3}$)

Ponka and Virtanen (1994)	Hospital admissions for chronic bronchitis or emphysema per 100,000 (all ages) per year	=	Low: Mid: High:	0.11 2.91 5.75	*	Change in annual SO_2 concentration (in $\mu g/m^3$)

should not be disregarded. The scope of uncertainty will surely be reduced as research in air pollution epidemiology continues.

3.2.4. *Interpretation of the health endpoints*

The likely severity of these effects is relevant both to general understanding and to economic valuation. For example, almost certainly the acute mortality effects occur predominantly in generally older people, with serious pre-existing ill-health conditions, though the precise mechanism of action has not (yet) been established. Length of life lost in those who die prematurely following higher pollution days is also unknown, but is likely to be very short relative to average life expectancy. Average reduced life expectancy among those who die prematurely from chronic disease is likely to be much greater.

Minor restricted activity days (MRADs) do not involve bed disability, or time off work or school. Any symptom, as in Krupnick *et al.* (1990), covers a very wide range of occurrences most of which would not be considered severe.

3.3. Valuing mortality impacts

The mortality approach in the valuation literature is based on the estimation of the willingness to pay for a change in the risk of death. This is converted into the 'value of a statistical life' (VOSL) by dividing the WTP by the change in risk. So, for example, if the estimated WTP is ECU 100 for a reduction in the risk of death of 1/10000, the value of a statistical life is estimated at 100*10000, which equals one million ECU. This way of conceptualising the willingness to pay for a change in the risk of death has many assumptions, primary among them being the 'linearity' between risk and payment. For example, a risk of death of 1/1000 would then be valued at ECU 1 mn./1000, or 1000 ECU using the VOSL approach. Within a small range of the risk of death at which the VOSL is established this may not be a bad assumption, but it is clearly indefensible for risk levels very different from the one used in obtaining the original estimate.

Estimates of the WTP for a reduction in risk or the WTA of an increase in risk have been made by three methods. First, there are studies that look at the increased compensation individuals need, other things being equal, to work in occupations where the risk of death at work is higher. This provides an estimate of the WTA. Second, there are studies based on the CVM method, where individuals are questioned about their WTP and WTA for measures that reduce the risk of death from certain activities (e.g. driving); or their WTA for measures that, conceivably, increase it (e.g. increased road traffic in a given area). Third, researchers have looked at actual voluntary expenditures on items that reduce death risk from certain activities, such as cigarette smoking, or purchasing air bags for cars.

Table 3.6 below summarizes the European studies from these three categories, and gives the central (and range if available) estimated value of a statistical life in ECU. All values have been converted into £(1995) prices and then into ECU. The European studies show a range of values for VOSL of ECU 0.8–7.2 mn. The mean of the range is approximately ECU 4.2 mn. It is worth noting that, on average, the highest values come from the CVM studies and the lowest from the consumer market studies where actual expenditures are involved. By comparison, Table 3.7 gives the summary results from the US studies, converted into ECU. The range (taking the average in each category) is ECU 2.4–3.1 mn. for the US studies, whereas it is ECU 3.0–5.3 mn. for the European studies. Given the higher per capita income in the US compared with that in the European countries where the studies were carried out (mainly the UK), this is somewhat surprising. It is principally the result of higher values in the European CVM studies, particularly the earlier Jones-Lee study, (1976) and the Frankel study. Eliminating these two studies would produce a range and an average for the CVM group of ECU 2.5–3.6 mn. and 3.1 mn. respectively. Taking those figures would give a mean value of life of ECU 3.1 mn.

Table 3.6. European empirical estimates of the value of a statistical life (VOSL)

Country	Author	Year	VOSL values (ECU mn. 1995)
Wage Risk Studies			
UK	Melinek	73	0.8
UK	Veljanovski	78	8.4–11.8
UK	Needleman	80	0.4
UK	Marin *et al.*	82	3.7–4.2
Average wage-risk			*3.4–4.3*
CVM Studies (including contingent ranking method)			
UK	Melinek	73	0.5
UK	Jones-Lee	76	15.5–19.2
UK	Maclean	79	5.2
UK	Frankel	79	5.2–21.0
UK	Jones-Lee	85	1.3–5.8
SWE	Persson	89	2.6–3.2
AU	Maier*et al.*	89	3.2
Average CVM			*4.7–8.3*
Consumer Market Studies			
UK	Melinek	73	0.4–0.8
UK	Ghosh	75	0.8
UK	Jones-Lee	76	1.0–11.0
UK	Blomquist	79	1.0–3.5
Average UK Market			*0.8–4.1*

Sources: See original studies.

Table 3.7. Summary table for values of a statistical life (ECU Mn. (1995))

	European	USA
Wage-Risk	3.4–4.2	4.2–6.6
CVM	4.9–7.6	1.7–3.0
Market	0.8–4.1	1.2–1.3
Average	3.0–5.3	2.4–3.6

Source: Adapted from Pearce *et. al.* (1992).

which is probably the best central estimate from the European studies. It is this value that has been used in this study.

The main issues that arise with the application of the value of a statistical life to the present study are the following:

(a) the validity of the methods used in estimating the value of a statistical life;
(b) the distinction between voluntary and involuntary risk;
(c) the treatment of age dependent mortality and years of life lost;

(d) the transfer of risk estimate from different probability ranges.

3.3.1. *Validity of methods of estimating a statistical life*

All three methods of valuing a statistical life have been subject to considerable criticism. The wage-risk method relies on the assumption that there is enough labour mobility to permit individuals to choose their occupations to reflect all their preferences, one of which is the preference for a level of risk. In economies suffering from long standing structural imbalances in the labour markets this is at best a questionable assumption. Second, it is difficult to distinguish between risks of mortality and morbidity. Third, the WTA will depend on perceived probabilities of death. Almost all studies, however, use a measure of the long-run frequency of death as a measure of risk. This makes the results quoted unsatisfactory. Fourth, the probabilities for which the risks are measured are generally higher than those faced by most airbourne pollutants. This point is returned to below, but a related factor is that the high risk occupations involve individuals whose WTA for an increase in the risk of death is not typical of the population at large (e.g., steeplejacks).[2] The net impact of all these factors is difficult to gauge but it is *likely* that the estimated WTA will be lower.

3.3.2. *Voluntary and involuntary risk*

There is strong evidence to suggest that individuals treat voluntary risk differently from involuntary risk, with the WTA for a voluntary risk being much lower than that for an involuntary risk. Starr (1976) has estimated, on a judgmental basis, the difference between the willingness to accept a voluntary increase in risk and an involuntary increase. He finds the latter to be around ten times as high as the former for probabilities of death in the range 10^{-6}–10^{-7}. Interestingly, for lower probabilities that are typical of the fuel cycles, estimates of the differences are not available. In another study of the difference (Litai, 1980), it has been argued that the difference could be as much as 100 times.

The CVM methods are subject to the criticism that the choices are hypothetical and that individuals are not familiar with the concepts of risk involved. Certainly, there have been serious difficulties in conveying the impact of different probability changes through questionnaire methods.

Finally, the consumer expenditure approach is subject to the difficulties that perceived probabilities are very different from objective probabilities, and that the effects of the expenditures are to reduce the risk of death as well as of

[2] This is probably one reason that the etimated value of life declines as the mean risk level in a group increases. From a theoretical perspective one would expect the opposite if the populations were homogeneous. The US Fuel Cycle Report (ORNL/RFF, 1994) cites data to show that the value of a statistical life more than doubles if one allows for self-selection in the wage risk studies.

Table 3.8. Variation of VOSL with age

Age	20	25	30	35	40	45	50	55	60	65	70	75
VOSL as% of age 40	68	79	88	95	100	103	104	102	99	94	86	77

Source: Jones-Lee *et al.* (1985).

illness following an accident. It is difficult to separate the two impacts in the studies.

For all these reasons the studies are likely to be biased, with the wage-risk studies producing values that are too low and the CVM studies values that are too high. Taking an average, as has been done here, is averaging unknown errors and one cannot say what the final impact will be. However, they are all that is available and one can draw some comfort from the fact that the values are, in broad terms, in a plausible range.

This issue is of great importance to the present study, where the public opposition to nuclear fuel will not be reflected in the costs as estimated by using a value of life of around ECU 3.1 mn. This may be an issue of perceived versus objective probabilities, but can only be a partial explanation. At this stage, there is no alternative to using this value. However, as part of the ongoing research the issue of voluntary risk needs to be addressed and a revised estimate of the value of life obtained.

3.3.3. *Age dependent mortality*

The problem of age dependent mortality arises because the value of ECU 3.1 mn. is based on studies in which the individuals involved came from relatively narrow age bands (around 25–55 years, with a concentration in the 35–45 age band). One would expect the VOSL to vary with age, with a possibly lower value for older people, but this does not appear to be supported by the little evidence that is available. Variations of estimated VOSL by age do exist but they do not demonstrate a clear pattern.

A study that provides clear evidence on age-dependence is Jones-Lee *et al.* (1985) (see Table 3.8). Other studies, such as Shephard and Zeckhauser (1982) address this question but in a purely theoretical context. Jones Lee *et al.* found that VOSL at age 20 is about 70% of VOSL at age 40. Between 20 and 40 it rises slightly, and between 55 and 75 it falls slightly, so that by the age of 75 it is around 77% of the value at age 40.

These are quite small adjustments to VOSL and, generally, within the margins of error of the estimates anyway. Furthermore, in those cases where we are going to use VOSL, there are other factors that are more important in determining the value that should be applied than this small age-dependence variation (such as latency, manner of death, years of life remaining etc.). For all these reasons the GARP team concluded not to adjust VOSL for reason of age dependent mortality.

3.3.4. *The concept of value of life years lost (VLYL)*

The basis for this concept is the fact that some of the air pollution impacts will be on individuals who could not be expected to have long to live. The VOSL estimates, however, are based on studies of individuals with normal life expectancies. Therefore, for certain kinds of mortality impacts, using VOSL and applying it to data on excess deaths (as estimated by dose–response functions) will result in an overestimation of air pollution damages. It is possible to estimate the value of a year of life lost from the estimates of the value of a statistical life, if one has data on the age of the reference group, and some way of estimating the discount factor applied to present versus future years of life.

An important study in this area is Moore and Viscusi (1988). This is a wage-risk study which considers how workers view fatality risks at work. With fatalities, a young person loses a much greater amount of lifetime utility than an older person, a source of variation that is not incorporated in VOSL approaches. This is incorporated by weighting the standard death risk measure by the remaining life of each sample member. More specifically, the information used includes expected lifetimes, worker age, a discount rate that is computed as part of the estimation process, and measures of death risk.

This makes it possible to calculate the worker's expected remaining life at risk, reflecting the fact that the worker's principal concern is not simply the probability of a fatal accident, but the discounted duration of life and the associated lifetime utility at risk on the job. This discounted duration of life at risk is referred to as the *quantity-adjusted death risk* and is used in calculating the *quantity-adjusted value of life*. The *quantity-adjusted value of life* differs from conventional (VOSL) estimates in that the trade off is not between wages and death risk probabilities, but between wages and death risks that have been weighted by the discounted number of potential life years lost.

The rates of discount for workers is in the range 10–12%. Since these values converted to nominal rates are bounded from below by the prevailing home mortgage interest rate and bounded from above by credit card interest rates, there is little evidence of intertemporal irrationality. The implicit values per additional expected year of life equate to ECU 180,000 to ECU 211,500 (1995 prices). These values are the average WTP for an additional life in present value terms. The corresponding VOSL estimate is equivalent to ECU 6.34 mn. (1995), which is reflective of traditional VOSL approaches in the US literature.

A further study is Harrison (1990), who indicates that the implied value of a year of life lost, when used to value average years of life lost from premature death caused by cancer, cardiovascular disease etc., is only around 30% of the value that would emerge from an application of the value of a statistical life.

Pursuing this line of reasoning, one can calculate the value of a life year lost as follows. Consider a 'prime age male' with 37 years of life expectancy, for whom the VOSL of ECU 3.1 mn. is correct. Applying a discount rate of 3% to the annual value of a life year results in a value for a life year lost (VLYL) of ECU 138,115.[3] Using this value will have an impact on estimated health costs that is dependent on the number of years lost. This is the approach proposed by the GARP team.

As Navrud (1997a) notes, the VLYL approach implies very strong assumptions about the relationship between time and utility a person derives from life, namely that it is a linear function of time. It is plausible that this might lead to a significant understatement of the WTP for elderly people if there is a value attributed to being alive that is independent of the amount of time one expects to live. Johannesson and Johanson (1996) using a CVM approach, found that adding a year of life is more highly valued among the elderly. The WTP increased at a constant rate of 1–4% per year; this falls in the range of the Moore and Viscusi study.

These results indicate that people do have a positive discount rate for VOSL and it is, therefore, appropriate to use the approach proposed. *Hence, the starting point for the valuation of health effects is a VLYL of ECU 138,000 and a corresponding VOSL of ECU 3.1 mn.*

3.3.5. *Chronic and acute mortality*

The dose–response functions used to estimate excess deaths are based on cohort data (different regions and one point in time), as well as time series data (the same region over time). In the literature, the interpretation commonly given is that the cohort studies reflect chronic effects and time series studies reflect acute effects. This may not, however, be completely correct (J. Evans, pers. communication). It is highly likely that the cohort studies are picking up *both* effects, and that adding the results of the two may be double counting. The mechanisms of acute and chronic effects have been detailed in section 3.2.1. With chronic effects one would expect a greater loss of life years.

For acute effects, the epidemiologists advising the GARP team concluded that the average VLYL should be 15 months, *equivalent to a value of an acute 'case' of ECU 172,500* with a range of 1 month to 3 years. Correspondingly, for chronic effects, the number of life years lost was estimated to be 3, giving a *value for a chronic 'case' of ECU 414,000*, with a range of 1 to 6 years. At this stage of the work we have also implemented exposure–response functions both for acute and for chronic mortality effects, but we realise that there may be an element of double counting here.

---

[3] The calculation is VOSL = $VLYL_i . (1.03)^{-i}$. If VLYL is independent of i this gives the figure reported above when the summation is over 37 years.

3.3.6. *Transfer of risk estimates from different probability ranges*

Finally, there is the issue of the probability range over which the estimation is carried out and over which it is applied. Typically one is dealing with much lower probabilities of death from the airborne pollutants (of the order of 10^{-6} and lower), whereas the studies on which the estimated value of a statistical life is based are dealing with probabilities of between 10^{-1} to 10^{-5}. Furthermore, as the survey by Fisher, Chestnut and Violette (1989) has pointed out, the results from studies at the higher end of the probability range are less reliable. As mentioned earlier, theoretical models would tend to predict that the WTA for lower risks should be lower but, if anything, the empirical literature shows the opposite. Partly this is due to the fact that the groups are not homogeneous. The issue remains unresolved and there is little that can be done about this problem at this stage. In the medium term, research on the theoretical and empirical aspects of this issue is needed.

3.4. Morbidity impacts

There is an enormous US literature on valuing morbidity effects, and a virtual absence of one in Europe. In this study, as in the EXTERNE study that preceded it, we have made maximum use of the excellent work carried out in this area by the US team, with modifications to their findings as and when appropriate.

The WTP for an illness is composed of the following parts: the value of the time lost because of the illness, the value of the lost utility because of the pain and suffering, and the costs of any expenditures on averting and/or mitigating the effects of the illness. The last category includes both expenditures on prophylactics, as well as on the treatment of the illness once it has occurred. To value these components researchers have estimated the costs of illness, and used CVM methods as well as models of avertive behaviour.

The costs of illness (COI) are the easiest to measure, based either on the actual expenditures associated with different illnesses, or on the expected frequency of the use of different services for different illnesses. The costs of lost time are typically valued at the post-tax wage rate (for the work time lost), and at the opportunity cost of leisure (for the leisure time lost). Typically the latter is between one half and one third of the post-tax wage. Complications arise when the worker can work but is not performing at his full capacity. In that case an estimate of the productivity loss has to be made. It is important to note that COI is only a component of the total cost. However, since the other components are difficult to measure, estimates have been made of the relationship between the total WTP and COI. A COI approach has been used to value non-fatal cancers. The COI plus foregone earnings were the only estimate available. Although not ideal, it is better than the alternative of using no value at all.

CVM is the only approach that can estimate the value of the pain and suffering. The difficulties are those generally associated with the use of CVM and, in addition, of allowing for the fact that it is difficult to know which of the many costs are included in the given responses. In general, respondents will not include those costs that are not borne by them as a result of the illness (e.g., medical insurance). In that event, such costs need to be added. In this category one should also include the cost, in terms of pain and suffering, that the illness causes to other people (the so-called altruistic cost).

Avertive behaviour is the most complex of the three to model. It involves the estimation of a health production function, from which one would be able to estimate the inputs used by the individual in different health states, and taking the difference in value between these to obtain the cost of moving from one health state to another. The difficulty is in estimating that function, where many 'inputs' provide more than one service (e.g., bottled water, air conditioners), and where the changes in consumption as a function of the state of illness are difficult to estimate.

The EXTERNE study provides an extensive list of the empirical literature on the costs of morbidity. The broad groups under which the estimates can be classified are as follows:

(a) estimation of restrictive activity days;
(b) cost of chronic illnesses;
(c) valuation of symptom days;
(d) estimation of altruistic costs.

3.4.1. *Restrictive activity days (RADs)*

A large number of studies, using COI as well as CVM methods, have been used to estimate several categories of RADs. These are differentiated by illness (respiratory RAD or RRAD, angina RAD etc.), and by severity of impact (minor RAD (MRAD) versus normal RAD). It is stated that these impacts are among the easier of the health impacts to value, as they relate to acute events, lasting a well defined period. The US study provides central or best estimates for these impacts which can, as a first approximation, be taken in the European study using a purchasing power parity (PPP) exchange rate. Although there may be grounds for arguing that medical costs are somewhat higher in the US (at the PPP rate), the errors involved in transferring the estimates are likely to be dwarfed by those arising from other sources. Until a corpus of European studies is available therefore, it is recommended that these values for RAD be taken from the US study. *This gives a value of ECU 75 per RAD.*

The other alternative is to take a COI approach for the country concerned, and to gross-up the value to get the total WTP by using the estimated factor of between 2.0 and 3.0. The COI can be valued in terms of the medical costs plus any loss of value of time. This approach has not been adopted here, although a check should be made of the WTP values against the European based COI values.

The US study points out that the central values provided for a RAD cannot be simply multiplied by the number of days lost, because one would expect that the value of each additional day declines as one loses more days, and indeed the empirical evidence supports that (the average value declines as the number of days lost increases). The conclusion that the estimated rate of decline in the value of an RAD with the number of days is too inaccurate to be of use is probably correct, making the use of a single value the best course to follow at this stage.

3.4.2. *Chronic illness*

The valuation of the chronic illness is largely in terms of the COI approach (although there are a few CVM studies). The COI approach includes the direct as well as the indirect costs of the illness (such as lost earnings and loss of leisure time). The CVM approach operates in terms similar to the value of statistical life (VOSL) – i.e., by asking what the willingness to pay to reduce the risk of contracting a chronic respiratory illness would be. The corresponding value is referred to as the 'value of a statistical case' (VSC). A US study by Krupnick and Cropper (1992) suggests a best estimate for the VSC for respiratory disease to be equivalent to ECU 1.2 mn. Attributing a full VSC to the extra (as distinct from the new) cases each year would have seriously over-valued the impacts. Therefore, this value has not been used. In this study we have valued chronic bronchitis only in terms of 'episodes' or acute attacks, which is, in effect, under-valuing the overall impacts. Some further work is necessary on this issue. Chronic morbidity impacts also arise occupationally. It is acknowledged that these are important, but they were not considered in this study.

3.4.3. *Symptom days (SDs)*

The US Fuel Cycle Study (ORNL/RFF, 1994) provides an extensive literature review on the valuation of symptom-days. These include CVM studies, as well as some that combine avertive behaviour and CVM. Although the work carried out is impressive, there are unfortunately still many difficulties to be resolved. The CVM studies have problems of low response rates and extreme bids that have to be discounted. There is also a difficulty in knowing the extent to which the responses include the use of avertive measures. Some of the results appear to indicate that the latter are not always allowed for. Some recent studies (Dickie *et al.* 1986, 1987) have included information on actual avertive behaviour and revealed responses but, as the EXTERNE report points out, "the results of this study need considerable refinement before they can be used with confidence in a morbidity benefit analysis. The limitations arise in the theory, data, statistical, and implementation phases of the study".

Nevertheless, EXTERNE took the view that this was an important end-point and should be valued. It adopted the estimate of *ECU 7.5 per symptom day*, based on the US study. We have taken the same value.

3.4.4. *Altruistic impacts*

As with symptom-days, estimates of the impact of an illness on the utility of others is not at a stage at which it can be used in a valuation exercise. One US study, Viscusi, Magat and Forrest (1988) came up with an altruistic value for each case of poisoning avoided of more than 5 times the private valuation. The experiment consisted of a CVM, in which individuals were asked their WTP for a TV campaign that would reduce poisoning resulting from poor handling of insecticides. However, the study had a relatively unsophisticated design and the results need to be confirmed in other studies. Work in the UK by Needleman (1980) and Jones-Lee *et al.* (1985) has suggested that the altruistic values are around 40–50% of the private total valuations. Again, however, these are isolated findings and need to be corroborated. In view of the current state of the art in this area it is not recommended that altruistic valuations be included in this study.

3.4.5. *Other endpoints*

There are a number of other endpoints that have been valued. These are included in Table 3.9 below which provides an overview of all the morbidity factors covered under the GARP study. All values have been converted into ECU 1995 prices.

3.5. Accidents

The final category of impacts is those related to environmentally related accidents. The cost to the public is in excess of the direct costs of repair etc., and often includes the costs in terms of deaths, injuries etc. In this evaluation we did not value any such impacts, although they could be important. This was due to a lack of data and time to develop the appropriate methodology. In some of the follow-up work we hope to address this issue.

3.6. Results of the studies

3.6.1. *Physical impacts*

Table 3.10 summarises the physical impacts of the four country analysis. The main points to note are:

Table 3.9. Summary of morbidity endpoints valued in GARP I

Endpoint	**Value (ECU)**	Source	Impacts assessed
Restrictive activity days (RADs)	75	US CVM studies	PM_{10}(acute), O_3(acute)
Symptom days (SDs)	7.5	US CVM studies	PM_{10}(acute), O_3(acute)
Chest discomfort day	7.5	US CVM studies	SO_2(acute)
Emergency room visits (ERVs) for various effects	223	US CVM studies	PM_{10}(acute), SO_2(acute) O_3(acute)
Hospital admissions for respiratory infections (RHAs)	7870	US CVM studies	PM_{10}(acute), O_3(acute)
Hospital admissions for chronic obstructive pulmonary disease (COPD)	7870	US CVM studies	PM_{10}(acute), O_3(acute)
Hospital admissions for chronic bronchitis	7870	US studies	SO_2(acute)
Episodes of various respiratory illnesses in adults and children	138	US studies	PM_{10}(chronic)
Asthma attacks	37	COI approach, adjusted w.r.t WTP	PM_{10}(acute), O_3(acute)
Episodes of chronic bronchitis in adults	138	US CVM studies	PM_{10}

Notes: CVM – Contingent Valuation Method.
WTP – Willingness-to-Pay.
COI – Cost of Illness.

a. Coverage has been relatively complete for Italy, Netherlands and the UK. In Germany, however, only SO_2 impacts have been estimated. This was due to time constraints and problems over the definition of background levels.
b. The mid estimates for excess deaths attributable to SO_2 are 4,900 in Germany, 6,700 in Italy, 900 in the Netherlands and 9,300 in the UK. A major factor influencing the magnitude of the figures are the background levels adopted, and these differed among countries. It should also be noted that the estimates for SO_2 are not additive to those for PM_{10}.
c. For PM_{10} estimated earlier deaths per annum from *acute* effects are: 15,000 in Italy, 4,300 in Netherlands and 10,100 in the UK. Excess deaths from *chronic* effects are: 52,200 in Italy, 14,600 in the Netherlands and 24,500 in the UK. It is likely that these two estimates are *not additive. The numbers of deaths for Italy is surprisingly high and the figures need to be reviewed. As the Italian report notes they account for more than 100% of all deaths from respiratory effects though the deaths in question are cardio-respiratory, not respiratory only.* For the UK the PM_{10} figures (chronic plus acute) account for 12% of all cardio-respiratory deaths. In all cases, however, we need to look more carefully at these figures.

d. For ozone impacts the figures in all countries are generally small, ranging from a mid value of around 90 cases in Italy, 220 in the Netherlands to around 2000 in the UK. For Italy only data from the Lombardy region was considered.
e. In the case of *morbidity* the impacts for PM_{10} seem to be broadly consistent across the three countries (Italy, Netherlands and UK), with the Italian figures being slightly higher than the other two. In all cases, however, the number of cases of Restrictive Activity Days (RADs) and Symptom Days (SDs) are improbably high (accounting for 3% of all days, for example, in Italy). For ozone there is a big discrepancy between Netherlands and the UK, with the latter showing around 37,000 excess attacks of asthma and 32,000 Emergency Room Visits (ERVs).
f. The range of values for the impacts can be quite large. For the UK, the lower bound for PM_{10} impacts is 20–30% of the central estimate and the upper bound is 80–100% higher. This reflects the different background conditions assumed. For the Netherlands, which also took the same set of background levels, the range is somewhat smaller. In Italy the PM_{10} range of background values taken is very small; hence the small range for the estimates of impacts.

3.6.2. *Monetary impacts*

Tables 3.11 and 3.12 summarise the monetary values of the health impacts. Table 3.11 is based on the value of life years lost (VLYL), whereas Table 3.12 is based on the Value of a Statistical Life (VOSL). This parameter is clearly dominant in determining the total health costs of environmental pollution. Using the VLYL concept we have mid-range total values of ECU 3.36 bn. for Germany, ECU 36.44 bn. for Italy, ECU 11.31 bn. for the Netherlands and ECU 25.32 bn. for the UK. In *per capita* terms for the year 1990 the figures are ECU 42 (Germany), ECU 632 (Italy), ECU 759 (Netherlands) and ECU 441 (UK). It must be emphasized that these figures are not comparable due the differences in impacts assessed.

In contrast, with the VOSL concept the corresponding mid-range values are: ECU 17.79 bn. (Germany), ECU 220.09 bn. (Italy), ECU 65.01 bn. (Netherlands) and ECU 121.47 bn. (UK). The VOSL figures are clearly implausibly high; they account for nearly 25% of GDP in Italy, for example. As we noted earlier, there may be some double counting which has to be accounted for. But most importantly, the use of VOSL really does look inappropriate for this kind of valuation.

The other findings on monetary valuation are:

a. Using VLYL, morbidity costs range from 29% of total costs in Italy to around 50% of total costs in the UK. Most significant morbidity costs arise from symptom days (SDs) and Restrictive Activity Days (RADs), both of which need to be investigated further. Other morbidity costs are not significant.

Table 3.10. Physical health impacts of air pollution in selected countries for 1990

Number of cases unless otherwise stated

	GERMANY (1)		ITALY		NETHERLANDS		UNITED KINGDOM	
	Mid value	Range	Mid value	Range	Mid value	Range	Mid value	Range
MORTALITY IMPACTS								
PM10 ACUTE	N.E.	N.E.	15,302	8653–21951	4,329	2329–6742	10,100	2200–18000
PM10 CHRONIC	N.E.	N.E.	52,191	39886–64495	14,645	10036–19667	24,500	8000–41000
SO2	4,928	2464–7392	6,773	6711–6835	912	N.E.	9,300	5600–13000
OZONE (2)	N.R.S.	N.R.S.	92	61–122	224	88–422	2,095	190–4000
TOTAL (3)	4,928	2464–7392	74,358	48539–94164	20,110	12453–26831	36,695	15990–63000
PER 1000 OF POPULATION	62	31–93	1,351	883–1713	1,350	836–1801	189	188–1145
MORBIDITY IMPACTS								
PM10 ACUTE								
Hosp. adm. for resp. inf.	N.E.	N.E.	2,304	1382–3226	1,041	565–1517	1,590	380–2800
Hosp. adm. for COPD	N.E.	N.E.	2,781	1795–3766	1,252	733–1771	1,850	500–3200
E.R.V. for COPD	N.E.	N.E.	8,762	6468–11055	3,920	2642–5197	5,650	1800–9500
E.R.V. for asthma	N.E.	N.E.	7,025	4460–9590	3,510	1822–5198	5,350	1200–9500
Hosp. visits for Croup	N.E.	N.E.	36,709	24311–49107	16,508	9929–23087	24,350	6700–42000
Adult RADs (000's)	N.E.	N.E.	68,060	35463–100657	25,269	11843–38695	48,500	10000–87000
Adult asthmatic attacks (000's)	N.E.	N.E.	9,722	546–18897	1,350	542–2158	6,050	1100–11000
Other symptom days (000's)	N.E.	N.E.	564,251	245343–883159	258,111	10170–415151	414,000	68000–760000
PM10 CHRONIC								
Adult chronic bronchitis (000's)	N.E.	N.E.	717	421–1013	317	168–465	460	110–810
Adult resp. illness (000's)	N.E.	N.E.	976	562–1390	430	223–637	620	140–1100
Child chronic bronchitis (000's)	N.E.	N.E.	323	151–495	166	71–262	325	59–590
Child chronic cough (000's)	N.E.	N.E.	417	184–650	215	86–345	426	72–780

OZONE ACUTE								
Hosp. adm. for resp. infection	N.R.S.	N.R.S.	548	411–653	1,556	683–2202	9,550	1100–18000
Hosp. adm. for COPD	N.R.S.	N.R.S.	399	250–548	1,132	415–2202	7,830	660–15000
Hosp. adm. for asthma	N.R.S.	N.R.S.	369	182–554	1,048	304–2228	7,740	480–15000
E.R.V. for asthma	N.R.S.	N.R.S.	1,700	1111–2262	4,827	1846–9090	32.450	2900–62000
Adult Minor RADs (000's)	N.R.S.	N.R.S.	1,009	0–3381	2,340	0–11104	46,000	0–92000
Adult asthmatic attacks (000's)	N.R.S.	N.R.S.	2,634	1652–3611	1,816	666–3523	37,050	3100–71000
Other symptom days (000's)	N.R.S.	N.R.S.	2,898	1476–4325	9,691	2888–20466	72,300	4600–140000
SO2 – ACUTE								
Hosp. adm. for chronic bronchitis	N.E.	N.E.	484	479–488	3,433	N.E.	26,400	800–52000
Adult chest discomfort days (000's)	18,500	9580–27090	N.E.	N.E.	N.E.	N.E.	N.E.	N.E.
E.R.Vs (000's)	10,750	6944–14349	767	753–779	413	N.E.	N.E.	N.E.

Notes:
NRS: Not reported separately.
NE: Not estimated or not available.
Mid values are not necessarily arithmetically calculated.

(1) Dose–response functions for Germany differ from those used in the other countries. For further details see the Germany country study.
(2) Ozone figures for Italy are for the Lombardy region only
(3) In adding up the impacts, the damages from SO2 are not added to those from PM10.

Table 3.11. Valuation of health impacts of air pollution in selected countries for 1990

Mortality impacts valued using value of life years lost (VLYL) (1)
ECU million unless otherwise stated (1995 prices)

		GERMANY			ITALY			NETHERLANDS			UNITED KINGDOM		
	Unit value	Mid value	Range		Mid value	Range		Mid value	Range		Mid value	Range	
MORTALITY IMPACTS													
PM10 ACUTE	172,500	N.E.	N.E.	N.E.	2,640	1493	3787	745	402	1163	1,742	380	3105
PM10 CHRONIC	414,000	N.E.	N.E.	N.E.	23,204	16513	29896	6,063	4155	8142	10,143	3312	16974
SO2 ACUTE	172,500	850	425	1275	1,130	1158	1101	157	N.E.	N.E.	1,604	966	2243
OZONE ACUTE (2)	172,500	N.R.S.	N.R.S.	N.R.S.	16	11	21	39	15	73	361	33	690
TOTAL (3)		850	425	1,275	25,860	18,016	33,704	6,847	4,572	9,378	12,247	3,724	20,769
AS % OF 1990 GNP	%	0.05%	0.03%	0.08%	2.90%	2.02%	3.78%	2.89%	1.93%	3.95%	1.44%	0.44%	2.44%
PER INHABITANT	ECU	11	5	16	448	312	584	460	307	629	213	65	362
MORBIDITY IMPACTS													
PM10 ACUTE													
Hosp. adm. for resp. inf.	7,870	N.E.	N.E.	N.E.	18	11	25	8	4	12	13	3	22
Hosp. adm. for COPD	7,870	N.E.	N.E.	N.E.	22	14	30	10	6	14	15	4	25
ERVs for COPD	223	N.E.	N.E.	N.E.	2	1	2	1	1	1	1	0	2
ERVs for asthma	223	N.E.	N.E.	N.E.	2	1	2	1	0	1	1	0	2
Hosp. visits for Croup	223	N.E.	N.E.	N.E.	8	5	11	4	2	5	5	1	9
Adult RADs	75	N.E.	N.E.	N.E.	5,105	2660	7549	1,895	888	2,902	3,638	750	6,525
Adult asthmatic attacks	37	N.E.	N.E.	N.E.	360	20	699	50	20	80	224	41	407
Other symptom days	8	N.E.	N.E.	N.E.	4,232	1840	6624	1,936	758	3,114	3,105	510	5,700
TOTAL		N.E.	N.E.	N.E.	9,748	4,553	14,943	3,904	1,680	6,129	7,001	1,310	12,693
PM10 CHRONIC													
Adult chronic bronchitis episodes	165	N.E.	N.E.	N.E.	118	70	167	52	28	77	76	18	134
Adult resp. illness episodes	165	N.E.	N.E.	N.E.	161	93	229	71	37	105	102	23	182
Child chronic bronchitis episodes	165	N.E.	N.E.	N.E.	53	25	82	27	12	43	54	10	97
Child chronic cough episodes	165	N.E.	N.E.	N.E.	69	30	107	35	14	57	70	12	129
TOTAL		N.E.	N.E.	N.E.	402	218	585	186	90	282	302	63	541

OZONE ACUTE													
Hosp. adm. for resp. inf.	7,870	N.R.S.	N.R.S.	N.R.S.	4	3	5	12	5	17	75	9	142
Hosp. adm. for COPD	7,870	N.R.S.	N.R.S.	N.R.S.	3	2	4	9	3	17	62	5	118
Hosp. adm. for asthma	7,870	N.R.S.	N.R.S.	N.R.S.	3	1	4	8	2	18	61	4	118
ERVs for asthma	223	N.R.S.	N.R.S.	N.R.S.	0	0	1	1	0	2	7	1	14
Adult Minor RADs	75	N.R.S.	N.R.S.	N.R.S.	127	0	254	176	0	833	3,450	0	6,900
Adult asthmatic attacks	37	N.R.S.	N.R.S.	N.R.S.	97	61	134	67	25	130	1,371	115	2,627
Other symptom days	8	N.R.S.	N.R.S.	N.R.S.	22	11	32	73	22	153	542	35	1,050
TOTAL		N.R.S.	N.R.S.	N.R.S.	257	79	434	346	58	1,171	5,568	167	10,969
SO2 – ACUTE													
Hosp. adm. for chronic bronchitis	7,870	N.E.	N.E.	N.E.	4	4	4	27	N.E.	N.E.	208	6	409
Adult chest discomfort days	8	138	72	203	N.E.	N.E.	N.E.	N.E.	N.E.	N.E.	N.E.	N.E.	N.E.
ERVs	223	2,374	1,549	3,200	171	168	174	0	N.E.	N.E.	N.E.	N.E.	N.E.
TOTAL		2,512	1,620	3,403	175	172	177	27	0	0	208	6	409
TOTAL FOR ALL MORBIDITY		2,512	1,620	3,403	10,580	5,021	16,140	4,463	1,828	7,582	13,079	1,546	24,612
AS % OF 1990 GNP		0.15%	0.10%	0.21%	1.19%	0.56%	1.81%	1.88%	0.77%	3.20%	1.54%	0.18%	2.90%
PER INHABITANT (ECU)		32	20	43	183	87	280	300	123	509	228	27	429
SUB TOTAL		3,362	2,045	4,678	36,440	23,037	49,843	11,310	6,400	16,960	25,326	5,271	45,381
AS % OF 1990 GNP		0.21%	0.13%	0.29%	4.08%	2.58%	5.58%	4.77%	2.70%	7.15%	2.98%	0.62%	5.34%
PER INHABITANT (ECU)		42	26	59	632	399	864	759	430	1,138	441	92	791

Notes:

NRS: Not reported separately, NE: Not estimated or not available.

Mid values are not necessarily aritmetically calculated (see country studies). GNP and population figures taken from World Development Report 1992. 1990 GNP converted into 1995 prices using EU Consumer Price Index and into ECU at a rate of $1 = ECU0.7608.

(1) For acute effects 1.25 years of life are assumed to be lost. For chronic effects 3 years are assumed lost. Each lost year is valued at ECU 138.000.

(2) Ozone figures for Italy are for the Lombardy region only.

(3) Damages from SO2 are not added to those from PM10.

Table 3.12. Valuation of health impacts of air pollution in selected countries for 1990

Mortality impacts valued using value of statistical life (VSL) (1)
ECU million unless otherwise stated (1995 prices)

		GERMANY (2)			ITALY			NETHERLANDS			UNITED KINGDOM		
	Unit value	Mid value	Range		Mid value	Range		Mid value	Range		Mid value	Range	
MORTALITY IMPACTS													
PM10 ACUTE	3,100,000	N.E.	N.E.	N.E.	47,436	26824	68048	14,060	7220	20900	31,310	6820	55800
PM10 CHRONIC	3,100,000	N.E.	N.E.	N.E.	161,791	123647	199935	46,040	31112	60968	75,950	24800	127100
SO2 ACUTE	3,100,000	15,277	7638	22915	20,299	20804	19794	2,827	N.E.	N.E.	28,830	17360	40300
OZONE ACUTE (3)	3,100,000	N.R.S.	N.R.S.	N.R.S.	284	189	378	791	273	1308	6.495	589	12400
TOTAL (4)		15,277	7,638	22,915	209,510	150,660	268,361	60,890	38,604	83,176	113,755	32,209	195,300
AS % OF 1990 GNP	%	0.94%	0.47%	1.40%	23.47%	16.88%	30.06%	25.67%	16.28%	35.07%	13.39%	3.79%	22.99%
PER INHABITANT	ECU	192	96	288	3,631	2,611	4,651	4,087	2,591	5,582	1,982	561	3,402
MORBIDITY IMPACTS													
PM10 ACUTE													
Hosp. adm. for resp. inf.	7.870	N.E.	N.E.	N.E.	18	11	25	8	4	12	13	3	22
Hosp. adm. for COPD	7,870	N.E.	N.E.	N.E.	22	14	30	10	6	14	15	4	25
ERVs for COPD	223	N.E.	N.E.	N.E.	2	1	2	1	1	1	1	0	2
ERVs for asthma	223	N.E.	N.E.	N.E.	2	1	2	1	0	1	1	0	2
Hosp. visits for Croup	223	N.E.	N.E.	N.E.	8	5	11	4	2	5	5	1	9
Adult RADs	75	N.E.	N.E.	N.E.	5,105	2660	7549	1,895	888	2,902	3,638	750	6,525
Adult asthmatic attacks	37	N.E.	N.E.	N.E.	360	20	699	50	20	80	224	41	407
Other symptom days	8	N.E.	N.E.	N.E.	4,232	1840	6624	1,936	758	3,114	3,105	510	5,700
TOTAL		N.E.	N.E.	N.E.	9,748	4,553	14,943	3,904	1,680	6,129	7,001	1,310	12,693
PM10 CHRONIC													
Adult chronic bronchitis episodes	165	N.E.	N.E.	N.E.	118	70	167	52	28	77	76	18	134
Adult resp. illness episodes	165	N.E.	N.E.	N.E.	161	93	229	71	37	105	102	23	182
Child chronic bronchitis episodes	165	N.E.	N.E.	N.E.	53	25	82	27	12	43	54	10	97
Child chronic cough episodes	165	N.E.	N.E.	N.E.	69	30	107	35	14	57	70	12	129
TOTAL		N.E.	N.E.	N.E.	402	218	585	186	90	282	302	63	541

OZONE ACUTE													
Hosp. adm. for resp. inf.	7,870	N.R.S.	N.R.S.	N.R.S.	4	3	5	12	5	17	75	9	142
Hosp. adm. for COPD	7,870	N.R.S.	N.R.S.	N.R.S.	3	2	4	9	3	17	62	5	118
Hosp. adm. for asthma	7,870	N.R.S.	N.R.S.	N.R.S.	3	1	4	8	2	18	61	4	118
ERVs for asthma	223	N.R.S.	N.R.S.	N.R.S.	0	0	1	1	0	2	7	1	14
Adult Minor RADs	75	N.R.S.	N.R.S.	N.R.S.	127	0	254	0	0	1	3	0	7
Adult asthmatic attacks	37	N.R.S.	N.R.S.	N.R.S.	97	61	134	0	0	0	1	0	3
Other symptom days	8	N.R.S.	N.R.S.	N.R.S.	22	11	32	0	0	0	1	0	1
TOTAL		N.R.S.	N.R.S.	N.R.S.	257	79	434	31	11	55	210	18	402
SO2 – ACUTE													
Hosp. adm. for chronic bronchitis	7,870	N.E.	N.E.	N.E.	4	4	4	N.E.	N.E.	N.E.	208	6	409
Adult chest discomfort days	8	138	72	203	N.E.	N.E.	N.E.	N.E.	N.E.	N.E.	N.E.	N.E.	N.E.
ERVs	223	2.380	1.560	3.200	171	169	173	N.E.	N.E.	N.E.	N.E.	N.E.	N.E.
TOTAL		2,517	1,632	3,403	175	173	176	N.E.	N.E.	N.E.	208	6	409
TOTAL FOR ALL MORBIDITY		2,517	1,632	3,403	10,582	5,023	16,138	4,121	1,782	6,466	7,721	1,397	14,045
AS % OF 1990 GNP		0.15%	0.10%	0.21%	1.18%	0.56%	1.81%	1.74%	0.75%	2.73%	0.91%	0.16%	1.65%
PER INHABITANT (ECU)		32	21	43	184	87	280	277	120	434	135	24	245
SUB TOTAL		17,794	9,270	26,318	220,092	155,683	284,499	65,011	40,386	89,642	121,476	33,606	209,345
AS % OF 1990 GNP		1.09%	0.57%	1.61%	24.75%	17.44%	31.87%	27.41%	17.03%	37.79%	14.30%	3.96%	24.64%
PER INHABITANT (ECU)		224	117	331	3814	2698	4931	4363	2710	6016	2116	585	3647

Notes:

NRS: Not reported separately.

NE: Not estimated or not available.

Mid values are not necessarily arithmetically calculated (see country studies). GNP and population figures taken from World Development Report 1992. 1990 GNP converted into 1995 prices using EU Consumer Price Index and into ECU at a rate of $1 = ECU 0.7608

(1) VSL taken as ECU 3.1 million.

(2) Dose–response functions for Germany differ from those used in the other countries. For further details see the Germany country study.

(3) Ozone figures for Italy are for the Lombardy region only.

(4) In adding up the impacts, the damages from SO2 are not added to those from PM10.

b. With VOSL, morbidity costs are around 5–6% of total health costs and therefore insignificant in the context of this kind of valuation.
c. We do not report the results of changes in the number of life years lost but, broadly, taking the upper bound of the range, will raise health costs 2.4 times for acute effects and 2 times for chronic effects. Taking the lower bound will lower costs to 7% of the reported values for acute effects and 33% for chronic effects.

3.7. Conclusions

This Chapter has reviewed the literature on the valuation of environmental health impacts. Such health impacts are probably the most important of all to value and also the most difficult conceptually. The Chapter began by looking at the methodological issues arising in the valuation. Impacts to be valued were divided into mortality, morbidity and accidents. For mortality impacts there are several issues that need to be resolved but, at the present time, a value of ECU 3.1 mn. in 1995 prices appears to be the best central estimate for VOSL. A range of problems that need to be addressed was identified. The most important is the fact that pollution related deaths involve a much smaller loss of life years than the deaths on which VOSL is based. Hence using VOSL results in a substantial overvaluation of health costs. The Value of Life Years Lost (VLYL) was discussed and an estimate of ECU 138,000 per year was proposed for this study. Pollution related deaths were divided into acute and chronic. It was further estimated that, for acute effects the loss of life years was about 1.25, whereas for chronic effects it was 3. Hence the central estimate of VLYL are ECU 172,500 for acute mortality impacts and ECU 414,000 for chronic mortality impacts.

For morbidity impacts heavy use of the US survey of this area was made, as the literature in Europe is very thin. Categories of impacts were divided into estimates of restrictive activity days (RAD) and variants of that; estimates of chronic illness, symptom-days, and altruistic impacts. For RAD and variants, US estimates were employed with conversion into ECU made using a purchasing power parity exchange rate. In addition, it was also suggested that a European cost of illness (COI) approach be used, with the COI value being grossed-up for items not captured by that method (i.e., pain and suffering). Estimates of chronic illness can be made via two routes. One is to use the COI and gross-up as indicated above. Another is to use an estimate of the value of statistical case, as estimated in the US and convert it in the manner indicated above. For symptom-days only a first attempt, based on US studies, of the estimates exists.

On morbidity, some initial valuation studies in Europe appear to support the figures taken from the US studies. There are, however, many gaps that need to be filled.

Environmental related accidents were not valued in this study, which is a gap that needs to be filled in future studies.

The results reported were fairly complete for the Netherlands and UK, with Italy having limited coverage of ozone but Germany having coverage of only SO_2 impacts. The results underscore the importance of shifting from VOSL to VLYL in the valuation of mortality. With the former, the estimates of damages are too high to be credible. With the latter they amount to 3–4.8% of GDP and ECU 441–759 *per capita*. Ranges for the estimates are from 92–399 (lower bound), to 864–1138 (upper bound).

In terms of morbidity further work is needed on chronic illness costs and estimates of chronic illness. At the same time the estimated cases of RAD and SD in all three studies (Italy, Netherlands and UK) looked very large. These should be checked carefully in future work.

References

Bates, D.V., Baker-Anderson, M. and Sizto, R. (1990) Asthma attack periodicity: a study of hospital emergency visits in Vancouver. Environmental Research, 51, 51–70.

Blomquist, G. (1979). Value of Life Saving: implications of Consumption Activity. Journal of Political Economy, Vol. 78, No. 3.

Burnett, R.T., Dales, R.E., Razienne, M.E., Krewski, D., Summers, P.W., Roberts, G.R., Raad-Young, M., Dann, T. and Brook, J. (1994) Effect of low ambient levels of ozone and sulfates on the frequency of respiratory admissions to Ontario hospitals. Environmental Research, 65, 172–194.

Cody, R.P., Weisel, C.P., Birnbaum, G. and Lioy, P.J. (1992) The effect of ozone associated with summertime photochemical smog on the frequency of asthma visits to hospital emergency departments. Environmental Research, 58, 184–194.

Dickie, M.S, Gerking, S., Schulze, W., Coulson, A. and Tashkin, D. (1986). Values of Symptoms of Ozone Exposure: An Application of the Averting Behavior Method. In: Improving Accuracy and Reducing Costs of Environmental Benefit Assessments. USEPA, Washington, DC.

Dickie, M.S., Gerking, S., Brookshire, D., Coursey, D., Schulze, W., Coulson, A. and Tashkin, D. (1987). Reconciling Averting Behavior and Contingent Valuation Benefit Estimates of Reducing Symptoms of Ozone Exposure Draft. In: Improving Accuracy and Reducing Costs of Environmental Benefit Assessments. USEPA, Washington, DC.

Dockery, D.W. et al. (1989) Effects of Inhalable Particles on Respiratory Health on Children. American Review of Respiratory Disease, 139, 587–594.

Dockery, D.W., Pope, C.A. III, Xu, X., Spengler, J.D., Ware, J.H., Fay, M.E., Ferris, Jr B.G. and Speizer, F.E. (1993) An Association Between Air Pollution and Mortality in Six US Cities. New England Journal of Medicine, 329, 1753–1759.

Dockery, D.W. and Pope, C.A. III (1994) Acute Respiratory Effects of Particulate Air Pollution. Annual Review of Public Health, 15, 107–132.

European Commission DGXII (1995a). Science, Research and Development Programme (JOULE). Externalities of Energy 'EXTERNE' Project. Summary Report EUR 16520 EN.

European Commission DGXII (1995b). Science, Research and Development Programme (JOULE). Externalities of Energy 'EXTERNE' Project. Methologology Report EUR 16521 EN.

Fisher, A., Chestnut, L.G. and Violette, D.M. (1989). The Value of Reducing Risks of Death: A Note on New Evidence. Journal of Policy Analysis and Management, 8 (1), pp. 88–100.

Frankel, M. (1979). Hazard, Opportunity and the Valuation of Life. Unpublished manuscript, Department of Economics, University of Illinois, Champaign-Urbana, IL.

Ghosh, D., Lees, D. and Seal, W. (1975). Optimal Motorway Speed and Some Valuations of Time and Life. Manchester School of Economic and Social Studies, Vol. 43, pp. 134–143.

Harrison, D. and Nichols, A.L. (1990). Benefits of the 1989 Air Quality Management Plan for the South Coast Air Basin: a reassessment. Prepared for the California Council of Environmental and Economic Balance by National Economic Associates Inc., Cambridge Mass.

Holguin, A.H., et al. (1984). The Effects of Ozone on Asthmatics in the Houston Area. In, Evaluation of the Scientific Basis for Ozone/Oxidants Standards (Lee, S.D., ed), Houston, Texas. Air Pollution Control Association, 1984, 262–280.

Johannesson, M. and Johansson, P-O. (1996). To Be or Not to Be; That is the Question: An Empirical Study of the WTP for an Increased Life Expectancy at an Advanced Age. Journal of Risk and Uncertainty, Vol. 13, pp. 163–174.

Jones-Lee, M.W. (1976). The Value of Life: An Economic Analysis. Chicago: University of Chicago Press.

Jones-Lee, M., Hammerton, M. and Philips, P. (1985). The Value of Transport Safety: Results of a National Sample Survey. The Economic Journal, 95, March, pp. 49–72.

Kinney, P.L., Ito, K. and Thurston, G.D. (1994) Daily mortality and air pollution in Los Angeles: an update (Abstract only). American Review of Respiratory and Critical Care Medicine; Supplement: p. A660.

Krupnick, A.J., Harrington, W. and Ostro, B. (1990): Ambient Ozone and Acute Health Effects: Evidence from Daily Data. Journal of Economic Management, 18, 1–18.

Krupnick, A.J. and Cropper, M.L. (1992). The Effect of Information on Health Risk Valuations. Journal of Risk and Uncertainty, 5 (1), pp. 29–48.

Litai, D. (1980). A Risk Comparison Methodology for the Assessment of Acceptable Risk. PhD Thesis, Massachussets Institute of Technology.

Litai, D., Lanning, D.D. and Rasmussen. N.C. (1983). The Public Perception of Risk, in V.T. Covello, W.G. Flamm, J.V. Rodricks, and R.G. Tardiff (eds.), The Analysis of Actual Versus Perceived Risks (New York, Plenum Press), pp. 213–224.

Maclean, A.D. (1979). The Value of Public Safety: Results of a Pilot Scale Survey. London Home Office Scientific Advisory Branch.

Maier, G., Gerking, S. and Weiss, P. (1989). The Economics of Traffic Accidents on Austrian Roads: Risk Lovers or Policy Deficit? Mimeo, Wirtschaftuniversitat, Vienna.

Marin, A. and Psacharopoulos, G. (1982). The Reward for Risk in the Labor Market: Evidence from the United Kingdom and a Reconciliation with Other Studies. Journal of Political Economy, Vol. 90, n. 4, pp. 827–853.

Melinek, S.J., Whoolley, S.K.D. and Baldwin, R. (1973). Analysis Questionnaire on Attitudes to Risk, Fire Research Note No. 962, Joint Fire Research Organization, Herts, England.

Moore, M.J. and Viscusi, W.K. (1988). The Quantity Adjusted Value of Life. Economic Inquiry, (26), pp 368–388.

Navrud, S. (1997a). Economic Valuation Estimates for the EXTERNE Transport Project. External Costs of Transport in EXTERNE. Task 4.4 Economic Valuation. European Commission – DGXII. Position Paper, March 1997.

Needleman, L. (1980). The Valuation of Changes in the Risk of Death by Those at Risk. University of Waterloo, Working Paper n. 103.

ORNL/RFF (1994). Estimating Fuel Cycle Externalities: Analytical Methods and Issues. Report No. 2 on the External Costs and Benefits of Fuel Cycles. A Study by the US Department of Energy and the Commission of the European Communitites.

Ostro, B.D. (1987) Air Pollution and Morbidity Revisited: a Specification Test. Journal of Environmental and Economic Management, 14, 7–98.

Ostro, B.D. and Rothschild, S. (1989) Air pollution and acute respiratory morbidity; an observational study of multiple pollutants. Environmental Research, 48, 238–247.

Ostro, B.D, Lipsett, M.J., Wiener, M.B. and Selner, J.C. (1991) Asthmatic Response to Acid Aerosols. American Journal of Public Health, 81, 694–702.

Pearce, D., Bann, C. and Georgiou, S. (1992). The Social Costs of Fuel Cycles. A report to the UK Department of Energy. The Center for Economic and Social Research into the Global Environment, (CSERGE), University College London and University of East Anglia.
Persson, U. (1989). The Value of Risk Reduction: Results of a Swedish Sample Survey. Mimeo. The Swedish Institute of Health Economics.
Pönkä, A. and Virtanen, M. (1994) Chronic bronchitis, emphysema, and low level air pollution in Helsinki, 1987–1989. Environmental Research, 65, 207–217.
Pope, C.A. III, Thun, M.J., Namboodri, M.M., Dockery, D.W., Evans, J.S., Speizer, F.E. and Heath, C.W. (1995) Particulate air pollution as a predictor of mortality in a prospective study of US adults. American Journal of Respiratory Critical Care Medicine, 151, 669–674.
Schwartz, J. (1993a) Air pollution and daily mortality in Birmingham, Alabhama. American Journal of Epidemiology, 137, 1136–1147.
Schwartz, J. (1993b) Particulate air pollution and chronic respiratory disease. Environmental Research, 62, 7–13.
Schwartz J. (1994) Air pollution and hospital admissions for the elderly in Detroit, Michigan. American Review of Respiratory and Critical Care Medicine, 150, 648–655.
Shephard, D.S. and Zeckhauser, R.J. (1982). Life-Cycle Consumption and the Willingnss to Pay for Increased Survival in M.W. Jones-Lee (ed.) (1982), The Value of Life and Safety, North Holland, Amsterdam.
Sunyer J., Sáez, M., Murillo, C., Castellsague, J., Martinez, F. and Antó, J.M. (1993) Air pollution and emergency room admissions for chronic obstructive pulmonary disease: a 5-year study. American Journal of Epidemiology, 137, 701–705.
Starr, C. (1976) General Philosophy of Risk Benefit Analysis. In Ashley, H., Rudman, R. and Whipple, C. (eds), Energy and the Environment: A Risk-Benefit Approach, Pergamon, Oxford.
Thurston, G.D., Kazuhiko, I., Hayer, C.G., Bates, D.V. and Lippmann, M. (1994). Respiratory Hospital Admissions and Summertime Haze Air Pollution in Toronto, Ontario: Consideration of the Role of Acid Aerosols. Environmental Research, 65, 271–290.
Touloumi, G., Pocock, S.J., Katsouyanni, K. and Trichopoulos, D. (1994) Short term effects of air pollution on daily mortality in Athens: a time-series analysis. International Journal of Epidemiology, 23, 957–967.
Veljanovski, C. (1978). The Economics of Job Safety Regulation: Theory and Evidence – Part 1 in the Market and Common Law. Mimeo. Center for Socio-Legal Studies.
Viscusi, W. Kip. (1989). Prospective Reference Theory: Toward an Explanation of the Paradoxes. Journal of Risk and Uncertainty, 2 (September), pp. 235–264.
Viscusi, W.K., Magat, W.A. and Forrest, A. (1988). Altruistic and Private Valuations of Risk Reductions. Journal of Policy Analysis and Management, Vol. 7, No. 2, pp. 227–245.

CHAPTER 4

NOISE

4.1. Introduction

There are several case studies regarding the impact of noise, covering primarily impacts from road and aircraft sources. There is also some discussion of industrial noise in Germany. The main European studies on noise valuation are:

- The Roskill Commission (1970)
- Pearce, Barde and Lambert (1984)
- Pennington *et al.* (1990)
- Oosterhuis and Van der Pligt (1985)
- Pommerehene (1986)
- Hoffman (1984)
- Larsen (1985)

4.2. Issues and methodology

The valuation of noise costs is largely based on the use of the hedonic pricing (HP) method, although some studies have also used the contingent valuation method (CVM). With HP, property prices are assumed to reflect the characteristics of the house as well as the neighbourhood, including factors such as noise. By carrying out an econometric analysis of house prices as a function of household characteristics (no of rooms, age size etc.), and as a function of neighbourhood and environmental characteristics, it is possible to estimate how much the price varies with the ambient level of noise.

In order for a study to be useful to the project, it must show results in terms of marginal noise changes (i.e., unit db(A) or Noise and Number Index (NNI) changes). The results are often presented in the form of a percentage reduction in housing value due to an NNI unit increase for aircraft, or a db(A) unit increase for road traffic. This will enable some 'transferability' to the final accounting framework, targeted by the project.

4.2.1. *Details of noise indexes and measures*

All sound measurements are expressed in decibels. Decibels are measured on a logarithmic scale to the base 10 where, say, 60 db is ten times as intense as

50 db (the psychological sensation of loudness roughly doubles with each 10 db increase in the intensity level). Ear sensitivity is greatest in the middle range of frequencies and so weighting scales are available that selectively discriminate against sounds of very high and very low frequency. The most commonly used weighting scale is the A-scale denoted by dBA {or db(A)}.

The subjective annoyance from noise is the result of the number of vehicles, vehicle type, road configuration and time of day. The cumulative noise distribution resulting from these various factors can be categorized as follows:

L_{90}: the background or residual noise level, the level in dbA that is exceeded 90% of the time.
L_{50}: the median or average noise level, the level in dbA that is exceeded 50% of the time.
L_{10}: the peak noise level, the level in dbA that is exceeded 10% of the time.

The background noise level during the daytime is approximately 45–55 dbA, with a 5 db reduction at night. A given noise is usually more annoying at night, partly because of the reduction in the reference noise level L_{90}. However the traffic noise indexes described below tend to be dominated by peak noise reflecting the logarithmic nature of the decibel unit.

For two residential properties that differ only in their level of exposure to noise, the absolute amount of housing depreciation per decibel can be defined as:

$$D = \frac{\text{difference in total value due to noise}}{\text{difference in noise exposure}}$$

Dividing D by the price of a basic house at a low or zero reference noise level gives the percentage rate of depreciation, or noise depreciation sensitivity index (NDSI), is defined as:

$$\begin{aligned}\text{NDSI} &= (D/\text{property value}) * 100 \\ &= \frac{\text{difference in total percentage depreciation}}{\text{difference in noise exposure}}\end{aligned}$$

4.3. Main results of past studies

4.3.1. *Aircraft noise*

The Roskill study (1970) was concerned with the impact of aircraft noise on property values It showed that the percentage depreciation in values varied according to the initial market value of the property, with larger properties showing a greater depreciation. This study also showed that if a property was rurally located, i.e., near Gatwick airport, some 30 miles from central London, it was more sensitive to noise than those properties located near to an urban airport, e.g., Heathrow which is about 15 miles from central London.

Property values affected by aircraft noise were observed in various noise contoured regions: 35–45 NNI contour, 45–55 NNI contour and those over 55 NNI contour (55 NNI is the highest noise contour and 30 NNI about the lowest for which any impact can be found). A sample of 20 real estate agents were asked to compare house price levels within different areas as specified on a large scale map of the areas in the vicinity of Heathrow and Gatwick airports. The Roskill Commission Research Team tested the real estate agents' data and carried out an analysis of actual house price transactions to find the relative rates of increase in prices of noisy and quiet property. The sample was chosen so that the properties had not undergone any substantial change from the time before the noise to the time after the noise shadow was imposed.

The study by Pearce, Barde and Lambert (1984) calculated 'total' house price depreciation in France for different noise levels in 1980. Their study was based on NDSI depreciation values as estimated by Nelson (1980), on the basis of HP studies in the US. Thus a complete transfer of estimated depreciation due to noise from the US to Europe was considered satisfactory for this purpose.

Following Nelson, the value of a property is affected by a noise level from 55 db through to 80 db (max). The total depreciation of the housing stock in 1980 due to traffic noise was estimated at FF 61.424 bn. (ECU 10.46 bn., 1980). At a 5% discount rate, this amounts to FF 0.92–1.85 bn. (ECU 156.7 mn.–315.2 mn., 1980) a year for 30 and 20 years respectively. At the time the annual expenditure on traffic noise abatement was FF 0.5 bn. (ECU 85.2 mn., 1980) per year. The authors note that this ratio between (potential) benefits and costs indicates very high social rates of return for noise control.

Details of European studies are given in Table 4.1. The NDSI (the percentage depreciation caused by a unit increase in the noise nuisance level) figure obtained by Pennington *et al.* (1990): 0.4–0.5% (first result) is very similar to those figures obtained in earlier American works, and reported in surveys carried out by Walters (1975), and Nelson (1980). Indeed a small but statistically significant noise effect of 0.4% per unit of aircraft noise measure is the average value found in the American literature.

The Pennington work is probably the strongest European study in terms of its large data set. Their data set is one of the largest ever used for a study of aircraft noise and included 3472 observations, all of which refer to actual market transactions. Information on housing attributes of Stockport was supplied by the Halifax Building Society covering the period April 1985 to March 1986. Noise contour maps, supplied by Manchester Airport, indicate the magnitude and annoyance of noise in different postcode areas. The contours of the map follow the Noise and Number Index (NNI), the most commonly used in the UK. For an area with a Noise Number Index figure of 40 or below, people

Table 4.1. Summary of European noise studies

Noise source	Noise level	Study/year	Method	Estimated property depreciation
Road/Industrial	55–65 db(A)	Oosterhuis & Van der Pligt	HP	D*fl* 400 per 1 db(A)
Road	30 db (A) 70 db(A)	Pommerehene (1986)	CVM	1% of rent 1.4% of rent
Aircraft	NNI unit increase	Pommerehene (1986)	CVM	0.2% of rent
Aircraft	NNI unit increase	Pennington *et al.* (1990)	HP	0.4–0.5% per db(A) 0.6% per db(A) in severe cases
Aircraft	35–45 NNI 55–65 NNI 55+ NNI	Roskill (1970)	HP	0–3% (Heathrow) 4.5–16.4% (Gatwick) 2.9–13.3% (Heathrow) 10.3–29.0% (Gatwick) 5.0–22.5% (Heathrow)
Aircraft	60 db(A)+	Hoffman (1984)	HP	1% per db(A)
Road	1000 ADT[1]	Larsen (1985)	HP	0.8% per db(A)

Notes:
ADT = average daily number of vehicles on the nearest road.
NNI: Noise and Number Index is defined as:
NNI = PN db + 15 log n − 80, where:
PN db = log average of peak noise levels of aircraft heard;
n = number of aircraft (in the UK on a summer day).
For details of the survey methods used in the design of the NNI, see McKennel (1963).

are 'moderately' affected by noise, whereas for the 45–50 NNI area (nearer the airport) people are 'very annoyed' by aircraft noise.

The study was split into two separate parts. The first involved research into five affected postcode areas; two of which are only slightly affected by aircraft noise due to their more rural location. Those properties most affected by noise command prices, on average, 3.7% lower than those elsewhere. A property in the worst affected area, was found to suffer a reduction in value of 6.09% compared with identical property in Stockport.

The second part of the work involved a very detailed analysis, where the number of bedrooms and living rooms entered as a series of zero/one dummy variables. The authors also used Market Analysis which involved assigning to each property an ACORN code value (A Classification Of Residential Neighborhoods). This coding is based upon some forty different census data variables and accounts for 38 neighborhood types of which 31 are represented in the subset of the author's Halifax data. The final results, including the very detailed ACORN analysis, imply that there is a "low negative, but weak and non-robust, relationship between aircraft noise and property value". They warn that any future research attempting to measure the effects of noise on house prices must include neighborhood and structural factors if estimates are to hold any kind of accuracy.

4.3.2. *Road traffic noise*

The Dutch study by Oosterhuis and Van der Pligt (1985), looked at house price depreciation due to road traffic noise and industrial noise, measured from 3 different levels, i.e., D*f* 50,000, D*f* 100,000 and D*f* 150,000 (because of difficulty of measuring exact prices). They estimated a reduction in property values of around D*f* 400 (ECU 188, 1993) per 1 db(A), which would indicate a depreciation of 0.4% per unit increase in noise for a house valued at D*f* 100,000 (ECU 46948, 1993).

Pommerehene (1986) examined the rent reductions due to excess noise in Basel, Switzerland. This study took account of air and road traffic noise affecting some 200 dwellings. A rent function was formulated, (DV = Rent), with 70 explanatory variables, including characteristics of the dwelling, the quarter and the environment. Two noise variables were considered, i.e., a road noise variable measure in decibels and a variable for aircraft noise measured in a noise and number index (NNI). The occupants of the dwellings were also asked for a WTP to have the present noise levels of their houses halved. The resemblance between the two sets of results was quite high although for aircraft noise, the 'observed' WTP was lower than the 'asked' WTP, which was contrary to expectations. For road noise depreciation was estimated at between 1 and 1.4% of the rent per unit increase in the noise level. Finally, Larsen (1985) estimated depreciation from road noise in Norway at 0.8% per db(A) increase. On the basis of the above studies, therefore, the defensible range of estimates appears to be 0.4%-1.4% property price depreciation per dB(A).

It is evident that most of the studies considered to date have a Northern European focus. To try and improve the estimates, Furlain & Jourdain (1995) conducted a survey of the post-1990 European literature, including three HP studies from Middle and Southern Europe. These new studies give values which are at the lower end of the above range. Vanio (1995) found that an increase in the traffic noise level of 1 dB(A) caused a 0.36% reduction in house prices in Helsinki, Finland.

One problem that should be noted about hedonic price studies, is that the values elicited for peoples' willingness to pay possibly includes other goods which are produced in conjunction with road traffic noise (air pollution, accident rates and road dust/soiling). The majority of HP studies fail to single out the effect of noise alone.

4.3.3. *Railroad noise*

The only known study in this area is by Vågnes and Strand (1996) who used distance from the railroad as a proxy for actual noise levels in Oslo, Norway. Using both the HP method and expert judgements they found a significant decrease in housing prices with greater proximity to the railroad line. The principal problem with such an approach is that noise levels do not only vary proportionally with distance but also depend on the location of dwellings and local topography (Navrud 1997a). In addition, estimates cannot be made in the conventional form (i.e.% property depreciation per db(A)) which limits their applicability in a study such as this. Hence, this category will not be considered further.

4.4. Contingent valuation studies

There is one CVM study of noise in Germany that is relevant to our work (Weinberger *et al.*, 1991). The study asked 6,500 households for their WTP for less noise, based on noise maps showing present and describing potential future levels of noise. The study found a significant WTP for noise levels above 43 dB(A), amounting to about ECU 0.84 per month, or 0.9% per dB(A). This is completely consistent with the hedonic literature, which is encouraging.

The difference, however, arose when the respondents were asked for WTP for an improvement to noise levels described as 'almost no noise' and 'little noise'. Since these levels are rather imprecisely specified, it is likely that responses will be biased due to 'commodity mis-specification'. A great deal of uncertainty is then introduced when these changes are then translated into dB(A) levels. The values obtained were significantly higher than those emerging from the HP studies, which carefully measure the benefits of reducing noise to some background level and apply that to the different categories of housing stock at different levels of noise exposure. The values generated by Weinburger *et al.* are approximately 20 times as large (see the German study for details). This difference warrants further analysis which has not been undertaken so far. There are two points which can explain some of the difference. First, there is the problem of the 'probability of provision bias'. People believe the noise levels will not be delivered so therefore state a high WTP. Second, the estimates probably also include people's WTP for the associated reductions in noise levels, air pollution and soiling that the reduction in road traffic noise (implicitly road traffic) would result in. In this study we report results based on both approaches for Germany.

Two other CVM studies have also been identified. Using the *annoyance cost approach*, Bye and Andersen (1996) found that residents in Oslo, Norway who were 'moderately' or 'highly' annoyed by noise elicited on average a WTP of ECU 425 and ECU 575 per household, per year respectively. As mentioned earlier, the WTP is not just for noise but other goods as well, therefore it is difficult to solely estimate the value changes for road traffic noise alone.

Narvud (1997a) carried out a survey to focus solely on road noise. This survey asked for people's WTP to obtain a noise absorbing road layer in the street where they live, which would reduce the noise level to a level where their would not be 'annoyed' by traffic noise inside their homes. It was stressed that other parameters (e.g. air pollution) would remain the same. Results from the study underline the fact that different peoples perception of noise level is subjective. In a random sample of the Norwegian population, 3% of the 997 respondents were 'highly' annoyed by road traffic noise indoors and 13.3% said they were 'somewhat' annoyed. The remaining 83.7% said they were not annoyed by the noise but more than ¼ of them said they could hear the traffic noise indoors. The average annual WTP for the total sample was ECU 14 per household. Correcting for protest 'zero' answers this increases to ECU 24.

4.5. Defensive expenditures

For noise, the defensive expenditures are a very important category. Unfortunately they have not been estimated for all countries. Detailed estimates of expenditures on noise protection windows, and increased insulation in buildings is available for Germany and the Netherlands but not for Italy or the UK. We expect to compile such data as part of GARP II for the latter two countries.

4.6. Values selected for the study

The most important source of noise from energy development and generation is undoubtedly road noise. It is concluded that a linear noise sensitivity index (NSDI) with a range of 0.4–1.4% per dB(A) (L_{Aeq}) and an average of around 0.9% are the best values for use in this study. The results are fairly consistent, both within Europe and including Europe and the US, which suggests that the transfer of benefits may well be feasible for noise valuation in the fuel cycles. For aircraft noise the European and US studies indicate a fairly robust value of around 0.45% per unit of NNI.

In this study, the actual values taken for the UK study noise depreciation were those given above. For the Netherlands, however, the team took the Dutch studies as being most relevant and applied a depreciation factor of 0.4% for road and rail noise per dB(A), a value it also applied for aircraft noise, having converted the aircraft noise index values back into dB(A). For Germany, the team reports the results of taking values between 0.5% and 1.26% per dB(A) for road and rail noise, and between 0.3% and 1.2% for aircraft noise.

As reported above, for Germany, the results from the hedonic study were reported alongside those from the CVM study.

In order to value the damages a baseline level has to be selected. For Germany this was taken as 45 dB(A). In the UK a range of values were taken, from 50 to 60 dB(A), with the best estimate based on 55 dB(A). The Netherlands team also took a value of 55 dB(A), but did not carry out a sensitivity analysis. As the results below show, lowering the threshold to 50 from 55 can increase

Table 4.2. Valuation of noise impacts in selected countries for 1990

ECU million unless otherwise stated (1995 prices)

		GERMANY			ITALY			NETHERLANDS			UNITED KINGDOM		
	Unit value	Mid Value	Range		Mid value	Range		Mid value (2)	Range		Mid value	Range	
ROAD NOISE (1)													
City Road								208					
Highway traffic								33					
TOTAL (3)		593	337	848	N.A.	N.A.	N.A.	241	N.A.	N.A.	4.833	1.208	12.566
AS % OF 1990 GNP	%	0.04%	0.02%	0.05%	N.A.	N.A.	N.A.	0.10%	N.A.	N.A.	0.57%	0.14%	1.48%
PER INHABITANT	ECU	7	4	11	N.A.	N.A.	N.A.	16	N.A.	N.A.	84	21	219
AIR TRAFFIC NOISE (4)													
		147	58.60255	236	N.A.	N.A.	N.A.	82	N.A.	N.A.	N.A.	N.A.	N.A.
AS % OF 1990 GNP	%	0.01%	0.00%	0.01%	N.A.	N.A.	N.A.	0.03%	N.A.	N.A.	N.A.	N.A.	N.A.
PER INHABITANT	ECU	2	1	3	N.A.	N.A.	N.A.	6	N.A.	N.A.	N.A.	N.A.	N.A.
RAIL TRAFFIC NOISE													
	N.A.	N.A.	N.A.	N.A.	N.A.	N.A.	31	N.A.	N.A.	N.A.	N.A.	N.A.	
AS % OF 1990 GNP	%	N.A.	N.A.	N.A.	N.A.	N.A.	N.A.	0.01%	N.A.	N.A.	N.A.	N.A.	N.A.
PER INHABITANT	ECU	N.A.	N.A.	N.A.	N.A.	N.A.	N.A.	2	N.A.	N.A.	N.A.	N.A.	N.A.
SUB TOTAL		740	396	1.084	N.A.	N.A.	N.A.	354	N.A.	N.A.	4.833	1.208	12.566
AS % OF 1990 GNP	%	0.05%	0.02%	0.07%	N.A.	N.A.	N.A.	0.15%	N.A.	N.A.	0.57%	0.14%	1.48%
PER INHABITANT	ECU	9	5	14	N.A.	N.A.	N.A.	24	N.A.	N.A.	84	21	219
CONTINGENT VALUATION METHOD		13.895	13.291	14500	N.A.	N.A.	N.A.	0	N.A.	N.A.	N.A.	N.A.	N.A.
AS % OF 1990 GNP	%	0.85%	0.81%	0.89%	N.A.	N.A.	N.A.	0	N.A.	N.A.	N.A.	N.A.	N.A.
PER INHABITANT	ECU	175	167	182	N.A.	N.A.	N.A.	0	N.A.	N.A.	N.A.	N.A.	N.A.
DEFENSIVE EXPENDITURES													
		4.369	3545.1522	5192	N.A.	N.A.	N.A.	1716	N.A.	N.A.	N.A.	N.A.	N.A.
AS % OF 1990 GNP	%	0.27%	0.22%	0.32%	N.A.	N.A.	N.A.	0.72%	N.A.	N.A.	N.A.	N.A.	N.A.
PER INHABITANT	ECU	55	45	65	N.A.	N.A.	N.A.	115	N.A.	N.A.	N.A.	N.A.	N.A.
ESTIMATED AVERAGE NOISE DAMAGE FROM ALL METHODS													
FINAL SUB TOTAL		11.686	10.389	12.984	N.A.	N.A.	N.A.	2.070	N.A.	N.A.	4.833	1.208	12.566
AS % OF 1990 GNP	%	0.72%	0.64%	0.80%	N.A.	N.A.	N.A.	0.87%	N.A.	N.A.	0.57%	0.14%	1.48%
PER INHABITANT	ECU	147	131	163	N.A.	N.A.	N.A.	139	N.A.	N.A.	84	21	219

Notes:

N.A. = Not available in this subclassification.

(1) Estimated damages are 0.5–1.26% Per dB(A) in Germany, 0.9% Per NDSI in UK and 0.4% Per NDSI in Netherlands.
(2) This was the single total damage estimate from the study used.
(3) For the UK the figure is not a straight average but one based on the central value of the background level.
(4) Hedonic 0.3–1.2% Per dB (Germany). Netherlands converts noise index KE into DB(A) and applies 0.4% per dB(A).

total costs 2.5 times, and raising it to 60 dB(A) can decrease it by a factor of 4. Further work is needed, therefore, on agreed baseline values.

4.7. Results from the studies

The results from the studies are given in Table 4.2 (page 71). Noise damages were only reported for Germany, Netherlands and the UK. The main findings are as follows:

i. Using the hedonic pricing method, mid estimates of depreciation per unit of noise from all sources considered range from 0.05% (Germany) to 0.57% (UK), as a percentage of 1990 GDP. Differences are partly due to different depreciation values but may also be due to coverage of the property affected. This issue needs to be looked at more closely. In *per capita* terms the corresponding range of costs are ECU 9–84.
ii. The German CVM study comes up with *much* larger values of noise damages based on the CVM method. Total damages amount to around ECU 13 bn., compared to only ECU 740 mn. using the hedonic a method. We have not been able to resolve the reasons for this difference. The CVM method may be better at picking up overall benefits of noise reduction (for example, the hedonic method cannot pick up the benefits of reduced noise in some public areas). In part, the underestimation in the hedonic pricing could also be rooted in an underestimation of the number of affected households (see also p. 51 of the German country report). In Table 4.2 we report both sets of results and, for Germany, we take the average of the two.
iii. Defensive expenditures are significant, ranging from ECU 55 per person in Germany to ECU 115 in the Netherlands. In the both countries they are around five or six times the estimates of hedonic expenditures (damages). Although it is correct to add them to the hedonic estimates, it may not be appropriate to add them to the CVM estimates
iv. The overall results suggest that we need to (a) resolve the discrepancy between the hedonic approach and the CVM approach, and (b) obtain more comprehensive estimates of defensive expenditures for all countries.

References

Bye and Andersen (1996). Economic Valuation of Impacts from Road Traffic. M.Sc thesis (in Norwegian). Departent of Economics and Social Sciences, Agricultural University of Norway.

Furlan, S. and Jourdain, E. (1996). Review of Some Recent European Empirical Studies on the Valuation of Externalities. External Costs of Transport in ExternE. Task 4.4 – Economic Valuation. Institut Francais du Petrole, Paris.

Hoffman (1984). In Navrud, S. (1991) op cit.

Larsen (1985). In Navrud, S. (1991) op cit.

McKennel, A.C. (1963). Aircraft Noise around London (Heathrow) Airport, C.O.I., London.
Navrud, S. (1991). Summary of Norwegian Valuation Studies Relevant for Economic Valuation of External Effects of Fuel Cycles. Noragric, Agricultural University of Norway.
Navrud, S. (1997a). Economic Valuation Estimates for the ExternE Transport Project. External Costs of Transport in ExternE. Task 4.4 Economic Valuation. European Commission – DGXII. Position Paper, March 1997.
Navrud, S. (1997b). A New Approach to Valuing Road Traffic Noise. Department of Economics and Social Sciences, Agricultural University of Norway. Note of May 1997.
Nelson, J.P. (1980). Airports and Property Values: A Survey of Recent Evidence. Journal of Transport Economics and Policy, 14: 37–51.
Oosterhuis, F. and Pligt, van der (1985). Kosten en Baten van de Wet Geluidshiner. Commissie Evaluatie Wet Geluidshiner. CW-AS-06.
Pearce, D., Barde, J-Ph. and Lambert, J. (1984). Estimating the Cost of Noise Pollution in France. Ambio, 13 (1), pp. 27–28.
Pennington, G., Topham, N. and Ward, R. (January 1990). Aircraft Noise and Residential Property Values Adjacent to Manchester International Airport. Journal of Transport Economics and Policy, pp. 49–59.
Pommerehene, W. (1986). Der Monetäre Wert Einer Flug und Strasseenlärmreduktion: Eine Empirische Analyse auf der Grundlage Individueller Präferenzen. In: Kosten der Umweltverschmutzung, Umweltbundesamt, Berichte 7/86.
Roskill Commission (1970). Report of the Commission on the Third London Airport. HMSO.
Walters, A.A. (1975). Noise and Prices. Oxford: Oxford University Press.
Weinberger, M., Thomassen, H.G. and Willeke, R. (1990). Kosten des Lärms in der Bundesrepublik Deutschland: Forschungsbericht 10103110/05 im Auftrag des Umweltbundesamtes, Berlin.

CHAPTER 5

CROPS

5.1. Introduction

In this impact category there are several European studies[1] which have established dose–response functions for the effect of air pollutants on crops. We do not discuss them in detail here but refer the reader to the individual country studies.

In this chapter we discuss the issues arising in the estimation of the crops damages. These are:

(i) Price changes and adaptation to different pollution levels by farmers.
(ii) Valuation of damages, choice of gross or net damages.
(iii) Selection of prices for the valuation.
(iv) Use of defensive expenditures.

5.2. Price changes and other adaptations

In valuing crop damages, the estimates will depend on how much prices are affected by the changes in concentrations. The earliest crop damage estimate studies simply multiplied the reduction in output attributable to the pollutant (according to the biological dose–response relationship) by the market price. This only gives a partial equilibrium analysis and can only be taken as a rough estimate. It ignores behavioral responses or price changes which would be included in a general equilibrium analysis.

Indeed, in the context of higher pollution levels, producers are likely to switch to those crop varieties which can give a better resistance to the air pollutants in the region. Alternatively, the farmers may alter the use of other inputs such as calcium carbonate, which mitigates the pollution impact. In relation to price changes, a fall in the yield as a result of pollution will generally mean less output at a higher market price. Consumers will lose when pollution

[1] Although the studies are cited above as 'European', they use several estimates of dose–response relationships that have been established in the US. This is particularly true for ozone-related damages, which are based on a Weibull function estimated as part of the US National Crop Loss Assessment Network (NCLAN) project.

increases if the market prices are at all sensitive to output. This was shown clearly in the European studies referred to above. In addition to these studies, US studies have also shown the importance of the price effects, where the percentage of the total loss attributable to consumers ranged from 50% (Adams, Hamilton and McCarl, 1986) to 100% (Adams and McCarl, 1985). Given that producers can sometimes gain from an increase in air pollution, not including the consumer impacts can overestimate total losses and misrepresent the distribution of welfare effects. Producers, on the other hand, will lose or gain depending on the price elasticities of their particular crops. As an approximation, if the percentage reduction in quantity is greater (less) than the percentage increase in price then the producer will lose (gain). Shortle, Dunn and Philips (1986) and Adams and McCarl (1985) found that substantial increases in ozone increased the economic rents of corn belt growers. An American study by Adams, Crocker and Thanavibulchai (1982) found that using the simple multiplication approach gave valuation estimates 20% higher than if price and behavioral changes were taken into account. The authors used mathematical programming (consistent within the framework of standard microeconomic theory concerning firm and market behavior), which is able to account more completely for the adaptations economic agents make to mitigate or exploit the opportunities of environmental damage.

As far as the present study is concerned the question of importance therefore is to what extent the prices will change as a result of the changes in concentrations of pollution? To answer this question, it would be necessary to construct a model of demand and supply for the different products and, based on elasticities, calculate the resulting damages from the present levels of air pollution. It was not possible undertake such a model development within the time available. Hence the results reported are for damages *assuming* that prices are unaffected by the pollution levels. This probably results in an overestimate of damages but, given the small amount of damages anyway, it did not seem important to give priority to this refinement at this stage. In subsequent work we will refine the analysis, time permitting.

5.3. Selection of gross or net values

The next issue is whether one should take the change in value, net of costs of production, or the gross value. Ideally what one wants to measure is the change in the value of the rents derived from the production. In the case where there is a change in an exogenous factor such as the deposition of pollutants, it can be shown that the change in the net income after deducting the costs of inputs is given by the *gross value* of the change in production. This conclusion is dependent on the assumptions that (a) the producers are price takers, (b) there is no change in the prices of the inputs or outputs and, related to that, (c) the changes in the quantities are marginal. The formal demonstration of this result is given in Annex 1 to this chapter. In the case of overall pollution damages, the assumption of changes being marginal may not always hold, although in most cases we are looking at changes in yields of only 1–3%.

5.4. Selection of prices

For many of the crops that are affected, the prices are not determined by market forces but are influenced by the Common Agricultural Policy. In such cases, it has been argued by many economists that the changes in the value of the output should be valued at international or border prices and not the actual prices that are prevailing in the market in question (see, for example, Squire and van der Tak, 1975). The view taken here is that it is appropriate to use border prices as long as the reason for the divergence between the domestic and international price is not the correction of some externality. For major crops produced in the EU there is no direct externality involved and it is therefore appropriate to use the international price in valuing the changes. The teams worked on the basis of international prices, but these appear to vary not insignificantly from country to country. For example the price of 'wheat' was taken as (ECU/ton): 107 (Italy), 150 (Netherlands), 96 (UK and Germany). Differences are partly due to different categories of wheat, and to differences in c.i.f. costs. Nevertheless, this issue needs further investigation.

5.5. Use of defensive expenditures

The analysis for the UK looked at 'defensive' expenditures for crops in two areas: the benefits of nitrogen oxide depositions, the damages from soil acidification. For the former, it concluded that there were benefits in the fertilization and these were valued in terms of the cost of fertilizer, taking the view that the application of nitrogen in fertilizer was reduced by that amount. At the margin this is not too bad an assumption to make. The benefits appear to be of the order of ECU 28 mn. for the UK.

For acidification the estimates were based on the additional calcium carbonate required to neutralize the deposition of H^+. This will result in an overestimate of damages in so far as full neutralization is not required, or lower cost adjustments are possible. Given the small cost involved, however, the use of neutralization costs is acceptable.

5.6. Main results

The results obtained reflect different coverage in terms of crops. In part this reflects the differences in what is important, but in part they reflect incomplete coverage. To the maximum extent possible, we will try and address this in GARP II. The main gaps in coverage are for ozone damages. In both Italy and Germany it was not possible to estimate damages, due to lack of concentrations data. From the data for other countries, damages in the range of ECU 1–14 per person were estimated.

The main results are summarised in Table 5.1. The following points should be noted:

i. Overall damages are small, even when coverage is fairly complete, as in the case of the UK. For that country they amount to 0.005% of GDP, or under 1 ECU per person. The estimates are much higher for the Netherlands (0.089% of GDP, or ECU 14 per person). The limited coverage in Italy and Germany results in lower total damage figures, but the SO_2 figures are high. The UK figures are, it should be noted, particularly low.

ii. There was wide agreement on the selection of dose response functions, although it was not complete. In assessing SO_2 impacts, all teams adopted the same dose response functions for wheat, barley and oats, peas and beans. For potatoes the UK and Italian dose response functions differed, and the Italian team looked at damages to tomatoes and apples, which were not considered by any other team. For ozone impacts the UK team relied on the NCLAN functions, whereas the Dutch team used those as well as other estimates from Swiss and Dutch studies. For ozone damages for the Netherlands, the team also relied on another study (RIVM, 1989) to fill the gaps on estimates it could not make.

iii. Compared to other studies, these figures are somewhat low but not completely out of line. For example, the study by van der Eerden (1988) for the Netherlands valued the effects of air pollution – i.e., ozone (O_3), sulphur dioxide (SO_2), and hydrogen fluoride (HF) – upon crop yields (14 species), and was based on 1983 data. All air quality parameters were considered to occur in sufficiently high concentrations to cause adverse effects upon crop plants. The dose–response functions were chosen so as to assess the nationwide damage most accurately; each crop species was identified as either tolerant, moderately sensitive or sensitive. The study found that if air pollution were reduced to baseline levels, producer surplus would only decrease by ECU 38 mn. implying that volume declines were offset by price changes. Consumers, however, could receive a net loss of as much ECU 365 mn. at the nationwide level, due to a 5% drop in yield of sensitive crops (all figures converted to ECU 1995 prices). Of the 5% drop in crop yield, ozone accounted for 3.4% of the loss, SO_2 for 1.2% and hydrogen fluoride for 0.4%. Thus, total damages would be ECU 403 mn., compared to ECU 212 mn. estimated in these calculations. This reflects partly the selection of more up to date dose response functions.

Table 5.1. Valuation of crop impacts in selected countries for 1990

ECU million unless otherwise stated (1995 prices) (1)

	GERMANY			ITALY			NETHERLANDS			UNITED KINGDOM		
	Mid value	Range		Mid value	Range		Mid value	Range		Mid value	Range	
Damages due to SO2 (2)												
Wheat	31.8	12.2	51.6	34.9	17.0	46.4	2.2	0.8	3.6	15.1	15.1	15.1
Barley	17.8	6.8	28.8	N.E.	N.E.	N.E.	0.4	0.1	0.7	5.9	5.9	5.9
Rye	14.2	5.4	23.0	N.E.	N.E.	N.E.	0.0	0.0	0.0	N.E.	N.E.	N.E.
Oats	2.4	0.9	3.9	N.E.	N.E.	N.E.	N.E.	N.E.	N.E.	0.5	0.5	0.5
Potato	N.E.	N.E.	N.E.	4.9	4.9	4.9	N.E.	N.E.	N.E.	N.E.	N.E.	N.E.
Tomato	N.E.	N.E.	N.E.	13.3	13.3	13.3	N.E.	N.E.	N.E.	N.E.	N.E.	N.E.
Green Peas and beans (3)	N.E.	N.E.	N.E.	N.E.	N.E.	N.E.	0.4	0.4	0.5	0.7	0.7	0.7
Apple	N.E.	N.E.	N.E.	9.7	9.7	9.7	N.E.	N.E.	N.E.	N.E.	N.E.	N.E.
TOTAL	66.2	25.3	107.3	62.8	44.9	74.3	3.0	1.3	4.8	22.2	22.2	22.2
AS % OF 1990 GNP	0.004%	0.002%	0.007%	0.007%	0.005%	0.008%	0.001%	0.001%	0.002%	0.003%	0.003%	0.003%
PER INHABITANT (ECU)	0.8	0.3	1.3	1.1	0.8	1.3	0.2	0.1	0.3	0.4	0.4	0.4
Damages due to Ozone (4)												
Winterwheat	N.E.	N.E.	N.E.	N.E.	N.E.	N.E.	6.0	3.0	9.0	14.1	14.1	14.1
Peas	N.E.	N.E.	N.E.	N.E.	N.E.	N.E.	0.7	0.2	1.1	0.8	0.8	0.8
Potatoes	N.E.	N.E.	N.E.	N.E.	N.E.	N.E.	41.1	41.1	41.1	25.1	25.1	25.1
Other crops (4)	N.E.	N.E.	N.E.	N.E.	N.E.	N.E.	161.2	164.7	157.8	6.3	6.3	6.3
TOTAL	N.E.	N.E.	N.E.	N.E.	N.E.	N.E.	209.0	209.0	209.0	46.4	46.4	46.4
AS % OF 1990 GNP	N.E.	N.E.	N.E.	N.E.	N.E.	N.E.	0.088%	0.088%	0.088%	0.005%	0.005%	0.005%
PER INHABITANT (ECU)	N.E.	N.E.	N.E.	N.E.	N.E.	N.E.	14.0	14.0	14.0	0.8	0.8	0.8
Defensive Expenditures (5)												
Value of Nitrogen Inputs via NOx	N.E.	N.E.	N.E.	N.E.	N.E.	N.E.	N.E.	N.E.	N.E.	−27.8	−27.8	−27.8
Calcium carbonates	N.E.	N.E.	N.E.	N.E.	N.E.	N.E.	N.E.	N.E.	N.E.	1.7	1.7	1.7
TOTAL	N.E.	N.E.	N.E.	N.E.	N.E.	N.E.	N.E.	N.E.	N.E.	−26.1	−26.1	−26.1
AS % OF 1990 GNP	N.E.	N.E.	N.E.	N.E.	N.E.	N.E.	N.E.	N.E.	N.E.	−0.003%	−0.003%	−0.003%
PER INHABITANT (ECU)	N.E.	N.E.	N.E.	N.E.	N.E.	N.E.	N.E.	N.E.	N.E.	−0.45	−0.45	−0.45
FINAL SUB TOTAL	66	25	107	63	45	74	212	210	214	43	43	43
AS % OF 1990 GNP	0.004%	0.002%	0.007%	0.007%	0.005%	0.008%	0.089%	0.089%	0.090%	0.005%	0.005%	0.005%
PER INHABITANT (ECU)	0.8	0.3	1.3	1.1	0.8	1.3	14.2	14.1	14.3	0.7	0.7	0.7

Notes:

(1) Valuation of damages has used international prices.

(2) Mid values for Germany and Italy are not averages but 'mid' values as given in the report.

(3) Figures are for peas in the Netherlands and beans in the UK.

(4) For the Netherlands, the RIVM study estimated total damage at ECU 209 mn. (1995 prices).

iv. Similarly, one can compare the German figures obtained this study with other estimates. Heinz (1980) in his estimate of agricultural damage in Germany due to air pollution assumed that the yield of most sensitive plants fell by 10% and less sensitive plants by only 5%. Highly resilient species were assumed to be unaffected. He estimates total agricultural damages to be DM 125 mn. (ECU 47.2 mn., 1977 prices). This is broadly comparable to the figures reported here, although we should note that our coverage is much less complete, suggesting that our total damages would be higher. Wicke, 1986 argues that Heinz has made a serious underestimation and puts agricultural damages at a lowest limit of DM 1 bn. (ECU 377.4 mn., 1977), which is justified by the fact that, contrary to Heinz's assumptions, vegetative damages are not restricted to polluted areas. Long range transport of pollutants is also damaging regions which are not directly influenced by industrial activity. Based on the Dutch estimates, it is possible that, including ozone damages, the German estimate could be as high as ECU 700 mn.

In conclusion therefore, we can say that, although crop damages have been only partially covered, the full estimates are not going to dominate the overall calculations of environmental damages. Improvements to be made include more consistent use of dose response functions, wider coverage of crops, and allowance for changes in prices resulting from the present pollution levels.

References

Adams, R.M. and McCarl, B. (1985). Assessing the Benefits of Alternative Oxidant Standards on Agriculture: The Role of Response Information. Journal of Environmental Economics and Management, 12: 264–276.

Adams, R.M., Crocker, T.D. and Thanavibulchai, N. (1982). An Economic Assessment of Air Pollution Damages to Selected Annual Crops in Southern California. Journal of Environmental Economics and Management, 9: 42–58.

Adams, R.M., Hamilton, S.A. and McCarl, B.A. (1985). An Assessment of the Economic Effects of Ozone on US Agriculture. Journal of the Air Pollution Control Association, 35: 938–943.

Heinz, I. (1980). in: Kosten der Umweltverschmutzung. Symposium Report. BMU (Bundesministerium für Umwelt, Naturschutz und Reaktorischereit) Bundesumweltministerium/Erich Schmidt Verlag, Berlin.

RIVM (1989). Concern for Tomorrow. Bilthoven. National Institute for Public Health and Environmental Protection (RIVM).

Shortle, S., Dunn, J.W. and Philips, M. (1986). Economic Assessment of Crop Damage Due to Air Pollution: The Role of Quality Effects. Staff Paper 118. Department of Agricultural Economics, Pennsylvania State University, State College, PA.

van der Eerden et al. (1988). Crop Loss Due to Air Pollution in the Netherlands. Environmental Pollution, 53: 365–376.

Wicke, L. (1986). Die Ökologiscen Milliarden. Das kostet die zerstörte umwelt – so können wir sie retten. München.

ANNEX TO CHAPTER 5

Valuing marginal damages derived from a dose–response function

If (a) a producer is a profit maximizer and price taker, (b) pollution changes result in marginal changes in outputs and inputs, and (c) prices are not affected by the changes in inputs and outputs, then the appropriate measure of the value of the change in damage following the change in pollution level is the change in value of gross output.

Suppose the production function for a single output y is given by

$$y = F(\mathbf{X}, Z)$$

Prices of inputs are given by the price vector P, and the price of output is given by s. The pollution level influences the output directly but not the inputs (labor, fertilizer etc.).

The profit function for the producer is given by

$$\Pi = \Phi(s, P, Z)$$

We are interested in the change in profit that results from the change in Z, i.e., in

$$\partial\Pi/\partial Z$$

By the definition of the profit function

$$\Pi = \max[s \cdot y - P \cdot \mathbf{X}] = \max[s \cdot F(\mathbf{X}, Z) - P \cdot \mathbf{X}]$$

which gives

$$\partial\Pi/\partial Z = s \cdot (\partial F/\partial Z) - P \cdot (\partial\mathbf{X}/\partial Z) = s \cdot (\partial F/\partial Z)$$

The last equation follows from the envelope theorem.

The last expression is of course the gross value of the change in yield. It should be noted that an implicit assumption is that inputs are not directly affected by the pollution.

CHAPTER 6

FORESTS

6.1. Introduction

Depositions of sulphur dioxide and nitrous oxides have become a major threat to the health and survival of forests. The valuation of this damage is, unfortunately, handicapped by uncertainties over the relative contribution of air pollution compared to other factors. There are many European studies of interest in this damage category which include the estimation of commercial timber loss due to air pollution and also the loss in the recreational value of forestry. The major studies include:

- Nilsson *et al.* (1992)
- Kuylenstierna and Chadwick (1994)
- NME (1988)
- Ewers *et al.* (1986)
- Linden and Oosterhuis (1988)
- Navrud (1990)
- The Austrian Ministry for the Environment (no date)
- Hoen and Winter (1991)
- Kristrom (1988)
- Grayson *et al.* (1975)
- Everett (1979)
- Willis and Benson (1989)

6.2. Review of existing studies and issues arising

There are several data limitations in this impact category, similar to those for crops. Despite the large number of European studies, virtually none of them contain details regarding SO_2 deposits in a localized scenario. There are many difficulties in actually translating the physical impacts into monetary damages. A forest has many functions including those of providing timber, other non-timber products, ensuring a habitat for animal and plant species and acting as a natural buffer against environmental shocks (e.g., global warming). It is thus not unreasonable to suggest that a significant portion of the total value of a forest is from non-market values.

Quantification of forest damage is further complicated by issues of transferability. It is unlikely that estimated valuations will be applicable or transferable to other locations since the blame and causation of forest damage tends to be very 'site specific'. All physical impacts and valuations will vary according to location. The categories of damages are commercial timber, recreation and non-use functions. In this section attention is concentrated on the first two, with non-use functions looked at in the next Chapter.

Commercial timber is valued by combining information on pollution-induced physical changes in the resource with details of market prices. The chief pollutants which concern forests are those carried in the air: SO_2, NO_x and ozone.

In the case of recreational damage, transferability to other regions or countries is much more limited than for many other categories of damages. The marginal value of forests as a recreational asset may be very different according to local demand and supply conditions. For example, a positive willingness to pay has been estimated in Sweden for the preservation of open agricultural land (open space) in what is a densely forested area; the marginal value of forests was therefore negative. Such a result would be very unlikely in say, the Netherlands, where forests are more scarce and their recreational values have been shown to be largely positive.

Almost all recreational damage valuations rely on the contingent valuation method, in which the marginal impact of pollution on recreation values is very rarely assessed. The valuations obtained are thus average values. Furthermore, there are good reasons to think that the answers given to such questions express a wider concern for the effects of air pollution, not just forest damage (i.e. they 'embed' other values). This makes it incorrect to attribute the damages of the air pollution to forests alone, and adding the 'forest' damages to other damages involves double counting.

6.3. Main results of other studies

Unfortunately, almost all the studies lack data on the baseline levels of pollutants and the land area of forest surveyed. Consequently, for the vast majority of studies no attempt can be made to establish an external marginal valuation for forestry based on per tonne of pollutant. NME (1988) calculated losses over 25–30 years from present production levels in Norway on the assumption that current ozone and acid depositions continue. Average timber losses were NOK 1 bn. (1988) or ECU 140,000 per year starting 25–30 years from now. This study also used expert judgements on the rate of loss of tree growth, a practice that is quite common in this area of damage.

Another Norwegian study is that by Hoen and Winter (1991) who established a total value (both use and non-use) of multiple-use forestry and preservation of virgin coniferous forests in Norway. They found a mean annual WTP per household (in different sub-samples totaling 1204 persons) of ECU 16–46 (for multiple-use forestry) and a median of ECU 6–12. For the preservation of virgin forests a mean value of ECU 26–36 was established with a median of ECU 12–15. Most of the estimated value was due to the recreational value of the forest. Results show that the WTP for multiple-use forestry is (significantly) dependent on the experimental design in different sub-samples. This is not the case for the preservation values and so the authors note that this may be because 'multiple-use forestry' is too unfamiliar and vague a concept to respondents.

Linden and Oosterhuis (1988) looked at the continued damage by acid rain to 2010 in the Netherlands. If no appropriate abatement policy is adopted they assumed that this will lead to 80% of forest becoming damaged and 90% of heath becoming grassy vegetation. On the basis of these assumptions, they estimated timber losses at DFL 13.1 mn. per year (ECU 5.62 mn., 1988). This is the expected loss if no measures against acid deposition–emission are taken. On the recreational side they estimated that there is a WTP per household per month of DFL 22.83 (ECU 10.2, 1990) to maintain current conditions and not allow the damages referred to above to take place. Extrapolated to the Dutch population this amounts to DFL 4.28 bn. (ECU 1.84 bn., 1990) per year (which is a large but plausible damage figure given other studies). Significant variables included income (level and changes), the perception of the acid rain problem, number of visits, age, education and social class.

Ewers *et al.* (1986) focused on the recreational value of a forest and examined the rate of forest loss in West Germany over the period 1984–2060 (using different air pollution reduction scenarios over time). The study established a WTP per visiting hour per person of DM 4.87 (ECU 2.55, 1993) instead of assessing a standard 'visitor' day. This could well lead to an overestimation of damages since recreational experience is believed not to increase at a constant rate with the number of hours spent. In addition, this cannot be associated with a measured pollutant deposition.

The figures are, however, valuable in estimating recreational loss of value due to forest damages. In overall terms he estimated damages resulting from recreation losses to be between ECU 1.15 bn. and ECU 3.15 bn., if damage is restricted by the achievement of the clean air targets as set out in 1986. Since these targets have been broadly met, damages can be said to be in the above range. The German team made use of these figures in estimating recreational losses from forestry.

Navrud (1990) looked at the recreational value of mountainous forest. Respondents were confronted with the proposal that the forest could be subject to 3 different management practices: clear cutting, selection forestry and no cutting (preservation). However, this does not relate to reduced recreational facilities resulting from air pollution.

The Austrian Ministry for the Environment made a national evaluation of forest decline attributable to air pollution. They established a total estimated forest loss at AS 4.5 bn. per year (ECU 0.22 bn., 1983). 85% of the forestry losses came from items other than commercial timber.

A Swedish study by Kriström (1988) studied the benefits of preserving selected regions of virgin forests, i.e., preventing commercial forestry. This established a WTP (open) of SEK 1014 (ECU 155.0, 1990) and a WTP (closed) of SEK 1005–2741 (ECU 153.6–419.0, 1990). Thus for 3 million Swedish households this gave a total WTP of SEK 3 bn.–8.2 bn. (ECU 458 mn.–1.25 bn., 1990). The WTP question was posed in a 'closed' format and an 'open' format. The closed format involves a model that describes the probability that the suggested bid is lower or higher than the person's actual WTP given in the conventional open format.

There are three major UK studies which estimated the recreational value of national forests. They show a consumer surplus per hectare ranging from around ECU 52/ha. to ECU 154/ha. (Willis and Benson, 1989). As none of these figures can be related to losses of recreation value resulting from increases in forest damage from air pollution, they are of no use in the context of this study.

One study which has been applied at the Europe-wide level is that by Nilsson *et al.* (1992). It was carried out at the International Institute For Applied Systems Analysis (IIASA). Further details of the models used in the analysis are given in the coal report. One of the critical sources of data was the IIASA database which gave a forest inventory for Europe. This was combined with the RAINS model, which provided data on emissions of acid rain in each country, including the source of the pollution (see below). Thus Nilsson was able to estimate the acid deposition, and to identify which forest areas and which species were affected. He combined this information with dose-response functions from an East German model, PEMU, to estimate the reduced yields in forest growth.

The estimates produced by Nilsson were reviewed by the ExternE team, as well as by the present team, and found to have some serious problems. Given its importance in the framework of the European studies, however, the results of applying it to the four countries have been reported in Table 6.1 below.

Finally, this study has looked at the model developed by Kuylenstierna and Chadwick (1994). The details are reported in the UK country study, but essentially it is based on a damage assessment of forests resulting from the increased aluminium concentrations that are caused by the acidification and that affect the ability of trees to take up nutrients and water. Its advantages are transparency and clarity, even though it has a number of gaps, a point recognized by the authors. The results of this study are also reported in this chapter.

6.4. Defensive expenditures

There are a number of defensive expenditures associated with forest damage. These have been investigated for Germany but unfortunately not for the other countries in this study. The German study (Kroth *et al.*, 1989) include damage

Table 6.1. Economic valuation of forest impacts in selected countries for 1990

ECU million unless otherwise stated (1995 prices)

	GERMANY			ITALY			NETHERLANDS			UNITED KINGDOM		
	Mid value	Range		Mid value	Range		Mid value	Range		Mid value	Range	
Valuation of Timber Loss (1)												
Based on IIASA Forest Study (2)	725.0	725.0	725.0	199.4	199.4	199.4	12.1	12.1	12.1	217.5	217.5	217.5
Based on KCM (3)	59.2	26.6	93.0	2.8	1.2	4.3	0.0	0.0	0.1	1.8	0.8	2.9
AS % OF 1990 GNP	0.004	0.002	0.006	0.000	0.000	0.000	0.000	0.000	0.000	0.000	0.000	0.000
PER INHABITANT (ECU)	0.74	0.33	1.17	0.05	0.02	0.08	0.00	0.00	0.00	0.03	0.01	0.05
Valuation of Recreational Loss (4)				N.A.	N.A.	N.A.				N.A.	N.A.	N.A.
Assuming clean air targets are met (DE)	2597.8	1389.5	3806.1	N.A.	N.A.	N.A.				N.A.	N.A.	N.A.
WTP to avoid further damage (NL)				N.A.	N.A.	N.A.	2227.3	2227.3	2227.3	N.A.	N.A.	N.A.
TOTAL	2597.8	1389.5	3806.1	N.A.	N.A.	N.A.	2227.3	2227.3	2227.3			
AS % OF 1990 GNP	0.159	0.156	1.605	N.A.	N.A.	N.A.	0.939	0.939	0.939	N.A.	N.A.	N.A.
PER INHABITANT (ECU)	32.7	17.5	47.9	N.A.	N.A.	N.A.	149.5	149.5	149.5	N.A.	N.A.	N.A.
Defensive Expenditures												
TOTAL	93.1	49.8	136.4	N.A.	N.A.	N.A.	N.A.	N.A.	N.A.	N.A.	N.A.	N.A.
AS % OF 1990 GNP	0.006	0.003	0.008	N.A.	N.A.	N.A.	N.A.	N.A.	N.A.	N.A.	N.A.	N.A.
PER INHABITANT (ECU)	1.2	0.6	1.7	N.A.	N.A.	N.A.	N.A.	N.A.	N.A.	N.A.	N.A.	N.A.
SUB TOTAL	2750.1	1465.9	4035.5	2.8	1.2	4.3	2227.3	2227.3	2227.4	1.8	0.8	2.9
AS % OF 1990 GNP	0.17	0.09	0.25	0.00	0.00	0.00	0.94	0.94	0.94	0.00	0.00	0.00
PER INHABITANT (ECU)	34.6	18.4	50.8	0.0	0.0	0.1	149.5	149.5	149.5	0.0	0.0	0.1

Notes:

(1) Figures for mid values are not averages but 'mid values' as given in the country studies.

(2) Nilsson *et al.*, 1992.

(3) Kuylenstierna and Chadwick Model (1994); estimates used in this study.

(4) CVM based values, assuming that clean air targets are realised.

German figures are from Ewers *et al.* (1986). Dutch figures are from Linden and Oosterhuis (1988).

protection through liming, reforestation of forest stands, cultivation of underwood, and other measures. The range of total costs for Germany is ECU 50 mn. to ECU 136 mn. These costs have to be added to the damage costs discussed above and further work is needed for Italy, Netherlands and UK to obtain these estimates.

6.5. Main results

The main results of the study on forestry can be seen in Table 6.1. There are several gaps, the most prominent being the lack of data on recreation losses for Italy and the UK. The figures for Germany and the Netherlands in this regard must be regarded as preliminary, as the CVM results need to be validated by more careful studies. Nevertheless the figures are of some interest. They suggest that:

(i) There is a big difference in the valuation of timber losses between the KCM and the Nilsson studies (about two orders of magnitude for Italy, Netherlands and the UK). It is large, but less so for Germany. The issue of which values to take has yet to be resolved but in our opinion, the KCM model is closer to the truth than the IIASA model.

(ii) Per capita damages from recreational–amenity losses have been estimated at ECU 33 for Germany and ECU 150 for the Netherlands, which is quite a wide range. More careful work is needed before we can understand why the range is so large. The Dutch figure was not included in the team report, but has been included here, to provide some idea of the range of values being obtained in related studies.

(iii) Defensive expenditures are available for Germany and amount to ECU 1.2 per person or ECU 10/ha of forest land. Similar estimates are needed for the other countries.

(iv) Total damages are dominated by losses of recreational value, and amount to about ECU 2.75 bn. for Germany (0.17% of 1990 GNP) or ECU 35 per capita. If the Dutch recreational costs are of the right order, costs there could be as high as ECU 150 per capita.

References

Everett, R.D. (1979). The Monetary Value of the Recreational Benefits of Wildlife. Journal of Environmental Management, 9:.203–213.

Ewers, H. et al. (1986). On the Monetarization of Forest Damages in the Federal Republic of Germany. In: Kosten der Umweltverschmutzung, Umweltbundesamt, Berichte, July, pp. 121–143.

Grayson, A.J., Sidaway, R.M. and Thompson, F.P. (1975). Some Aspects of Recreational Planning in the Forestry Commission, in Searle, G.A.C. (ed.), Recreational Economics and Analysis. London: Longman.

Hoehn, H.F. and Winter, G. (1991). Attitudes to Willingness To Pay for Multiple Use Forestry and

Preservation of Coniferous Forests in Norway. Draft, Department of Forestry, Agricultural University of Norway.

Kriström, B. (1988). On the Benefits of Preserving Virgin Forests. Paper prepared for the Conference on Multiple Use Forests, Economics and Policy, Oslo.

Kroth, W., Grottker, Th. and Bartelheimer, P. (1989) Forstwirtschaftliche Massnahmen bei neuartigen Waldschäden – Nutzen – Kosten – Untersuchung forstwirtschaftlicher Massnahmen in durch Luftschadstoffe geschädigten oder gefährdeten Waldbeständen. Schriften reihe des Bundesministers für Ernährung, Landwirtschaft und Forsten, Reihe A: Ange Wandte Wissenschaft, Heft 368, Münster-Hiltrup: Landwirtschaftsverlag.

Kuylenstierna, J. and Chadwick, M. (1994). Research Report on Modelling Forest Damage for Air Pollution. mimeo, Submitted to the ExternE project.

Linden, J. Van Der, and Oosterhuis, F. (1988). De Maatschappelijke Waardering voor de Vitaiteit van Bos en Heide. Ministry of Public Housing, Physical Planning and Environmental Management: Leidschendam.

Navrud, S. (1990) Cost-Benefit Analysis of Liming River Audna: An Extended Analysis, Report from the Directorate for Nature Management. In Press.

Nilsson, S., Sallnas, O. and Duinker, P. (1992). Forest Potentials and Policy Implications: A Summary of a Study of Eastern and Western European Forests by the International Institute for Applied Systems Analysis. Executive Report 17, IIASA, Laxenberg, Austria.

NME (Norwegian Ministry of Environment) (1988). Forest Damage Scenarios: How Pollution Might Affect Norwegian Forests in a 25–30 Years Perspective. Report T-907, Norwegian Ministry of Environment (NME), Oslo.

Willis, K. and Benson, J. (1989). Values of User Benefits of Forest Recreation: Some Further Site Surveys. Report to the Forestry Commission, Department of Town and Country Planning, University of Newcastle, Newcastle.

CHAPTER 7

ECOSYSTEMS

7.1. Introduction

This chapter covers the impacts of economic activity on natural terrestrial ecosystems, and on water related environments. Natural terrestrial ecosystems are affected by gaseous pollutants SO_2, NO_X, NH_3 and O_3. The changes in vegetation can have consequent effects on the associated vertebrate and invertebrate faunas and on soils. Indirect effects result from the deposited load of the pollutants S and N and their compounds. Acidic deposition resulting from the burning of fossil fuels can impact indirectly on various water bodies. Water chemistry, which is affected by direct deposition to the surface of the body, is one factor which influences the status of fish stocks. The relationship between deposition and water chemistry, and between water chemistry and fish survival have been explored by the UK teams in this study.

The assessment of the impacts in these areas have all been in physical terms. The main problem has been the establishment of reliable and generalizable dose–response functions linking gaseous pollutants to the 'endpoints' that we would wish to value, such as plants, fauna etc. The approach adopted has been to establish critical values for specific pollutants (see below). These critical levels tend to be applied to broad natural vegetation groups rather than to individual species or plant communities. They are being refined and revised as knowledge increases. The impact of current, ambient levels of the various pollutants can be assessed by overlaying spatially disaggregated data of pollutant concentrations on spatially disaggregated information regarding the occurrence of relevant receptors to determine the proportion of the area of the stock at risk for which the critical level is exceeded. Damage is assumed to occur in areas where the critical load for a given receptor is exceeded. This methodology does not, however, provide any quantification of the magnitude of the impact on, for example, growth or flowering. It should also be noted that there is little or no data currently available for valuation of air pollution impacts on ecosystems (European Commission, 1995).

For fisheries, linkages have been established between stream classes and fish stocks, now and in pre-industrial times. This allows one to estimate the loss of fish stocks but the valuation of this loss has not been undertaken as yet. The reasons for this are discussed in greater detail below.

This chapter begins by discussing the critical loads data and its usefulness in the context of an environmental accounting exercise such as this. It then goes on to review the evidence on valuation of natural and terrestrial ecosystems and water ecosystems. Finally it concludes with a discussion of the future directions for impact valuation in this area.

7.2. Critical loads assessment

The UK study established pollutant data on one and five km. grids for SO_2, NO_X, NH_3 (mean annual concentration) and for 20 km grids for O_3 (total ppb hours above 40 ppb). The stock at risk was defined using the ITE Land Cover map, based on satellite imagery and included the area of semi-natural vegetation, defined as semi-natural grassland, herbaceous semi-natural vegetation and deciduous forest. Critical levels for SO_2, NO_X, NH_3 were taken from the UN-ECE Manual on 'Methodologies and Criteria for Mapping Critical Levels/Loads and Geographical Areas Where They are Exceeded' (Federal Environment Agency, 1993) For ozone they were taken from Fuhrer and Ackerman (1994). Combining the two sources of information allowed the team to establish maps giving the areas where critical loads for different pollutants were exceeded, and how much of the land in each area consists of semi-natural vegetation. For further details see the UK report.

Having established areas where critical loads are exceeded the problem is one of establishing the *impact* of the exceedence on the eco-systems, and thereby establishing monetary values for these impacts. So far this has not proved possible. Progress on valuation can be made in three ways. One is to establish the impacts in physical terms and then go on to the valuation stage; the second is to value the excedance directly, in terms of individual WTP; and the third is to estimate the costs of remediation. All these options are considered further in Sections 7.3.–7.5.

In the Italian study, there is an assessment of pollutant concentrations in key areas, and of the prevailing vegetation. No attempt has been made to establish areas where critical loads are exceeded. Preliminary data on forest areas is being collected and forest decay is documented but there is no assessment of what the pollutant loadings in the different areas are. The Italian study also provides lists of endangered fauna and flora.

The Netherlands study does not cover ecosystems damages.

The German study also supports the critical loads approach but provides no details of areas where ecosystems are suffering from critical load excedance. Like the Italian study it provides lists of fauna and flora which are endangered (with 4 categories of endangerment).

7.3. Valuation of impacts on ecosystems: use of dose response functions combined with CVM methods

As noted above the valuation of ecosystem impacts can proceed either by establishing a dose–response function linking the pollutant to a physical endpoint, and then valuing the damage to the endpoint; or by valuing an acknowledged damage to some endpoint using the CVM method, and then attributing part of that damage to the pollutants of interest; or by estimating the costs of remediation. We begin by considering valuation of damages to natural and semi-natural eco-systems based on the first approach, and then go on to consider the application of, and the potential for, the other two approaches.

The main areas where dose response functions have been used are: valuation of commercial fisheries, and valuation of recreational fisheries. Each of these is discussed further below.

7.3.1. *Commercial fisheries*

The following studies here have a limited use for estimating marginal damages for commercial fisheries:

- Ewers and Schulz (1982)
- Rasmussen *et al.* (1991)

Ewers and Schulz (1982), in a study on the Lake Tegeler, West Germany (see also non-ecological use of surface water) measured the level of commercial catch as affected by reduced phosphorus emissions. The pollution levels used in the study made no reference to actual physical measures of phosphorus. The authors only mentioned an annual producer surplus increase of DM 12,400–DM 44,000 which they said was gained if pollution improved from level 4 up to level 1 (cleanest). Thus, the link between the phosphorous levels and the water quality indicators needs to be established. However, such a link is not difficult to make.

The study by Rasmussen *et al.* (1991), includes the effects on deep sea fishing, coastal fishing in the North and Baltic Seas, river fishing, lake fishing and aquaculture (ponds) in West Germany over the period 1950–1982. They estimated an annual loss of DFL 130 mn. (ECU 56 mn., 1987). Total social costs were estimated. This involved taking into account the fact that marginal costs decrease with increased harvests and so there is an additional penalty imposed because of the smaller catch. The estimated damages came to DFL 245 mn. (ECU 105 mn., 1987). Thus, this figure is found by taking the total lost income, adding the additional cost mentioned above and subtracting the reduction in total expenditures. Again the study gives no indication of pollution levels and, in this case does not even identify levels of improvement in water quality. Hence its transferability to this project is even more limited.

Table 7.1. Maximum annual WTP of Norwegian households for an improvement in the freshwater fish populations [WTP-fish] from reduced European sulphur emissions (1986 NOK)

Subsample number (emission reductions)	No. observations	Response rate fish/public goods	First bid	WTP–Fish/HH/year Mean	Med	Std dev
1	288	98/97	200	278	200	338
2	270	97/95	500	455	200	826
3	238	99/96	1000	603	300	772
4	239	97/99	0–10000 (p.c)	335	100	514
5	206	99/96	200	366	200	660
6	206	98/98	500	578	200	1036
7	204	97/97	1000	597	200	796
8	192	98/98	0–10000 (p.c)	291	100	492
9	189	98/97	0–10000	387	200	648

Sub-samples 1–4 were asked for a WTP to reduce sulphur by 30%.
Sub-samples 5–8 were asked for a WTP to reduce sulphur by 50%.
Sub-sample no. 9 was asked for a WTP to reduce sulphur by 70%.

(p.c) = payment card, where 3 of the sub-samples were shown amounts ranging from 0 to NOK 10,100 and asked to pick the value that reflected their WTP.

Furthermore, the historical nature of data raises difficulties in separating effects of pollution from effects of fishing techniques and policies.

Both these studies represent one form of valuation which needs to be pursued if monetary damage estimation in the framework of environmental accounting is to advance. To date, all water related damages have not been covered due to a lack of time.

7.3.2. *Non-commercial fisheries*

One study that makes use of dose response functions in relation to non-commercial fisheries is that by Navrud (1989) in Norway. He examined the water quality affecting fish (Brown Trout stock) populations from 1980. Respondents were asked to reveal their WTP for various intensities of lime application to achieve 30%, 50% and 70% reductions of sulphur effects at all sites. Respondents were also questioned regarding quality of all public goods affected by acid rain. This was used to control against 'constant budget bias', but the results are themselves useful in indicating the WTP for an improvement in water quality that has been damaged by acid rain.

The results are summarised in Table 7.1. This study could be of potential value in this study as it contains marginal valuations of sulphur emissions, both in terms of NOK/fish/year as well as NOK/HH/year. It also provided careful estimates of changes in fish stocks, based on dose–response functions, something which other studies do not always do. One problem with this study is that, in spite of this careful estimation of impacts, he finds, as with the Travers Morgan study (see below), very little increase in WTP for increased sulphur reductions. Given the large expected physical impacts of such reductions, it suggests that there is still a problem in eliciting marginal valuations.

Such a valuation method can be applied to the countries in this study but it does need considerable data collection. First we would need to know the pollution doses in different water bodies. Second, in the case of non-commercial fisheries, we would have to estimate the WTP for reductions in the pollutant. The method does require the respondent to understand the impacts of the pollutants on the aspects of the environment that matter: fish stocks, for example. The Navrud study says little about how much is known about this linkage. The values obtained here may also be transferable to other countries in the EU. This question also needs further investigation, although there is some evidence to suggest that transferability in this area may be limited (see European Commission, 1995).

7.4. Valuation of impacts on ecosystems: direct use CVM and other methods

Contingent valuation (CVM) is a potentially important tool for valuing ecosystem damage or loss of biodiversity, although there are many problems that need to be addressed. The main one is that we have no way to validate the results of an experimentally-created market. Some techniques are available to validate the results of a CVM approach (comparing the results, for example, with travel cost or hedonic approaches) but there is no obvious way to apply them to the estimated willingness to pay for conservation.

CVM methods have been used to value environmental amenities and damages in the following areas:

(a) Loss of species
(b) Loss of amenity (not water related)
(c) Loss of water related recreational benefits
(d) Loss of non-use benefits in relation to water bodies.

7.4.1. *Loss of species*

Table 7.2 below, adapted from Pearce *et al.* (1991), shows the results of CVM surveys of studies valuing endangered or rare species. The results are interesting because of their broad consistency. Most of the valuations tend to cluster around ECU 7, (1990 prices) if we exclude the relatively high value for humpback whales. The range is ECU 1.4–15.4 excluding humpback whales and ECU 1.4–42.0 including humpback whales. The results suggest that:

Table 7.2. Per capita preference valuations for endangered species and prized habitats

	(US 1990 $ per annum per person) (ECU 1990 per annum per person)	$ value	ECU value
SPECIES			
Norway	Brown bear, wolf and wolverine	15.0	12.3
USA	Bald eagle	12.4	10.2
	Emerald shiner	4.5	3.6
	Grizzly bear	18.5	15.3
	Bighorn sheep	8.6	7.0
	Whooping crane	1.2	1.0
	Blue whale	7.5	6.2
	Bottlenose dolphin	5.4	4.5
	California sea otter	6.0	4.9
	Northern elephant seal	8.1	6.7
	Humpback whales[1]	40–48	32.9–39.5
		49–64	40.4–52.7
HABITAT			
USA	Grand Canyon (visibility)	27.0	22.3
	Colorado wilderness	2.7–6.0	2.2–4.9
Australia	Nadgee Nature Reserve NSW	28.1	23.1
	Kakadu conservation	40.0	32.9
	Zone, NT[2]	93.0	76.6
UK	Nature reserves[3]	40.0	32.9
	Flow country	28.6	23.5
Norway	Conservation of rivers against hydroelectric development	59.0–107.0	48.6–88.1

Source: Pearce *et al.* (1991), converted to ECU at £1 = 1.4 ECU.

(a) they are not large proportions of respondent income, and
(b) habitat appears to be more highly valued than species, which is to be expected since a wider array of benefits is being secured through conservation of habitat than through targeting individual species.

The transferability of per capita values of this kind is limited. The values in one country are very much a function of local factors and it is inappropriate to take them out of context. This issue is discussed further below where some systematic studies of transferability are reported. However, the conclusion that *per capita* values for different species and different locations are not generally transferable remains. The relevance to our study is limited, because it is difficult to establish a link between various socio-economic activities and damage to species. Nevertheless it is something to be pursued in future work.

7.4.2. *CVM studies of loss of amenity*

Table 7.2 also reports some studies that have valued habitant in special places, such as the Grand Canyon, nature reserves etc. Such values are only useful in the context of our study if we can establish a link to the destruction of the habitat from specific pollutants. In many cases that is not possible. Furthermore, the transferability of the values (in terms of damages per person) is extremely limited.

More directly, studies are now being initiated, which look at the WTP for damages due to acid deposition. Ecotec (1994) is one such contingent valuation study, which was performed to value a reduced threat of acid deposition damage to upland vegetation in the UK. Results are specific to the exact case in question, and the degree to which they can be extrapolated more generally is being investigated. The work does demonstrate, however, that valuation work on pollution impacts to natural ecosystems is possible.

Valuation of loss of habitat due to project development

One related area of valuation that could be relevant for our purposes is that of loss of habitat due to development projects. One such study is that by Hervik *et al.* (1987), who interviewed people regarding the conservation of rivers against the development of hydroelectric power plants. They found a WTP of US$ 59–107 (ECU 53.2–96.5, 1990) per person per annum. This latter result could be of use in valuing the loss of habitat and eco-system function from hydropower development. Although that is not currently being looked at it is an area for further work.

Another study that looks at loss of habitat in this context is that by Travers Morgan (1991). Using CVM methods, it estimated non-user values for wildlife and habitat in the Mersey Estuary. The construction of a barrage to capture tidal energy in the estuary would have adversely affected surrounding wildlife. Some non-user values were estimated by questioning people in Bristol and Sheffield, which were considered to be far enough away not to represent use values. A WTP was elicited from a sample of 300 which, if extrapolated across the UK, would give a value of ECU 146 mn. The extension of the results across the UK must be problematic and the embedding problem (i.e. that people are expressing a general value for loss of wildlife) remains. Finally, as noted earlier, such a study is not transferable outside this context and the use of CVM for non-use values is still subject to much criticism.

7.4.3. *Loss of water related recreational benefits*

Water related benefits that have been valued using the CVM method, and that could be of relevance to environmental accounting are provided in the following studies:

- Hjalte *et al.* (1982)
- Kanerva and Matikainen (1972)

- Green and Tunstall (1990a/b); WRC/FHRC (1989)
- Magnussen and Navrud (1991) / Magnussen (1991)
- Hervik, Risnes and Strand (1987); Strand and Wenstop (1991).
- Mäntymaa (1991)
- Aarskog (1988)
- Heiberg and Hem (1988)
- Dalgard (1989)

Hjalte *et al.* (1982), in Lake Vombsjon, Southern Sweden (drinking water source of the town called Malmo) estimated the recreational value per visitor per year at SEK 4 (1982), on average for all recreational activities. WTP = SEK 4.50 for angling alone. (1 SEK = ECU 0.13). Variables measuring water quality were in terms of angling, bathing, boating, bird watching, walking and lakeside view. This study used a theoretically acceptable travel cost model, e.g. taking account of substitute sites, varying on-site costs and not over stating time costs (i.e. use 15% of the average industrial working salary). Three different water quality levels were examined in qualitative terms. In these circumstances the damage function would have to be discontinuous at the three points indicated by the three levels of the quality. If a pollution source were to cause the water to fall from one quality level to another, the estimated damages would be measured in terms of the loss of that specific function. There is no reason why data of this type should not be used in an environmental accounting assessment, but it will require specific estimates of the impact of pollutants on the local water quality.

Magnussen and Navrud (1991)/Magnussen (1991) is probably the most comprehensive Norwegian study on user and non-user benefits from water quality changes because it considers both the linkages to physical damage functions and the constructed CVM models. Water quality was found to be mainly affected by Nitrogen and Phosphorus emissions and the study looked at the benefits of a 50% reduction in emissions. The average WTP per household for the improvement was estimated at 600–5000 1991 NOK (72–600 ECU, 1990). Again, this study only measures large changes in water quality so that the valuation procedure in which it is used would have to be adapted accordingly.

Another study which could be relevant to this study in that it considers the effect of diminished water quality is Mäntymaa (1991). Again, it operates in terms of 5 quality levels. The task of linking specific pollutants to these quality levels remains, however, incomplete, or at least not available from the published version.

The unit values estimated in Green and Tunstall (1990a/b); WRC/FHRC (1989), can only act as a rough indicator of the valuations in the UK that people place on water quality improvements. Problems in using these studies stem from the fact that people do not perceive the difference between the present water quality levels and those water quality improvements that are

Table 7.3. Mean annual willingness to pay per local household for improved water quality in three Norwegian fjords

		Mean WTP per household per year		
Fjord	Author[1]	Users	Non-users	Weighted average of users and Non Users
Kristiansand	Heiberg and Hem (1987)	–	–	430 (62.5)
Inner Oslo	Aarskog (1988)/Heiberg and Hem (1988)	906 (123.4)	598 (81.4)	837 (114.0)
Drammen	Dalgard (1989)	849 (110.5)	416 (54.1)	563 (73.3)

Notes: 1. All studies have used contingent valuation mail surveys using colour coded maps and detailed verbal descriptions of different pollutants before and after the improvement. All figures are in NOK; except the WTP figures in brackets which are in ECU, 1990.

Source: Navrud and Strand (1992).

being valued. It is thus very difficult to place much reliance on the results and to transfer these results to the present study.

Table 7.3 below shows the findings of three main Norwegian studies. The physical changes measured in these studies (i.e. oxygen concentrations, eutrophication, heavy metal and chlorinated hydrocarbon concentrations and surface pollution of oil and litter) were 'translated' into perceivable measures of water quality but the water quality improvements were rather large, from the prevailing state to a 'nearly unpolluted' state. Thus, although the improvement is very large, both steps required to relate the results to an environmental valuation study are available – i.e. the link between the pollutants and the measures of water quality, and the relationship between water quality and WTP.

For recreational use values, a useful way of making use of the results is as the value of a recreational day, which has been provided by Navrud (1994). He estimates, from a range of Scandinavian and US studies the following values for different recreational days: Atlantic salmon/sea trout fishing (ECU 16.3); brown trout fishing (ECU 7.6); big game hunting (ECU 20.6); small game hunting (ECU 16.35); hiking and cross country skiing (ECU 5.4); picking mushrooms (ECU 8.1); and swimming (ECU 4.3). To test the transferability of these estimates he examined an EIA for a proposed hydropower site in Sauda Fjord in South West Norway. Losses of fish and wildlife stocks had been estimated in the EIA, and 'expert opinion' was used to convert these into lost recreational days. Taking the above values for the different recreational days he obtained an estimate of total loss due to the hydropower development of ECU 142,000/year. He then conducted a direct CVM of the potential users, focusing on the loss of recreational value in the area. By this means he obtained an estimate six times larger (ECU 809,000/year).

Navrud's analysis is interesting but not conclusive. It strongly suggests that marginal valuations made up of benefit transfer and expert assessments are misleading. However, it does not tell how transferable a 'proper' marginal valuation study would be. For example, could the results of the specific study be convertible into value per 1000 loss of fish population and then be transferred to a new situation? If that were the case, they could be used in the valuation of fish stock changes, as have been estimated in the UK study. Certainly such an issue is worth exploring in future work.

7.4.4. *Non-use of surface water*

Non-use benefits of surface water are divided into those relating to fisheries and those relating to non-fisheries.

For non-fisheries the following studies are relevant:

- Kyber (1981)
- Magnussen and Navrud (1991) / Magnussen (1991)
- Hervik, Risnes and Strand (1987); Strand and Wenstop (1991).
- Mäntymaa (1991)
- Green and Tunstall (1990a/b) ; WRC/FHRC (1989)
- Aarskog (1988)
- Heiberg and Hem (1988)
- Dalgard (1989)

Kyber (1981), made a hedonic study of the water quality in Valkaekosia, Finland affecting average value per square metre of shore area with summer cottages and permanent dwellings. He used the National Board of Waters' quality levels where quality level II indicates 'good' quality; level III (satisfactory); IV (passing) and level V (poor). Valuations were found but no details were given regarding the levels of physical pollutants and the equivalent measures used by the national water board. Since water quality in this sense is only weakly related to pollution levels, a link between the two might be very difficult to establish. A similar comment applies to the study by Mäntymaa (1991) of Lake Oulujarvi, Finland.

Green and Tunstall (1990a/b); WRC/FHRC (1989), in a UK CVM survey of 1500 people on user and non-user recreational and amenity value, found people were better at recognising higher pollution than they were at recognising cleaner water. This indicates that peoples' valuations of reduced water quality have a high correlation with physical damage than their valuation of the opposite change. Among those not resident in the immediate area, 56% of users had an annual WTP of £19.56 (ECU 27.4) per person. 45% of non-users had an annual £13.60 per person (ECU 19.1, 1990). (£UK figures in 1987 prices, £1 = 1.40 ECU in 1990). The authors made a more detailed description of water quality involving an index of perceived water quality relating to biochemical water quality indices. The three levels of water quality were as follows:

Class 1. 'To be safe for children to swim in'.
Class 2. 'To support many fish including trout, dragon flies and many plants'.
Class 3. 'For water birds to use water'.

(The definitions are equivalent to the official classification ordering).

The mean WTP per person per annum was: ECU 814 for class 1, ECU 786 for class 2 and ECU 737 for class 3. The small differences in values are remarkable – even for non-users one would expect bigger values and for users, US studies suggest much bigger differences. The finding was, however, supported by a survey of all user benefits where the recreational values per visit were ECU 0.17, ECU 0.84 and ECU 0.73 for water quality levels, 3, 2 and 1 respectively. The typical present value of the benefits of improving the water quality level from one class to the next is as follows:

Class 3 to Class 2 = £450,000 (for a country park)
Class 2 to Class 1 = £90,000 (for a local park site)

These results are too general to be of direct use in the present study. If one could establish that the operation of a particular pollutant resulted in a change in quality as identified above, however, it might be possible to use this and similar studies to value the impact.

Hervik, Risnes and Strand (1987); Strand and Wenstop (1991), established a WTP of 550–1490 NOK (1990) through a CVM survey for the preservation of the River Rauma-Ulvaa, West Norway. The study was based on the most cost effective way of ranking rivers for hydro-electric development. This may be relevant to the external costs of hydro power in valuing the non-use value of the site prior to development.

Methodologically, it is difficult to separate the use and non-use values as clearly as one might like. Where genuine non-use values are involved, there will always be a problem of relating such valuation to one or more pollutants. By their very nature non-use values are not closely or quantitatively linked with pollution levels. If preservation is threatened, the relevant loss of value is that associated with the affected habitat or environment and some of the preservation valuations that have been carried out may be of relevance, although it will be important to ensure that the results are broadly transferable. In other cases, such as those where individuals enjoy the visual impacts of a clean lake or object to the idea that a water body is being polluted even when they will make no use of it, valuations *may* involve a relationship between the level of pollution and the WTP. However, this is very difficult to measure and the best that can be hoped for is a specific CVM study when the main source of the pollution damage is a specific power plant. Naturally such a study would be specific to that plant, although it may be possible to transfer some of its results to other plants in similar circumstances.

Fisheries non-use
Similar remarks may apply to the valuation of non-use benefits related to fisheries. There are studies, however, that have estimated such benefits and linked them to specific pollutants.

The following studies are of potential relevance:

- Navrud (1989)
- Navrud (1990a/b)
- Amundsen (1987)
- Carlsen (1985)
- Strand (1981)
- Silvander (1991)

The results are summarised in Table 7.4. All the studies employ the CVM to elicit WTP. With the exception of Silvander (1991), which is Swedish, they are all Norwegian. The Silvander study is also the only one to use mail surveys whereas the others conduct in-person interviews (which should give better results). Four of the studies give the total WTP (use and non-use): Navrud (1989), Amundsen (1987), Carlsen (1985) and Strang (1981). This is important to avoid double counting. Strand (1981) and Navrud (1989) have divided the use and non-use values, however, the method for doing this is rather ad hoc. Hence, one cannot place much reliance on the separate numbers.

The results in Table 7.4 are of direct relevance to the estimation of the costs of acid depositions (see also Chapter 9 on Forests). It is from detailed studies such as those carried out by Navrud that valuations of acid depositions will be obtained and fed into the proper valuation of acid rain damages.

7.5. Mitigation costs

Mitigation costs have been used in valuing damages to surface and ground water. The following studies are relevant:

- Baan (1983)
- SPCA (1991)
- Winje *et al.* (1991)
- Ewers and Schultz (1982)

Baan (1983) studied the social benefit in the Netherlands resulting from higher surface water quality. The SPCA (State Pollution Control Authority) (1991) made a study in the North Sea as to the cost effectiveness of different measures to achieve reductions of pollutants. This involved a 50% reduction in national emissions of Nitrogen and Phosphorus to the North Sea. The discounted costs (discount rate = 7%) of different measures for reducing emissions of Nitrogen and Phosphorus by 1 ton were 0.02, 0.15, 6.95 and 30.90

Table 7.4. Non-use values and fisheries

The change in fish stocks valued	Cause of change	Author	WTP per year (1990 NOK)
Detailed description of the increased number of trout lakes and salmon rivers with restored stocks on southern Norway (national)	Reduced acidification due to 30-70% reduction in European SO_2 emissions	Navrud (1989)	$390^{\#}$ (46.8 ECU) (246–343 [29.5–41.1 ECU] is non-use value the rest is use value).
Avoiding the extinction of the current salmon and sea trout stocks in River Audna (local)	Stop liming, i.e. the neutralization of acid depositions	Navrud (1991a)	115* (13.8 ECU)
Avoiding the extinction of current trout stocks in the Gjerstadskog lakes (local)	Stop liming, i.e. the neutralization of acid depositions	Navrud (1991b)	46* (5.5 ECU)
Avoiding an unspecified 'reduction' of the current trout stocks in the Oslomarka lakes (local)	Not start liming to neutralize acid depositions	Amundsen (1987)	$360^{\#}$ (43.2 ECU) (both use and non-use values.
Avoiding 'some' and 'considerable' reductions in the salmon stock in river Numedalslågen (local)	Different operation schemes of the hydro-power dams	Carlsen (1985)	$41–85^{\#}$ (4.9–10.2 ECU) (only 24–25% of the households were WTP, 165–340 [19.8–40.8] per household) (both use and non-use values)
Avoiding extinction of all freshwater fish stocks in Norway (national)	Acidification	Strand (1981)	1650–2650* [197.8–317.7 ECU] (1000–1600 [120–191.8 ECU] is non-use value; the rest is use value)
[1] Avoiding extinction of the species most popular for consumption, in all Swedish saltwater areas (national)	Eutrophication due to leaching of nitrogen	Silvander (1991)	310* (37.2 ECU)

Notes: 1: Saltwater study; #: WTP per household *: WTP per individual.
All ECU figures are 1990 where 1 NOK = 0.125 ECU.

Source: Navrud *et al.* (1992).

(1990 NOK) respectively (1 NOK = 0.125 ECU). The basic data on the relationship between costs and reductions of pollutants could be of relevance in deriving a response to any estimated damages from the pollutants but cannot in general be a substitute for estimating those damages in the first place.

In a similar vein, Winje *et al.* (1991) in West Germany looked at the costs of treating different pollutants in 3 sectors including public drinking water supply, private and industrial water supply. Pesticide residuals (PPP), chlorinated Hydrocarbons (CHC) and nitrate concentrations were investigated. Treatment costs for industry = DM 120 mn. per annum (1983) and public supply = DM 780 mn. per annum (1983). 60% of these additional future costs are due to nitrate concentrations in ground water.

Ewers and Schultz (1982), similarly value the benefits of improved water quality in terms of the reduced cost of treating it. They look at the value of improved water quality (reduced eutrophication due to reduced emissions of phosphorous) in Lake Tegeler, Germany. For drinking water quality improvements (see later for the same study valuing commercial fisheries) from the present level of quality (level 4) up to the highest level (level 1) were valued in the range DM 4.7–6.9 million annually. For catering firms the corresponding estimate was DM 0.4–4 million. The drinking water benefits were valued by the associated cost reductions of improved water quality.

7.6. Conclusions

From a monetary valuation viewpoint, the treatment of damages to terrestrial and aquatic ecosystems is inadequate. The present study has reported damages in terms of areas with critical load excedence, numbers of species under various levels of threat etc. It has not sought to make any monetary valuation of these impacts, primarily because it could not establish the damages done to the ecosystems. In the case of fisheries, the UK team did establish loss of stocks resulting from various pollutants, but did not value them owing to a lack of valuation data.

In fact there is a body of valuation studies that are of varying degrees of usefulness for this purpose, and which can be used to value damages in the context of an environmental accounting exercise. We have reviewed these, dividing them into studies that start from dose response functions and then estimate damages (using CVM and other studies); studies that seek to establish damages by going directly to the CVM approach, and studies that rely on mitigation costs.

The dose response function has been used for commercial and recreational fisheries and could be adapted to other countries and situations, although the scope for, and the conditions under which, such transfer can be made are a matter of some discussion. Some of this will be taken up in the next stage of this research.

The 'direct CVM' approach is less suitable for the environmental accounting exercise. Applications were divided into species valuations, amenity valuations, water related recreational benefits, and non-use benefits related to water. The valuations of species have to be linked to damages to species based on anthropogenic activities. This may be feasible in some cases. But that is not enough, we also need valuations of losses that are relevant to the countries and situations. Transferability of species values is more limited than is sometimes suggested.

Amenity valuations are generally very site specific and not often related to pollution emissions. Hence they are of limited use in the valuation exercise. They may be of importance, however, in relation to losses of amenity resulting from investments. This is something we have not looked at in this study.

Water related recreational benefits' studies are becoming more relevant as they relate to the damages caused by specific pollutants, or relate to water quality indicators that can be related to specific pollutants. It is possible to make use of these studies in the context of this Green Accounting exercise, and is something we will undertake in the follow up work.

Non-use benefits are generally regarded as difficult to measure using the CVM method. The applications reported here must, therefore, be treated with caution. Nevertheless, some of the research in this area has been impressive and the results may be applicable to the valuation of damages to water bodies suffered by persons who never make use of them.

Finally we have the use of mitigation costs. These have been used particularly in connection with the impacts of pollution on general surface water quality, as well as surface and ground water for drinking. Although such a procedure is not generally valid as an estimate of the damages from a polluting source, it can be a measure of the benefits if it can be established that the water treatment measures would have to be taken on grounds of public health, or if the damages would be so high that the treatment costs would be much less than any damages. Since this is not known in general, the use of mitigation data for valuing such benefits needs, therefore, further investigation.

References

Aarskog, E. (1988). Willingness To Pay for Cleaning up the Inner Oslo Fjord. MSc. Thesis, Department of Economics, University of Oslo, Report No. 871013-2, March, Center for Industrial Research, Oslo.

Amundsen, B.T. (1987). Recreational Value of the Fish Populations in the Recreational Area Oslomarka. MSc. Thesis, Agricultural University of Norway.

Baan, P.J. (1985). Inventory Studies of the Benefits of Recovery and Protection of Soil Quality. Delft Hydraulics Laboratory, n. R2182/S618-D85/004, December.

Baan, P.J. (1983). Inventory Studies of the Benefits of an Improved Water Quality. Delft Hydraulics Laboratory, n. R1853 R83/001, July.

Carlsen, A.J. (1985) Economic Valuation of Hydroelectric Power Production and Salmon Fishing. In Carlsen, A.J. (ed), Proceedings – UNESCO Symposium on Decision Making in Water Resource Planning, 5–7 May 1986, Oslo, pp. 173–182.

Dahle, L., et al. (1987). Attitudes Toward and WTP for Brown Bear, Wolverine and Wolf in Norway. Department of Forest Economics, Agricultural University of Norway. Report May.

Dalgard, M. (1989). Willingness To Pay for Regulatory Actions Towards Water Pollution in the Drammen Fjord. MSc. Thesis, Department of Economics, University of Oslo. Report n. 881108-2, August, Center for Industrial Research, Oslo.

ECOTEC (1994). An Evaluation of the Benefits of Reduced Sulphur Dioxide Emissions: Annex 3; A Contingent Valuation Study of Upland Vegetation. Report to the Department of the Environment, London.

European Commission, DGXII, Science, Research and Development, JOULE (1995). Externalities of Energy – 'ExternE' Project, Report Number 2, Methodology.

Everett, R.D. (1979). The Monetary Value of the Recreational Benefits of Wildlife. Journal of Environmental Management, 9: 203–213.

Ewers, H. and Schulz, W. (1982). The Monetary Benefits of Measures for Water Quality Improvement: A Case Study of Lake Tegeler in Berlin. Umweltbundesamt, Berichte 3/82: Erich Schmidt Verlag, Berlin.

Federal Environment Agency (1993) Manual on Methodologies and Criteria for Mapping Critical Levels/Loads and Geographical Areas where they are Exceeded. Berlin: Federal Environment Agency.

Fuhrer, J. and Ackermann, B. (Eds.) (1994) Critical levels for Ozone; a UN-ECE workshop report. FAC Report no. 16, Swiss Federal Research Station for Agricultural Chemistry and Environmental Hygiene, Liebefeld-Bern, 1994.

Green, C. and Tunstall, S. (1990a). The Amenity and Environmental Value of River Corridors. Flood Hazard Research Center, Publication n. 171, Middlesex Polytechnic.

Green, C. and Tunstall, S. (1990b). The Benefits of River Water Quality Improvement. Report from the Flood Hazard Research Center, Middlesex Polytechnic (in press, Applied Economics).

Heiberg, A. and Hem, K.G. (1988). Regulatory Impact Analysis for the Inner Oslo Fjord: A Comparison of Three Different Methods. Report 880105-1, September, Center For Industrial Research.

Heiberg, A. and Hem, K.G. (1987). Use of Formal Methods in Evaluating Countermeasures to Coastal Water Pollution: A Case Study of the Kristiansand Fjord, Southern Norway. Center For Industrial Research. In Seip. H and Heiberg (eds.) (1989). Risk Management Of Chemicals in the Environment. London: Plenum Press.

Hervik, A. and Strand, J. (1991). In Navrud, S. (1991). Review and Evaluation of European Valuation Studies in the Impact Categories: Crops, Forests, Fisheries and Water, pp. 35.

Hervik, A., Risnes, M. and Strand, J. (1987). Implicit Costs and Willingness to Pay for Development of Water Resources. In Carlsen, A.J. (ed.), Proceedings – UNESCO Symposium on Decision Making in Water Resource Planning, 5–7 May 1986, Oslo, pp. 195–202.

Hjalte, et al. (1982). Economic Consequences of Water Quality Changes in Lakes. Report, March. TEM University of Lund.

Kanerva, V. and Matikainen (1972). The Economic Losses Caused by Water Pollution in Lake Saimaa. Rakennustalouden laboratorio 7, Otaniemi marraskuu, report from The Technical Research Center, Finland.

Kyber, O. (1981). Harmful Effects of Water Pollution on Recreation and their Estimation at an Inspection. Research Notes 23/1981. Technical Research Center of Finland.

Magnussen, K. and Navrud, S. (1991). Valuing Reduced Pollution of the North Sea. Report to the Ministry of Environment, Noragric/Agricultural University of Norway and Norwegian Agricultural Economics Research Institute.

Magnussen, K. (1991). Valuation of Reduced Water Pollution Using the Contingent Valuation Method: Methodology and Empirical Results. Norwegian Agricultural Economics Research Institute. Paper presented at the second annual meeting of EAERE, 10–14th June, Stockholm.

Mäntymaa, E. (1991). Some New Ideas and Preliminary Results for Using the CVM in Measuring the Environmental Benefits of a Lake. University of Oulu, Research Institute of Northern Finland.

Navrud, S. (1988). Recreational Value of Atlantic Salmon and Sea Trout Angling in the River Vikedalselv in 1987 – before regular Liming. Report from the Directorate for Nature Management.

Navrud, S. (1989). Estimating Social Benefits of Environmental Improvements from Reduced Acid Depositions: A Contingent Valuation Survey. In Folmer, H. and Ierland, E. van der (eds.), Valuation Methods and Policy Making in Environmental Economics. Studies in Environmental Science, 36. Amsterdam: Elsevier Science Publishers, pp. 69–102.

Navrud, S. (1990a). Cost–Benefit Analysis of Liming Lakes: Two Case Studies. Report from the Directorate for Nature Management.

Navrud, S. (1990b). Cost–Benefit Analysis of Liming River Audna: An Extended Analysis. Report from the Directorate for Nature Management. In Press.

Navrud S. (1991). Review and Evaluation of European Valuation Studies in the Impact Categories: Crops, Forests, Fisheries and Water, Report to the ExternE Project.

Navrud, S. (1994). Economic Valuation of External Costs of Fuel Cycles: Testing the Benefit Transfer Approach. In: A.T. de Almeida (ed.). Models for Integrated Electricity Resource Planning. Kluwer Academic Publishers.

Navrud, S. and Strand, J (1992): Norway. Chapter 6 in S. Navrud (ed.) 1992: Pricing the European Environment, Scandinavian University Press/ Oxford University, Oslo, Oxford, New York.

Pearce, D., Bann, C. and Georgiou, S. (1992). The Social Costs of Fuel Cycles. A Report to the UK Department of Energy. The Center for Economic and Social Research into the Global Environment, (CSERGE), University College London and University of East Anglia. (also referenced in an earlier version as 1991).

Rasmussen, T. et al. (1991). Lost Income to the Fishing Industry in the FRG Resulting from Environmental Pollution and Other Anthropogenically Related Impacts. Umweltbundesamt.

Silvander, U. (1991). The Willingness To Pay for Angling and Ground Water in Sweden. Dissertations, No. 2. Swedish University of Agricultural Sciences, Department of Economics.

Singsaas, T. (1991). Estimating the Economic Value of Recreational Fishing for Atlantic Salmon and Sea Trout in the River Gaula. MSc. Thesis, Agricultural University of Norway.

SPCA (1991) The North Sea Declaration. Nutrients. A proposed regulatory plan to reach the North Sea Declaration's goal of nutrient emission reduction, combined with the largest possible improvement in local water quality. Report from the State Pollution Control Authority. September 1991.

Strand, J. (1981). Valuing Benefits from Recreational Fishing in the River Gaula. Memo, November 18, Department of Economics, University of Oslo. English (short) version: Valuing Benefits of Recreational Fishing in Norway: the Gaula Case. In Carlsen, A.J. (ed.). Proceedings–UNESCO Symposium on Decision Making in Water Resource Planning, 5–7 May 1986, Oslo, pp. 245–278.

Strand, J. and Wenstop, A. (1991). Implicit Willingness to Pay to Avoid Negative Consequences of Hydroelectric Development in the 'Master Plan for Water Resources. In: Environmental Costs and the National Economy. Report to the Norwegian Water Resources and Energy Administration (NVE).

Travers Morgan Economics (1991). Mersey Barrage Feasibility Study – Economic Valuation: Waders and Wildfowl. Contingent Valuation Method. London: Travers Morgan.

Winje, D., Homann, H., Luhr, H.P. and Butow, E. (1991). The Influence of Pollution in the Aquatic Environment on the Costs of Supplying Water in the Federal Republic of Germany. Report No.2/91, Federal Environmental Agency, Berlin, UFOPLAN No 101 03 110–08.

WRC (Water Research Center) and FHRC (Flood Hazard Research Center) (1989). Investment Appraisal for Sewage Schemes: The Assessment of Social Costs. Project Report, WRC and FRHC, WRC, Swindon.

CHAPTER 8

MATERIALS

8.1. Introduction

The studies in this valuation category have been separated into three sections.

1. Materials
2. Residential and commercial buildings
3. Historic buildings and monuments.

The Chapter begins by discussing the general issues of valuation for this set of impacts, and then goes on to survey the European studies of damage. There are serious concerns that previous damage estimates are out by an order of magnitude and the reasons for this are discussed. The Chapter concludes by comparing preliminary results from the EXTERNE study some of these earlier studies, and by making recommendations for future work.

There are ten previous European studies which have made estimates relevant to these three sections

- Building Effects Review Group (BERG) (1990)
- BMU (1986)
- ECOTEC (1986)
- Feenstra (1984)
- Glomsrod and Rosland (1988)
- Heinz (1980)
- Isecke *et al.* (1991)
- Fenger *et al.* (no date)
- National Environmental Program of the Netherlands (1990)
- Lee *et al.* (1995)

8.2. Issues and choice of dose response functions

Acidic deposition from the burning of fossil fuels is the process most widely associated with the erosion of buildings and materials, primarily stone and steel structures. Details of sulphur emissions and depositions are available for some of the studies in this impact category which should, in principle, make it possible to calculate an external marginal value. The main pollutants considered are acidity, SO_2, particulate matter and chlorides. NO_x is also considered to be involved indirectly (through the formation of ozone), and some preliminary results for this pollutant are reported in the UK study. Types of damage included are primarily rubber goods.

For normal residential and commercial buildings, the best method of valuation is to estimate physical damage using a dose response function, and then to apply values to the damages, as in the case of crops. For special buildings of historic and cultural interest, however, such a method is inappropriate. It is not the physical damage that is critical there but the fact that damages are effectively irreplaceable. For such buildings, use values may be estimated by CVM or travel cost methods. In addition information on defensive expenditures is relevant (this also applies to non historic buildings). Non-use values can only be obtained from CVM methods, but the success of using the methodology for this purpose is limited.

The studies reviewed place raise many problems, and there is some uncertainty as to their validity as far as the monetary estimates are concerned. The most serious problems that arise are with respect to the estimation of the dose response functions. The EXTERNE study reviewed the literature on these and found that the forms used in many widely quoted studies are scientifically poor, out-of-date and sometimes both. The section below reviews the literature on the dose response functions. Given that an acceptable estimate of the function has been obtained, the next step is to value the damages. Again, experience from US and other studies has shown that there is considerable scope for judgment in linking the physical damages to monetary damages. One key issue is the extent to which individuals can limit damages by suitable choice of materials, e.g. by using galvanized guttering instead of non-galvanized guttering. Another is the extent to which deteriorating stone and paint results in replacement of materials or merely in a poorer appearance and upkeep. Both these factors relate to behavioural aspects and cannot be addressed outside of a model of human decision-making.

If the physical and social aspects are not integrated, the consequences are similar to those discussed in Chapter 5 for crops and partly in Chapter 3 for health. First, the dose–response function itself is not immune to the linkage. If damages have been estimated in areas where more protective materials have been used *and this quality change has not been allowed for*, the dose response function will underestimate the damages of an increase in depositions in a previously unpolluted area, and overestimate them in a highly polluted area (where individuals will be able to switch to less costly materials). These impacts tend to be ignored. Second, the US studies found that actual maintenance in polluted areas was generally less than the estimated increase in the frequency of maintenance, based on physical dose response functions. This would suggest that costs based on the latter alone would be *overestimates*.

In all the European studies reviewed below, neither of these issues have been addressed. It is difficult to know how much the estimates are affected but it does add to the general weakness of the monetary estimates in this area.

Dose response functions for materials damages

The EXTERNE Study reviewed the literature on the estimates in some depth. It concluded that reliable estimates of damages from acid deposition could only be made for stone, mortar, concrete, paint, steel, zinc and aluminum. Details can be found in the UK study, where they have been further updated.

8.3. Conclusions on dose response functions

In the view of the EXTERNE team the dose response functions used in the UK study are the best that can be used, given the data available. It should be noted that the studies on which they are based are multi-national, with extensive use of meta analysis. A large amount of US experience is reflected in these. Such data can and need to be used if a proper estimate of damages is to be made. However, it also needs to be applied with caution, recognizing that some of the materials used are different. In some previous studies carried out in Europe, using US dose response functions, that does not appear to have been done.

Given the dose response functions, for estimates of monetary damage an inventory of building materials affected and an estimate of the damage is needed. Inventories are generally based on extrapolating from a small sample of detailed surveys of materials. As a procedure this is acceptable as long as the initial surveys are representative enough of the areas to which they are applied. Clearly, however, such a procedure cannot be satisfactorily applied to special buildings, such as those of historical interest. For them a case by case approach has to be followed.

Finally there are the monetary costs used to convert the physical damages into monetary ones. As explained at the outset, adjustments in the choice of materials and in the frequency of maintenance need to be studied as independent important factors. In general this has not been done but some judgmental estimates have been made of the changes in the frequency of painting/replacing of materials with and without the corrosion caused by atmospheric pollution. The present study has taken the same approach, recognizing its shortcomings but aware that, at this stage, there was no alternative. Damages are estimated in market prices, net of taxes, as there is no reason to believe that the prices that prevail for these materials are significantly different from the social costs of using them.

8.4. Main results from other European studies

The results of other European studies are taken from the studies cited above.

8.4.1. *Materials in buildings (other than historic and cultural)*

Fenger *et al.* (no date) estimated damages from acid deposition (mainly sulphur) in Denmark at ECU 68 mn. Pearce *et al.* (1992) report that in the late 1980s sulphurous depositions in Denmark were 121,800 tons. This would suggest damages of ECU 562 per ton of SO_2, or ECU 70 per capita. Since emissions affecting materials also come from outside the country, the use of domestic emissions is not, however, a good normalising factor.

The National Environment Program of the Netherlands, 1990 gives the following annual material damage estimates:

	Millions
Steel and zinc sheet	Dfl 40
Metals	Dfl 113
Concrete damage	Dfl 175–350
Facade cleaning	Dfl 12–25
Total	*Dfl 340–528 (ECU152–237)*

Again, taking depositions of SO_2 as given in Pearce *et al.*, 1992 for the Netherlands as 172,600 tons for 1988, this yields damage estimates of ECU882.6–1,371.0 per ton of SO_2 or 10–15 per head.

Glomsrod and Rosland (1988) estimate the direct and indirect costs of SO_2 in Norway. The direct costs are estimates as in the above studies (i.e. using dose response functions). The indirect costs are estimated by looking at how the materials damages affect growth in the economy. The materials examined include galvanized steel, paint on steel, wood and stone and stain on wood.

Local and long range SO_2 emissions result in an additional cost of approximately:

Direct costs = NOK 420 mn. 1990 (ECU 52 mn. per annum).

Indirect costs = NOK 136 mn. 1990 (ECU 17 mn. per annum).

This makes an annual total of ECU 69 mn. The study concludes that 27% of these costs are due to long-range sources. Since depositions of sulphur oxides were 302,400 tons in 1990, then damage works out at ECU 158 per ton or ECU 11 per head.

The Glomsrod and Rosland (1988) study has lower figures than the other two studies cited. This may be due to its having considered fewer materials. On the other hand, the fact that it allowed for very limited mitigation measures should have resulted in higher estimates.

The UK conducted a study on the effects of acid rain on building materials

(Building Effects Review Group (BERG), 1989). It reported a benefit of ECU 4.7 per capita (1990 prices) for a 30% reduction in SO_2 in Birmingham. This figure cannot be converted into a benefit per ton of sulphur dioxide without data on the depositions in that city. The per capita figure is not dissimilar to that from another UK study (ECOTEC, 1986) which is described below and which used the same basic data. The BERG report supports the conclusions reached in the EXTERNE study, in recognizing the poor quality of the dose response functions, including the effects where pollutants act synergistically.

Heinz (1980), reported the additional costs for maintenance, replacement and cleaning in the more polluted area as opposed to clean areas of West Germany. The categories under consideration were: buildings, steel, and windows (cleaning). However, as Heinz himself has pointed out, there are several shortcomings in the Study, whose results can only be regarded as first approximations. First, the inventory of damaged materials is not complete, e.g. no account is taken of damage to paints and plastics. Second, estimates neglect indirect costs e.g. costs incurred due to closure of a bridge for maintenance activities. Third, as pointed out above, extra maintenance expected can only be a reliable damage estimate if a direct relationship with the preferences of the victims can be estimated. The correct method is to include only those maintenance costs which are actually paid by the victims of environmental damage.

The total monetary damage in Heinz's study was estimated at DM 2.3 bn. (ECU 1.0 bn., 1983), with painting of residential buildings accounting for the largest single component. The impacts assessed are the result of all air pollutants, not just sulphur. Allowing for inflation, the figure amounts to about ECU 20 per head in 1990 prices.

A similar study was carried out by Isecke *et al.* (1991), and which has been used as the basis of the figures reported in this study. This looked at the additional damages from air pollution in high deposition areas, as opposed to low deposition areas. The damages to galvanized steel, stone, paint, concrete and brickwork were estimated on the basis of the depositions of SO_2 that occurred in 1987. They obtained an estimate of between ECU 0.9 bn. and ECU 1.6 bn. for the highly polluted areas. This is a wider range than the Heinz study, with the upper bound being about 60% greater. Converting the figures to damages per ton is problematic as damages in the so called clean areas are not included. Furthermore damages referred to include several pollutants. Nevertheless it is interesting to note that the estimated damages are very large. Taking the 1987–88 depositions of around 1.1 mn. tons, gives damages of about ECU 2360/ton from the Isecke study. The Heinz study gives an estimate of around ECU 1,100/ton. In per capita terms, damages are around ECU 40/head in the Isecke study.

ECOTEC (1990) estimate the benefits of reduced building damage from a 30% reduction in SO_2 in England. The building stock in the UK was estimated from aerial surveys and converted into a range of typical buildings (houses, schools, offices, shops, government buildings, factories and warehouses). For each of these the amount of different types of materials exposed was estimated. Then dose response functions were used to calculate the damages. The benefit of a 30% reduction in sulphur dioxide came to £8 bn., or ECU 14.4 bn. Taking the total depositions of sulphur dioxide in the UK in 1990 as 1.167 mn. tons, gives the damages from 30% of that SO_2 as:

$$(14.4/(1.167*0.3) = \text{ECU } 41{,}262/\text{ton}.$$

(A conversion factor of 1 UK ton = 1.018 metric tons has been used).

This is a lifetime cost of the SO_2. ECOTEC assume a discount rate of 5% and a lifetime of around 30 years. That would give an annualised cost of ECU 2556/ton, which is considerably higher than the other studies have produced. The corresponding value in per capita terms is about ECU52. The study was reviewed in detail by the EXTERNE team and found to have many shortcomings. The criticisms of the study were:

a. the 30% reduction in SO_2 on which the benefits were based assumed that, in the absence of a reduction, levels would increase uniformly across the country. This is very unlikely to be the case;
b. the dose–response functions used are not explicitly declared. It appears that all materials damages are based on three, out-of-date functions. Three categories of materials were included: metals, minerals and paint. Damages for metals were based on an unreferenced Norwegian study on galvanised steel. However, such a study is not valid for uncoated steel and aluminum. Because the corrosion rates for the latter two are much lower, the study overestimates the damages as a result of this error;
c. the 19% change in benefits for minerals is much higher than is indicated by dose response functions in the recent literature. Based on the functions recommended in this study, a better estimate would be around 6%;
d. a 7% estimate is made for painted woodwork, and a 30% estimate for joinery replacement. This seems unreasonable as joinery replacement would take place when the painted woodwork had to be replaced or repainted. Hence a damage of 7% would be more reasonable;
e. the Report estimates benefits from cleaning of paintwork. However, SO_2 would not in general affect the frequency of such cleaning.

For all these reasons the estimates of damages are likely to be seriously biased upwards. This is also confirmed by looking at the implied benefits in ECU/Kg, which work out at around ECU 2.7/Kg (excluding historic buildings). The range of values in the studies reviewed so far has been between ECU 0.16 and 1.37/kg, with a mean of ECU 0.74/kg.

The above studies have little evidence on ozone damages. These have been reviewed in the UK study, based on Lee at al, 1995. Damages are believed to occur to rubber goods, paint and textiles. For rubber preliminary estimates have been made, based on damages as well as protection costs. They amount to ECU 221–450 mn., which is significant, but smaller than the SO_2 damages. The authors stress the very preliminary nature of these estimates. For paint initial figures suggest damages in the range ECU 0–78 mn., and for textiles there are not estimates available. Ozone damages are very much, therefore, an area for further research.

A recent alternative approach to the assessment of soiling costs has been taken by Pons and Curtiss (1995). They conclude that the amenity loss is equal to the repair costs at the time of cleaning. Thus, total damage to buildings equals is equal to twice the level of cleaning or repair costs. It is not known whether this approach could be used on a consistent basis in the context of the GARP framework.

8.4.2. *Historic and cultural buildings and monuments*

The user values of a historic monument could be estimated via the travel cost method by looking at visiting patterns for different sites and relating them to the characteristics of that site. However, such a procedure would demand a large quantity of data and, moreover, would not capture the non-use values, which, in the case of historic buildings, can be very high. The latter would require the use of CVM analysis, which has been discussed earlier in this report (see Chapter 2). Even in this case, it is necessary to proceed by asking people to value various states of repair and disrepair through questionnaires using photographs of monuments, buildings, works of art and books, etc. Dose–response functions then become relevant, in linking the states of damage to the depositions of pollutants. Finally, one can look at damages in terms of costs of restoration. Such costs are only a part of total costs, as it can generally be assumed that restoration (a) does not achieve complete replacement of historic monument losses and (b) the WTP for damages is in excess of the restoration costs.

The first attempt to estimate damage to cultural property on a comprehensive basis is Feenstra (1984), who adopted the restoration cost approach and looked at data on restoration and prevention costs in the Netherlands. He included in his study the damages done by air pollution to monuments, objects of art, archives and buildings. He estimated that the restoration costs of monuments in the Netherlands amounted to about ECU 50–80 mn. (ECU 312–519) per ton of SO_2.

In his comment on the quantitative estimates, the author noted that not all of the properties at risk were evaluated. Furthermore, the proportion of damage that an be attributed to air pollution is difficult to estimate. Costs of urgent restoration and prevention measures which have not been carried out were included. However, the most important criticism of such a study is that it does not base the values of historic monuments on the WTP for their aesthetic and

Table 8.1. Economic valuation of materials impacts in selected countries for 1990

ECU million unless otherwise stated (1995 prices) (1)

	GERMANY (1)			ITALY (7)			NETHERLANDS			UNITED KINGDOM		
	Mid value	Range		Mid value	Range		Mid value	Range		Mid value	Range	
Damages from SO2 (2) (3)												
Stone refacing	N.A.	N.A.	N.A.	N.A.	N.A.	N.A.	N.A.	N.A.	N.A.	3.6	1.1	12.1
Plaster and facades	397.0	199.0	595.0	N.A.	N.A.	N.A.	32.4	6.5	13.5	N.A.	N.A.	N.A.
Brick	77.3	20.5	131.7	N.A.	N.A.	N.A.	N.A.	N.A.	N.A.	77.3	20.5	131.7
Paint	170.5	86.1	254.9	N.A.	N.A.	N.A.	N.A.	N.A.	N.A.	606.6	95.5	1213.1
Galvanised steel (painted and unpainted)	963.6	488.2	1439.1	N.A.	N.A.	N.A.	27.6	27.6	27.6	1139.4	488.2	1439.1
Industrial damages (4)	19.3	19.3	19.3	N.A.	N.A.	N.A.	N.A.	N.A.	N.A.	N.A.	N.A.	N.A.
Other (5) (6)	983.1	685.9	1280.4	N.A.	N.A.	N.A.	227.3	180.0	274.6	N.A.	N.A.	N.A.
TOTAL	2610.9	1499.0	3720.5	N.A.	N.A.	N.A.	287.3	214.0	315.6	1826.9	605.2	2796.0
As% of GDP in 1990	0.16%	0.09%	0.23%	N.A.	N.A.	N.A.	0.12%	0.09%	0.13%	0.22%	0.07%	0.33%
PER INHABITANT (ECU)	33	19	47	N.A.	N.A.	N.A.	19	14	21	32	11	49
Soiling costs												
Cleaning costs	222.3	112.4	332.3	N.A.	N.A.	N.A.	N.A.	N.A.	N.A.	135.3	135.3	135.3
Amenity costs	0.0	0.0	0.0	N.A.	N.A.	N.A.	N.A.	N.A.	N.A.	135.3	135.3	135.3
TOTAL	490.5	247.9	733.0	N.A.	N.A.	N.A.	N.A.	N.A.	N.A.	270.7	270.7	270.7
As% of GDP in 1990	0.03%	0.02%	0.04%	N.A.	N.A.	N.A.	N.A.	N.A.	N.A.	0.03%	0.03%	0.03%
PER INHABITANT (ECU)	6	3	9	N.A.	N.A.	N.A.	N.A.	N.A.	N.A.	5	5	5
Damages from ozone												
Damage to rubber goods	N.A.	N.A.	N.A.	N.A.	N.A.	N.A.	N.A.	N.A.	N.A.	236.2	227.2	245.3
Protection of goods	N.A.	N.A.	N.A.	N.A.	N.A.	N.A.	N.A.	N.A.	N.A.	169.2	39.9	298.5
Damages to paint	N.A.	N.A.	N.A.	N.A.	N.A.	N.A.	N.A.	N.A.	N.A.	47.1	0.0	94.2
Tyres	N.A.	N.A.	N.A.	N.A.	N.A.	N.A.	N.A.	N.A.	N.A.	3.1	0.0	6.3
TOTAL	N.A.	N.A.	N.A.	N.A.	N.A.	N.A.	N.A.	N.A.	N.A.	455.6	267.0	644.3
As% of GDP in 1990	N.A.	N.A.	N.A.	N.A.	N.A.	N.A.	N.A.	N.A.	N.A.	0.05%	0.03%	0.08%
PER INHABITANT (ECU)	N.A.	N.A.	N.A.	N.A.	N.A.	N.A.	N.A.	N.A.	N.A.	7.9	4.7	11.2

Cultural assets												
Defensive expenditures	26.3	8.8	45.7	N.A.	N.A.	N.A.	N.A.	65.1	108.3	N.A.	N.A.	N.A.
FINAL SUB TOTAL	3127.7	1755.7	4499.2	N.A.	N.A.	N.A.	287.3	279.1	423.9	2553.3	1142.9	3710.9
As % of GDP in 1990	0.19%	0.11%	0.28%	N.A.	N.A.	N.A.	0.12%	0.12%	0.18%	0.30%	0.13%	0.44%
PER INHABITANT (ECU)	39.3	22.1	56.6	N.A.	N.A.	N.A.	19.3	18.7	28.5	44.5	19.9	64.7

Notes:
N.A. = Not assessed.
(1) Figures for Germany refer only to West Germany and may include some ozone related damages.
(2) Figures for mid values are not averages but 'mid values' as given in the country studies.
(3) Individual figures for Germany refer to residential buildings only. Industrial buildings are given separately below.
(4) Industrial costs include transmission pylons, railway contacts, bridges, oil depots and sludge towers.
(5) For Germany 'Other' includes balconies, gutters, roofs, chimneys and windows.
(6) For Netherlands 'other' includes greenhouse heating pipes, greenhouse pallisades, concrete and other metals.
(7) The only figure reported for Italy is that of defensive expenditures for households. These amounted to 0.05% of GDP and about ECU 8 per head.

cultural values, but on costs of restoration. This is precisely the same methodological issue as was discussed in connection with water damages; one cannot value the damages by looking at the costs of correcting the damages. The method does not value the existence benefits or the benefits from visits as such.

Using a similar methodology, Altshuller *et al.* (1983), estimated damages for Germany that are reported in the German study. The figures here are somewhat lower than for the Netherlands and with a wider range (ECU 7–35 mn.). This may be due to a more limited coverage

Glomsrod and Rosland (1988) in Norway have examined the repair and replacement costs of additional, direct and indirect, material costs of corrosion due to SO_2 (NO_x effect is small according to Norwegian studies). They look chiefly at the effects on galvanized steel, paint on steel, wood and stone and stain on wood. Damages to other materials are not included due to lack of dose–response functions.

The only study that appeared to estimate WTP for cultural monuments is that by Navrud (1990). He estimates the existence and bequest value of one specific historic building, the stone church of Nidarosdomen. A survey is used to elicit the WTP to preserve the *original* church instead of preserving it through restoration. By this means one can get the additional value associated with a cultural monument, other than what can be captured through restoration. However, it is likely to be very specific and transferability will be a problem.

8.5. Results of this study

The main results of this study are summarised in Table 8.1. Only three countries reported any significant results, with the Italy team not being able to collect the relevant data on the stock of properties at risk. The Italian team did, however, report some limited data on defensive expenditures.

For the other teams it is important to note that the classification of materials and the use of dose–response functions *vary widely*. This is due, in part, to the way the data on materials is collected, and in part to differences in the agreed dose response functions. The latter applies in particular to the the Dutch and UK studies, which are the only two to have undertaken primary estimation of damages. Both these issues need further discussion, to see what standardisation is possible.

Because of the relatively smaller coverage in the Netherlands study, we have taken external estimates of damages for those categories not covered in the present study and used them to complement the figures obtained from the application of dose response models to the recently collected concentrations data.

The main results obtained as follows:

i. Estimates of total damages for the impacts assessed range from 0.12% of GDP for the Netherlands to 0.30% of GDP for the UK. In per capita terms, the range is ECU 19 (Netherlands) to ECU 45 (UK). These differences are largely the result of differences in coverage, and of differences in dose response functions.
ii. For Germany total costs of ECU 3.12 bn. are based on a review of previous studies. The range is quite wide (+/− 50% of the central value). Gaps to be filled are ozone damages and more complete coverage of cultural monuments.
iii. For the Netherlands, the present study only looked at some of the impacts. We have completed the picture by taking the best results available from other studies. This procedure of blending the two sources does, however, need to be confirmed.For the UK, coverage is wide. As with the German study the range is about +/− 50%. Damages to cultural monuments are a major gap and it is noted that the ozone damage figures are preliminary.
iv. Compared to earlier studies, which can really only be done for the UK, the figures are not out of line, but are a little lower. The ECOTEC study estimated per capita damages at around ECU 63 (1995 prices), compared to our mid-estimate figure of ECU 45.

References

Altshuller, A.P., Lindhorst, R.A., Nadler, J.S., Niemeyer, L.E., McFee, W.W. (1983). Acidic Deposition Phenomenon and Its Effects, Critical Assessment Review Papers, Volume 1: Atmospheric Science, EPA-600/8, PB 84–171644, Raleigh, NC, USA.

BERG (1990). The Effects of Acid Deposition on Buildings and Building Materials. Ukn Building Effects Review Group, HMSO, London.

BMU (Bundesministerium für Umwelt, Naturschutz und Reaktorischereit) (1986). Kosten der Umweltverschmutzung Symposium Report. Bundesumweltministerium/Erich Schmidt Verlag, Berlin.

ECOTEC (1986). An Assessment of the Effects of Air Pollution on the Building Materials in the UK: An Evaluation of Techniques for Grossing Up Stock at Risk. Department of Environment, UK.

European Commission DGXII (1995a). Science, Research and Development Programme (JOULE). Externalities of Energy 'EXTERNE' Project. Summary Report EUR 16520 EN.

European Commission DGXII (1995b). Science, Research and Development Programme (JOULE). Externalities of Energy 'EXTERNE' Project. Methologology Report EUR 16521 EN.

Feenstra, J. (1984). Cultural Property and Air Pollution. Ministry of Housing, Physical Planning and Environment: Leidschendam.

Fenger, J., Jensen, T. and Brinch Madsen, H. (no date) Forsuringsprojektet: Materialskader (Acidification Project: Material Damage). Miljoministeriet, Copenhagen.

Glomsrod, S. and Rosland, A. (1988). Air Pollution and Material Damages: Social Costs. Report 88/31. Norway: Central Bureau of Statistics.

Heinz, I. (1980) See BMU (1986).

Isecke, B., Weltchev, M., and Heinz, I. (1991). Volkwirtschaftliche Verluste durch Unweltverschmutzungbedingte Materialschäden in der Bundesrepublik Deutschland. Texts of the Federal Environmental Agency 36/91, Berlin.

Lee, D.S., Holland, M.R. and Falla, N. (1995). An assessment of the Potential Damages to Materials in the UK from Tropospheric Ozone, Atmospheric Environment, forthcoming.
Ministry of Housing and Environment, Netherlands (1986). The Benefits of Environmental Policy in the Netherlands. Journal of Political Economy: 90, 4.
Pearce, D., Bann, C. and Georgiou, S. (1992) The Social Costs of Fuel Cycles. A Report to the UK Department of Energy. The Center for Economic and Social Research into the Global Environment, (CSERGE), University College London and University of East Anglia. (also referenced in an earlier version as 1991).

CHAPTER 9

GLOBAL WARMING

9.1. Introduction

The treatment of climate change is somewhat different from the other categories of damages that have been assessed. One reason is the extreme uncertainty surrounding the figures. Although the numbers presented here have relatively small ranges, there are researchers who take the view that the values are extremely high, and others who argue that we really do not know what the damages will be. The second is that we are dealing with impacts far into the future. Hence valuation methods have to look at long time periods. Third, although damages are global, and much of the research has been on estimating global damages, our concern is at the national level. Hence we need to look at the factors that affect local environments more closely.

In contrast to the other types of damages covered in the GARP *study, the issues surrounding global warming for the four countries under study are all contained in this section.*

9.2. Background and methodology

Anthropogenic increases in the concentrations of greenhouse gases in the atmosphere are believed to warm-up the Earth's atmosphere, potentially causing economic damage through a number of different impact pathways, e.g., warming, sea level rise, changes in rainfall patterns, increases in the incidence and severity of storms and hurricanes.

The latest scientific assessment predicts that a doubling of pre-industrial greenhouse gas concentrations in the atmosphere will result in a 1.5–4.5°C rise in temperature, which translates to a global mean surface temperature rise of 0.25°C per decade over the next century. However, there may be large regional differences. Warming in the Northern hemisphere, with greater land mass relative to oceans, may be in the order of 0.5°C per decade. This warming is likely to be accompanied by high regional climate variability and 'surprises'. The central estimate for sea level rise is 45 cm by the year 2100 as compared to 1990 (Pearce *et al.*, 1994).

There are a number of gases contributing to the enhanced greenhouse effect. They include CH_4, N_2O and CFCs. By far the most important, however, is carbon dioxide (CO_2) emissions. Anthropogenic increases in the concentration of CO_2 in the atmosphere mainly result from the burning of fossil fuels and deforestation. It has been estimated that global emissions of CO_2 in 1990 amounted to 6.9 Giga tons of carbon (GtC), of which 5.6 GtC were caused by fossil fuel burning and 1.3 GtC were caused by deforestation (IPCC, 1992a).

Climate change fits somewhat uneasily in the accounting framework used in the present study. Apart from the large measure of uncertainly related to its causes and effects, the major problem lies in the substantial time-lag between greenhouse gas emissions and the resulting warming effect. Moreover, unlike most other air pollutants, greenhouse gases are so-called stock pollutants. Global warming damage is not caused by the flow of emissions as such, but by their accumulation in the atmosphere (Fankhauser, 1994). This resembles another stock pollutant in our study: acid deposition. In acidification it is also not the flow of acid deposition, but rather the stock of acid that causes damage. For acidification we have, in our accounting framework, assessed the *current* (1990) damages, which are, in part, caused by historical deposition patterns. In theory, this approach could also be used for climate change. This would entail the assessment of current (1990) damage due to the increase in greenhouse gas concentrations in the atmosphere from pre-industrial levels (about 280 ppm) to current levels (about 350 ppm). However, little is known about current damage, neither about its incidence nor its severity. Research focuses on the assessment of future damages, especially the damage associated with a doubling of greenhouse gas concentrations in the atmosphere (the so-called $2*CO_2$ case). However, we are not interested in a point estimate of damage in a particular year, but rather with the flow of damages over time which can be attributed to an amount of CO_2 emissions in a particular period. *To remain as close as possible to the 1990 accounting year and the overall methodology of this study, we will calculate the present value of expected future climate change damage in the four case-study countries due to global carbon emissions in the year 1990.*

9.3. Estimation of physical damages

The literature has identified a large number of possible impacts of global warming on the economy. The impacts range from positive to negative (benefits and costs), from adaptation costs to unmitigated damage, and from market to non-market impacts. Some of these impacts have been monetized and others have not. For a full discussion of these issues the reader is referred to IPCC (1996) covering all aspects of the economic and social dimensions of climate change.

Table 9.1. Estimates of yield changes from global warming in sample countries

Country	Crop	Yield change (%)
Germany	Wheat, spelt	−5
Italy	Wheat, spelt	−10
Netherlands	Wheat, spelt	−1

Source: Kane, Reilly and Bucklin (1989).

Table 9.2. Estimates of global warming impacts ($2{*}CO_2$) on agriculture in Germany, Italy and the Netherlands

	1988 GDP $ billion			Implied value change	
Country	Total	Agriculture	Yield change in %	$ab = 0.75$	$ab = -1.25$
Germany	1130	16.5	−5	−0.6	−1.0
Italy	765	31.1	−10	−2.3	−3.8
Netherlands	215	8.5	1	0.1	0.2
Total	2110	56.1	−14	−2.8	−4.6

Based on Parry (1990).

9.3.1. *Agriculture*

Global warming is likely to have mixed effects on agriculture. The main damage will arise from heat stress and decreased soil moisture. Kane, Reilly and Bucklin (1989) estimate that in Northern latitudes [near 70° N] agricultural prospects will improve. In the mid latitudes [near 40° N and S] increases in temperature and drying of continental interiors is expected to lead to reductions in productivity. Projected coastal inundation of rice growing regions in parts of Southeast Asia, combined with the projected movement of the Asian monsoon away from the Indian sub-continent could also lead to reduced agricultural production. IPPC (1990c) draw the conclusion that food production can be maintained at essentially the same level as without climate change but the costs of achieving this are unclear. A restriction on such an optimistic view is that the yield increase in nature due to carbon dioxide fertilization will not be as high as laboratory results have suggested. Yield estimates for most of the countries under study are given in Table 9.1.

Table 9.2 applies the estimates of Table 9.1 to the value of the agricultural base in each country in order to obtain an impression of agricultural effects globally. The final two column.s apply the central percentage yield change to the initial agricultural GDP base with a multiple ranging from $ab = 0.75$ to $ab = 1.25$, where a is the ratio of the output reduction after low-cost adaptational changes to the initial yield reduction, and b is the ratio of total change in consumer and producer surplus to volume output reduction at base period prices.

The figures in Table 9.2 do not take account of trade patterns. All consumer and producer surplus effects are treated as proportionate to the production base. Taking international trade into consideration, consumer surplus losses would tend to be relatively greater for food importers, and producer surplus effects relatively greater for food exporters. Worldwide damage for the normal benchmark of $2*CO_2$ equivalent warming (2.5°C) would be $40 billion. The corresponding estimate for very long-term warming would be about $212 billion.

9.3.2. *Forests*

Global warming is supposed to cause a poleward migration of forests and a change in forest composition. Boreal forests would decline by 40%, temperate forests by 1.3%, whereas tropical forests would rise by 12% (OECD, 1992). US forests may lose 23–54% of standing biomass in the Great Lakes region and 40% in Western forests over the next 100 years (EPA, 1989). On this basis, one can estimate that global warming from $2*CO_2$ could cause a 40% loss for US forests.

For Europe the effects of climate change on forests are not yet analyzed in full detail. Research on regional climate change which is important in order to predict the effects of $2*CO_2$ equivalents on forest ecosystems is still lacking. In addition, climatic factors show an interrelationship with other factors (e.g. insects) which influence the state of forests.

9.3.3. *Sea level rise*

The extent of sea level rise is fiercely disputed. The USEPA dispute the IPCC estimates becasue of the implicit assumption that the Antarctic region would accumulate more ice than be a source of melting over the next 100 years – a point of view not generally accepted by glaciologists. The areas in Europe likely to be affected include the German beaches along the North Sea and nearby islands, the Po delta in Italy and low-lying land areas in the Netherlands and Belgium. Sea level rise will result in three types of costs: capitals costs of protective constructions (dikes, sea walls), the costs of foregone land services, and the costs of increased flooding hazards. There may also be migratory effects.

9.3.4. *Other effects*

Human dis-amenity damages will depend on location and income level. For high latitude countries the effects may be favourable, for mid and low latitude countries dis-amenity will play a major role. A warmer climate could aggravate urban pollution problems and some vector-borne diseases. If one assumes that human amenity is an income-elastic service, valuation tends to be higher in high-income countries (also relative to GNP). A predicted reduction of water supply in some areas could not only affect agriculture, but also households and industry.

Water supplies are also likely to be affected. Water balance models for Europe suggest a reduced precipitation and runoff in the south (Spain, Portugal, Greece) and possible decreases in runoff in the central region (this applies to Germany). The North (the UK, the Netherlands, and Belgium), however, will experience significant increases.

Finally, an increase in the incidence and severity of extreme weather events can cause damage to infrastructure and human lives.

9.4. Monetary estimation of the physical impacts

Several studies have assessed global costs and benefits of climate change in the event of a doubling of pre-industrial greenhouse gas concentrations ($2*CO_2$). The studies predict a net damage of 1.5%–2.5% of current world GNP. In absolute terms the $2*CO_2$ equilibrium warming leads to net damage costs of about US$ 270 bn. (Fankhauser, cited in Pearce *et al.*, 1994) to US$ 316 (Tol, ibid.) *in the year in which the doubling of CO_2 is achieved.*

These point estimates, however, tell us little about the social damage costs of a ton of carbon emitted today. There are considerable difficulties in calculating marginal costs of CO_2 emissions, especially because of the non-linearity of the climate system and damages. Moreover, because of the long-term impact of a unit of carbon dioxide in the atmosphere, calculations heavily depend on the discount rate used to compare damages which occur in different time periods. Estimates of marginal costs of present CO_2 emissions range from US$ 5 to US$ 45 per ton of carbon. A sort of central estimate is Fankhauser's US$ 20 per ton. This estimate has the additional advantage of being derived from a model which seems to be somewhat more appropriate for our purposes than some of the other models.[1]

Accepting the Fankhauser estimate of US$ 20 per ton of carbon, the present value of global damages due to global 1990 carbon emissions is 6.9 GtC $\times$ 20 US$/tC $\times 10^9 = 138$ bn. US$. Is it acceptable to multiply total carbon emissions in 1990 with the marginal costs of one ton of carbon, i.e., do we not violate marginality conditions? In defence of our linearity approximation is the fact that the global carbon emissions of one year only add a small percentage (about 0.9%[2]) to the atmospheric concentration.

[1] The distinction referred to is that Fankhauser's model is not an optimal control model to assess the optimal carbon tax. An optimal control model calculates the shadow price of carbon to keep carbon emissions on the optimal trajectory. Fankhauser, by contrast, does not assume an optimal trajectory of future emissions.

[2] 1 GtC = 0.47 ppm, thus 6.9 GtC is 3.2 ppm. Current concentration is about 350 ppm, thus the increase is ($3.2/350*100 = 0.9\%$).

Table 9.3. Damage costs of global warming to EU and four EU countries

	ECU Bn. 1989		% EU total	
	1°C	4°C	1°C	4°C
Germany	7.2	18.2	13.7	8.4
Italy	9.2	44.3	17.5	20.5
Netherlands	1.0	4.5	0.2	0.2
United Kingdom	3.2	13.5	6.0	6.2
EU total	52.5	216.1	100.0	100.0

We are, however, not interested in global damages as such, but in damages to the four case study countries. To this end, we need to introduce two additional, arguably controversial steps. The first step is to assess the ratio between world damages and EU damages; the second step is to assess the ratio between EU damages and specific EU member country damages.

For the first step we use regional estimates of $2*CO_2$ damages as reported in Pearce *et al.*, 1994. Fankhauser has specifically estimated EU damages in the $2*CO_2$ case to be US\$ 63.6 bn., compared to global damages of US\$ 269.6 bn. The EU/World ratio of damages is thus $63.6/269.6 = 0.24$. Tol gives figures for OECD-Europe and World, respectively. OECD-Europe is larger than EU, but Tol's figure of US\$ 56.5 bn. is less than the EU figure of Fankhauser. Tol's World estimate is US\$ 315.7 bn. The EU/World ratio according to Tol is therefore $56.5/315.7 = 0.18$ (or somewhat less, strictly speaking). Using these EU/World ratios, damages to the EU because of global 1990 emissions of carbon can be calculated as 0.18 to 0.24 times US\$ 138 bn., is US\$ 24.8 bn. to US\$ 33.1 bn.

For the second step we make use of a monetary assessment of climate change damages to the EU and EU member countries (CRU/ERL, 1992). Economic damage in this report is presented per degree of warming. Specifically, numbers are presented for temperature rises or 1°C and 4°C. The numbers are presented in Table 9.3 overleaf.

The CRU/ERL study distinguishes between five impact indicators and six economic sectors. The impact indicators are: sea level rise (including the effects of increased storm surges); run off (water supply); area above comfort threshold;[3] heating and cooling degree days; and agricultural yields.

Note that no health effects have been assessed. The economic sectors are labelled: agriculture; mining; manufacturing; electricity, gas, water; tourism; transport; services.

[3] The 'comfort index' is an index using temperature, sunshine and rainfall as indicators of the suitability of climates for leisure activities; changes of the index have impacts on the tourist industry. It can be interpreted as an human amenity index.

Table 9.4. Damages and benefits of 4°C temperature rise per impact indicator[1]

	Bn 1989 ECU		Comfort threshold	Heating and cooling	
	Sea level rise	Run-off	deg/days	deg/days	Agricultural
Germany	43.9	3.7	(15.4)	(11.0)	(3.0)
Italy	37.2	5.4	2.9	(0.6)	(0.8)
Netherlands	10.7	0.0	(1.4)	(2.9)	(1.8)
UK	26.6	0.0	(0.6)	(11.4)	(1.1)
EU Total	272.2	18.4	(27.1)	(35.5)	(12.2)

[1] Benefits are between brackets.

At a temperature rise of 4°C the damage as a percentage of 1990 GDP ranges from 2% (Germany) to 8.8% (Italy). The major share of the damage is due to sea level rise. For the Northern countries (Germany, Netherlands, and UK) the net impact would be positive without sea level rise. The impacts on the agricultural sector show a North–South divide: positive for Northern countries, negative for Southern countries Table 9.4 presents an overview of costs and benefits for the four case-study countries per impact indicator.

The before-mentioned IPCC assessment reviewed the CRU/ERL study. It concluded that CRU/ERL may have, on the one hand, overstated sea level costs (through augmenting sea level rise costs by a factor 2.7 to account for storm surges), while, on the other hand, it may have understated the damages of non-sea level rise impacts.

If we take the arithmetic mean of the damages corresponding to 1°C and 4°C temperature rise (Table 9.3), country damages as a fraction of EU damages are: Germany 0.11, Italy 0.19, Netherlands 0.02, UK 0.06. Because of the likely overestimation of sea level rise costs and the likely underestimation of the other damage costs the fractions between country damage costs and EU damage costs may be less correct, i.e. biased against those EU countries with a long coast line (e.g.Italy). However, because of a lack of better estimates we nevertheless use the CRU/ERL figures.

Multiplying these fractions with total EU damage due to 1990 global emissions (derived above) gives the country damages presented in Table 9.5. *These have been converted into 1995 prices as this is the standard reference year adopted throughout the synthesis report.*

Notice that the damage costs in Table 9.5 are present values of *national* damage costs of *global* emissions. The mid-range estimates are in the range 0.2 to 0.7% of GDP, and from ECU 37 to ECU 115 per capita. The range is quite small, about +/− 10–20% of the mid value. This narrow range is misleading, however, as regards the true uncertainty about these estimates, which arise from differences about physical impacts and valuation of those impacts, both of which have been relatively constrained in this analysis.

Table 9.5. Present values of global warming damages due to global 1990 carbon emissions in 1990 (ECU Million). Figures updated to 1995 prices.

	Germany			Italy			Netherlands			UK		
	Mid	Low	High	Mid	Low	High	Mid	Low	High	Mid	Low	High
	3870	3260	4470	6640	5670	7610	660	600	720	2110	1810	2410
As % of 1990 GDP	0.23	0.20	0.27	0.74	0.63	0.85	0.27	0.25	0.30	0.24	0.21	0.28
Per capita (ECU)	49	41	56	115	98	132	44	40	48	37	32	42

References

CRU/ERL (1992), Development of a Framework for the Evaluation of Policy Options to deal with the Greenhouse Effect, A Report for the European Commission (Directorate General for Environment, Nuclear Safety and Civil Protection), prepared by Climatic Research Unit, University of East Anglia, and Environmental Resources Limited, London.

Environmental Protection Agency (EPA) (1989). The Potential Effects of Global Climate Change on the United States. J. B Smith and d. Tirpak (eds.), Washington.

Fankhauser, S. (1994), Evaluating the Social Costs of Greenhouse Gas Emissions, CSERGE Working Paper GEC 94–1, London.

IPPC (1990c). IPCC First Assessment Report, Vol. 1: Overview and Summaries. New York: WMO and UNEP.

IPCC (1992a). Climate Change 1992. The Supplementary Report to the IPCC Scientific Assessment. Cambridge University Press, Cambridge.

IPCC (1996). Climate Change 1995: Economic and Social Dimensions of Climate Change. Contribution of WGIII to the Second Assessment Report of the IPCC. Cambridge University Press, New York.

Kane, S. Reilly, J. and Bucklin, R. (1989). Implications of the Greenhouse Effect for World Agricultural Commodity Markets. Washington: USDA.

OECD (1992). Global Warming: The Benefits of Emission Abatement. OECD, Paris.

Parry, M. (1990). Climate Change and World Agriculture. Earthscan Publications, London

Pearce, D.W., W.R. Cline, A.N. Achanta, S. Fankhauser, R.K. Pachauri, R.S.J. Tol, P. Vellinga (1994, 4th draft), Chapter 6: The Social Costs of Climate Change: Greenhouse Damage and the Benefits of Control. IPCC Working Group III.

CHAPTER 10

CONCLUSIONS AND RECOMMENDATIONS

The study had the objective of investigating the credibility of estimating environmental damages for the EU, within a consistent, coherent framework. To this end, it carried out four national case studies: for Germany, Italy, Netherlands and the UK. The approach taken was receptor based – i.e. to look at environmental pollution concentrations in a spatially disaggregated fashion, and to estimate the damages associated with these pollutants.

The study did not cover all pollutants; in particular it focused on air emissions. Furthermore it did not address the question of whether or not the estimated damages were 'internalised' in the GDP figures – i.e. whether GDP was already lowered because of the damages. On account of its receptor based approach, the study did not look at damages outside national boundaries.

At the outset the teams tried to adopt as consistent an approach as possible. This meant agreeing on selection of dose response functions, valuation methodologies and selection of data sets. Inevitably, the coverage and approach were not fully consistent. The degree of spatial disaggregation varied by country, as did the availability of data on pollution concentrations and estimates of stocks at risk. These differences account for a significant part of the differences in the resulting damage estimates, as has been pointed out in the report.

Monetary estimation of the impacts of pollutants on health, materials and crops; and of economic activities on noise amenity have been established, with some credibility. Some further work is required to take these estimates to a level where they can be reported in a set of satellite environmental accounts. This work should not, however, be beyond the scope of the follow-on project. In addition, the teams found the *physical* data of environmental impacts to be of considerable value and it is recommended that the latter be presented alongside the monetary data.

The teams were less successful in attaining reasonable coverage of the monetary impacts of damages to forests and ecosystems. Some improvements in these areas are possible but it is highly unlikely that comprehensive estimates can be developed. We recommend therefore a hybrid approach, where monetary values and physical impacts are presented, with the latter being in the form that is closest to the 'end point' that affects human welfare.

The estimates of global warming damages reflect the consensus that currently exists among important sections of the scientific research community. But it is important to note that there is a significant group of researchers who challenge the assumptions on which the figures are based. We recommend national damage estimates of global warming be reported as a separate category, indicating that they are highly uncertain and possibly subject to major revisions.

Although there are many gaps in the numbers, we have summarised the monetary findings in Table 10.1 overleaf. This provides an overview of the data. The numbers are *not* comparable across countries as there are gaps in the data. They are most complete for Netherlands and UK, where damages amount equate to around 7.1% and 4.1% of 1990 GDP respectively. The big gap in the UK compared to the Netherlands is the absence of forestry recreational damage estimates.

The programme of research raises the question of how (and to what extent) these estimates can be integrated into a system of national accounts. In their present form, it is the teams' view that the data are more useful presented *alongside* the national income accounts rather than deducted from GDP. One reason is that we do not know the extent to which damages have been internalized. Subtracting them would result in double counting. The second is that by far the greatest benefit in these data is to guide policy on the economic and environmental interface, and the aggregation of damages is of little value in this regard.

To develop the research further we need to extend coverage within each country so that it is as uniform as possible. That is important for policy-making within the EC, where cross-country resource allocation and regulatory decisions have to be taken. Second, we need to attribute damages to sources. This is an essential linkage if the framework is to be of policy relevance. Third we should extend coverage to pollutants that affect water resources, as other studies indicate this to be of importance in terms of environmental priorities.

What is the scope for updating the results obtained here on a regular basis? Such updating is necessary if we are to use the data for policy purposes. We believe this can be done at reasonable cost. First it is not necessary to review all the components that go into the valuation. There are, broadly, three components: data on pollutants and their concentrations, data on stocks at risk and data on valuation of end points. The first will need the most frequent updating, to reflect physical changes in the environmental situation. The second will need to be revised as frequently as similar data is revised in other sections of the national accounts (e.g. data on stocks of buildings, persons resident in different areas, forest stocks etc.). The third need be revised less frequently. Changes in estimates of VLYL or property depreciation per unit of noise etc. can be made, for example, every 3–5 years. Similarly, new pollutants will need to be considered. When such changes are made, it will be necessary to re-estimate previous damages so as to maintain a consistent time series. This should not, however, be too costly an exercise.

This study has made a promising start in an important area. Much remains to the done but we are well on the way.

Table 10.1. Summary of economic valuation for selected environmental damages in four countries of the EU

ECU million unless otherwise stated (1995 prices)

	GERMANY (1)			ITALY (2)			NETHERLANDS (3)			UNITED KINGDOM (4)		
	Mid value	Range		Mid value	Range		Mid value	Range		Mid value	Range	
Health damages (5)												
Mortality	850.1	425.0	1275.1	25859.9	18016.0	33703.8	6847.0	4571.8	9377.9	12246.6	3724.3	20769.0
Morbidity	2511.7	1620.4	3403.0	10580.5	5021.3	16139.6	4463.4	1827.9	7581.9	13079.1	1546.4	24611.8
TOTAL	3361.8	2045.4	4678.1	36440.3	23037.3	49843.3	11310.4	6399.7	16959.8	25325.8	5270.7	45380.8
As % of GDP in 1990	0.21%	0.13%	0.29%	4.08%	2.58%	5.58%	4.77%	2.70%	7.15%	2.98%	0.62%	5.34%
Per inhabitant	42	26	59	632	399	864	759	430	1138	441	92	791
Noise damages												
TOTAL	11686.2	10388.7	12983.8	N.A.	N.A.	N.A.	2069.6	N.A.	N.A.	4833.2	1208.3	12566.3
As % of GDP in 1990	0.72%	0.64%	0.80%	N.A.	N.A.	N.A.	0.87%	N.A.	N.A.	0.569%	0.142%	1.479%
Per inhabitant	147	131	163	N.A.	N.A.	N.A.	139	N.A.	N.A.	84	21	219
Crop damages												
TOTAL	66.2	25.3	107.3	62.8	44.9	74.3	212.0	210.3	213.8	42.5	42.5	42.5
As % of GDP in 1990	0.00%	0.00%	0.01%	0.01%	0.01%	0.01%	0.09%	0.09%	0.09%	0.01%	0.01%	0.01%
Per inhabitant	1	0	1	1	1	1	14	14	14	1	1	1
Forest damages												
TOTAL	2750.1	1465.9	4035.5	2.8	1.2	4.3	2227.3	2227.3	2227.4	1.8	0.8	2.9
As % of GDP in 1990	0.17%	0.09%	0.25%	0.00%	0.00%	0.00%	0.94%	0.94%	0.94%	0.00%	0.00%	0.00%
Per inhabitant	35	18	51	0	0	0	149	149	149	0	0	0
Materials damages												
TOTAL	3127.7	1755.7	4499.2	N.A.	N.A.	N.A.	287.3	279.1	423.9	2553.3	1142.9	3710.9
As % of GDP in 1990	0.19%	0.11%	0.28%	N.A.	N.A.	N.A.	0.12%	0.12%	0.18%	0.30%	0.13%	0.44%
Per inhabitant	39	22	57	N.A.	N.A.	N.A.	19	19	28	44	20	65
Global warming damages												
TOTAL	3870.0	3260.0	4470.0	6640.0	5670.0	7610.0	660.0	600.0	720.0	2110.0	1810.0	2410.0
As % of GDP in 1990	0.23%	0.20%	0.27%	0.74%	0.63%	0.85%	0.27%	0.25%	0.30%	0.24%	0.21%	0.28%
Per inhabitant	49	41	56	115	98	132	44	40	48	37	32	42
FINAL SUB TOTAL	24862.0	18941.0	30773.9	43145.9	28753.5	57532.0	16766.7	9716.5	20544.9	34866.6	9475.3	64113.5
As % of GDP in 1990	1.52%	1.16%	1.89%	4.83%	3.22%	6.44%	7.07%	4.10%	8.66%	4.10%	1.12%	7.55%
Per inhabitant	313	238	387	748	498	997	1125	652	1379	607	165	1117

Notes:

(1) Figures for Germany refer only to West Germany in most cases. Health damages are very limited as PM10 and ozone are excluded.

(2) Figures for Italy include ozone damages to health for Lombardy only and lack data for noise and materials. Data on forests are very incomplete.

(3) Figures for Netherlands have included forest non-timber damages from other studies.

(4) Figures for the UK do not include non-timber forestry damages.

(5) Health damages for mortality are based on Value of Life Years Lost (VLYL) approach.

(6) 'Mid Values' are not always the average of the range, but a value selected as the mid figure by the country team.

N.A.: Not assessed, data not available, not possible to estimate separately

DUTCH CASE STUDY

Prepared by IVM* in collaboration with the other country teams.

* Instituut voor Milieuvraagstukken (IVM), Vrije Universiteit, De Boelelaan 1115, 1081 HV Amsterdam, The Netherlands. Tel: +(31) 20 4449555; Fax: +(31) 209 4449553; E-mail: onno.kuik@ivm.vu.nl

CHAPTER 1

INTRODUCTION AND DATA ISSUES

1.1. Introduction

The coverage of the Dutch assessment under the first phase of the Green Accounting Research Project (GARP) is presented in Table 1.1 below. The effects of SO_2 have been considered on the widest range of *receptor categories.* Due to severe limitations on both data and resources, the assessment was not able to consider the effects of atmospheric pollutants on forests or ecosystems. Considerable progress was made, however, in the assessment of noise related damages.

The section below assesses the data required on pollutant concentrations for the Netherlands on a detailed geographical level. As mentioned previously, the study benefitted from the methodology that was developed under the ExternE Project (see European Commission, 1995a–f). However, for the specific purposes of this study many additional data were acquired and processed. The assessment is based, as much as possible, on dose–response functions and methodologies presented. When specific assumptions have been made this is clearly indicated in the text.

1.2. Air quality data

1.2.1. *Data availability*

Air quality in the Netherlands is monitored by the National Institute of Public Health and Environmental Protection (RIVM) in the National Air Quality

Table 1.1. Pollutants and damage categories addressed in the Dutch assessment

	PM_{10}	SO_2	NO_x	O_3	Noise
Health	×	×		×	
Materials		×			
Crops		×		×	
Forests					
Ecosystems					
Amenity					×

Monitoring Network (LML). The monitoring stations of the LML are divided into regional stations, city background stations, and street stations. The regional stations are considered representative for the air quality in a ray of 5–50 km around the station, and are used by RIVM to calculate the overall-distribution of air pollutant concentrations by means of interpolation. The city background stations are located in cities with 40,000–800,000 inhabitants and are considered representative for the air quality in comparable cities. Finally, street stations are located in urban streets with a high traffic intensity (> 10,000 motorized vehicles per day) and are considered representative for the air quality in comparable streets (estimated at about 100). An overview of air pollutants monitored in the LML which are relevant for this study and the number of measuring stations in 1990 is presented in Table 1.2. Black smoke (BS) is also measured outside the LML by other organizations than the RIVM. The measurements of these stations, which are mainly located in urban and industrial areas, are also included in the Annual Air Quality Survey. Including these stations there are a total of 34 stations for BS.

The results of the LML are published annually by the RIVM in two publications. The Annual Air Quality Survey (Dutch: Jaaroverzicht Luchtkwaliteit) provides a broad overview of the air quality and the concentrations of a number of pollutants. The publication National Air Quality Monitoring Network: Results of the Measurements (Dutch: Landelijk Meetnet Luchtkwaliteit: Meetresultaten) gives the results of the indivual stations. This last publication consists of 4 volumes: 3 for the regional stations and 1 for the city background and street stations. Table 1.3 gives an overview of the availability of data for each pollutant. An X in a cell means that data for a certain pollutant are available in that specific form.

1.2.2. *Particulate matter*

The measure of particulate matter that was chosen for this study is PM_{10}. However, the measure of particulate matter which is monitored in the LML is

Table 1.2. Air pollutants measured in the LML and number of stations for 1990

		Number of stations			
Pollutant	Unit	Total	Regional	City	Street
SO_2	$\mu g/m^{-3}$	83	77	6	–
NO_x	ppb	43	25	5	13
–NO	$\mu g/m^{-3}$	43	25	5	13
$-NO_2$	$\mu g/m^{-3}$	43	25	5	13
O_3	$\mu g/m^{-3}$	36	20	4	12
Oxidant	ppb	36	20	4	12
Particulates: black smoke	$\mu g/m^{-3}$	20	14	–	6

Table 1.3. Pollutant data available for each monitoring station in the LML in 1990

	SO_2	BS[6]	NO_x	NO	NO_2	O_3
1-hour percentiles (50, 60, 70, 80, 90, 95, 98)[1]	×		×	×	×[5]	×
24-hour percentiles (50, 60, 70, 80, 90, 95, 98)[1]	×	×	×	×	×	×
8 highest 1-h measurements[1]	×		×	×	×	×
8 highest 24-h measurements[1]	×	×	×	×	×	×
Annual average[1]	×	×	×	×	×	×
Average concentration by windsector[2]	×		×	×	×	×
Daily averages[3]	×	×	×	×	×	×
Monthly averages[4]	×	×	×	×	×	×

[1] For periods: year 1990, summer 1990 (April–September [incl.]), year April 89/April 90 (April–March [incl.]) and winter 89/90 (October–March [incl.]).
[2] For periods: year 1990, summer 1990 and winter 89/90.
[3] Calculated from hourly measurements.
[4] Calculated from daily averages.
[5] Also a 99.9 percentile is given.
[6] BS = black smoke.

Black Smoke (BS). The maximum diameter of particulates measured with the BS method is approximately 5 μm, whereas PM_{10} consists of particulates with a maximum diameter of 10 μm. There are no measurements of PM_{10} available for 1990; PM_{10} is measured in the LML only since 1992. Also the number of stations measuring PM_{10} is less than the number of stations measuring BS, and most stations measuring PM_{10} have not been operating during all of 1992. Also the method used to measure PM_{10} systematically underestimates the true concentration by about 25%.

There are measurements available of TSP for 1990 but the stations that measure TSP are largely located in an urban or industrial environment and can therefore not be considered representative for the overall situation. Also these measurements are performed on a less frequent basis than the BS measurements. So to keep in line with the common base year 1990 and considering that BS is measured on a larger and more frequent scale than TSP, BS will be used as a measure of particulate matter in the Dutch implementation of this study.

Since the dose–response relationships use PM_{10} we have tried to establish a relationship between BS and PM_{10} using the daily mean concentrations from stations which measured BS as well as PM_{10} in 1992. Figure 1.1 shows a plot of the measured PM_{10} concentrations against the measured BS concentrations (left graph), and a plot of the ratio of BS to PM_{10} against the BS values (right graph) for one of these stations. From these graphs it becomes clear that the ratio of BS to PM_{10} is not constant; it increases with increasing BS concentrations. This means that BS concentrations cannot be converted to PM_{10} concentrations using a constant factor. Therefore a regression equation was

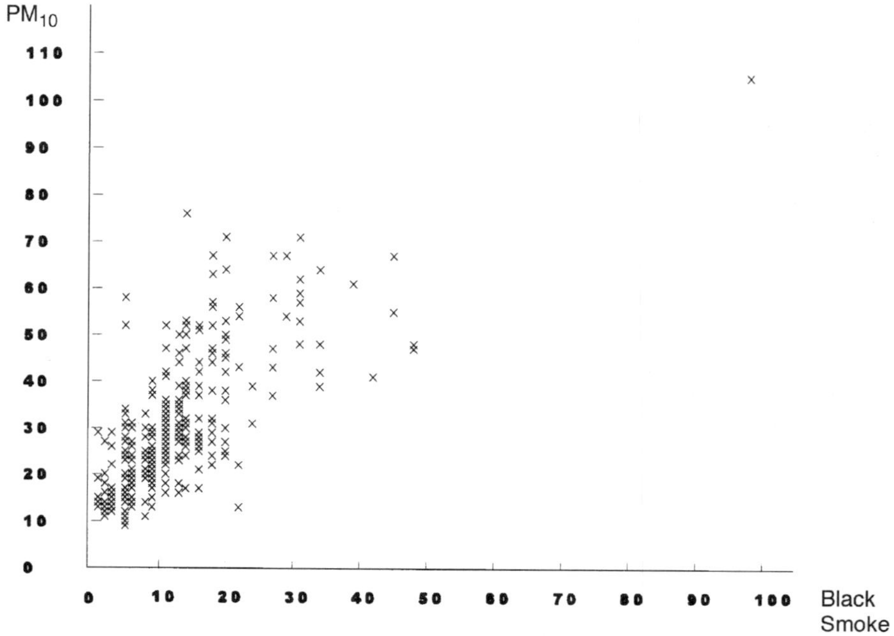

Fig. 1.1. Scatterplot of PM_{10} concentrations against black smoke concentrations (daily averages). Data are from station 133 in the LML 1992.

estimated which gives PM_{10} concentrations as a (linear) function of BS concentrations.

It was found that the best results could be obtained by estimating this equation for each station separately. Results of the regressions are shown in Table 1.4. From applying these relationships directly to annual mean concentrations it appeared that they might give a slight overestimate of the true annual mean PM_{10} concentration. Furthermore it has to be kept in mind that these equations give the relationships between *measured* concentrations; results of the equations have to be divided by 0.75 to correct for the systematic underestimation of PM_{10} concentrations by approximately 25%.

1.2.3. *Spatial distribution of air pollutant concentrations*

The spatial distribution of air pollutant concentrations is constructed by linear interpolation of the results of the regional stations. This means that no use is made of modelled data. Experience within the RIVM, which also uses interpolation methods to construct spatial distribution of pollutant concentrations, has shown that interpolation produces credible results. Although the RIVM uses a specially developed interpolation method, comparison of different interpolation methods has shown that results of different methods are within a range

Table 1.4. Results of the regression equations

	$PM_{10} = a + b * BS$				
Station	a[1]	b[1]	n	r	R^2
131	18.688	1.749	258	0.694	0.482
133	16.618	1.098	307	0.725	0.525
230	19.540	1.978	287	0.722	0.521
318	17.151	1.261	285	0.757	0.574
437	18.299	1.221	320	0.737	0.543
538	20.537	1.703	276	0.636	0.405
540	21.947	1.059	299	0.719	0.517
722	18.453	1.571	273	0.633	0.401
724	18.738	1.415	339	0.809	0.654
928	19.978	1.360	326	0.763	0.583
All	19.563	1.286	2970	0.713	0.508

[1] All coefficients statistically significant at the 0.1% level.

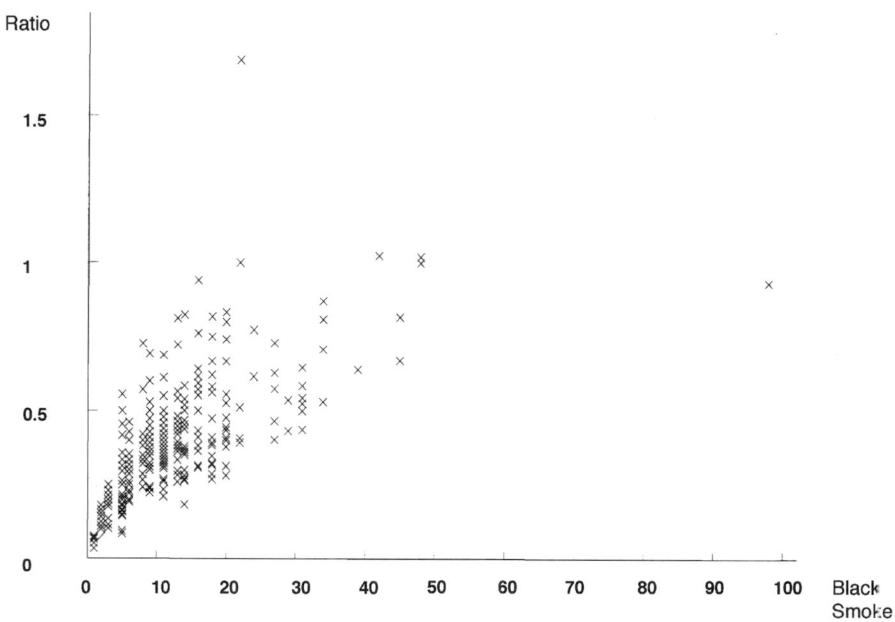

Fig. 1.2. Scatterplot of the ratio of PM_{10} concentrations against BS concentrations (daily averages). Data are from station 133 in the LML 1992.

of 10% of each other (personal communication with members of RIVM's Air Quality Department). Also, models are not available for all pollutants (e.g. O_3).

A modelling exercise for PM_{10} for 1992 using the TREND-model shows a considerable underestimation of actual concentrations. The estimated annual mean concentration was approximately 30 $\mu g/m^{-3}$ (including a 10% correction for wind-blown dust, sea salt aerosol etc.), with a range of 25–35 $\mu g/m^{-3}$, whereas the (for underestimation corrected) measured annual mean concentration was 44 $\mu g/m^{-3}$ (average of the regional stations) (RIVM, 1994). Given these considerations we believe that relying on monitored data and interpolation can be justified for the Dutch situation.

Linear interpolation was performed using the SPANS Geographical Information System using the nearest neighbourhood method (SPANS, 1993). Calculations were performed on a $5*5\,km^2$ grid surface. Since most of the other data for the calculation of impacts (population distribution, yields of agricultural crops etc.) were available on the municipality level the results of the interpolation were used to calculate an area-weighted concentration per municipality.

CHAPTER 2

DAMAGE TO HEALTH

2.1. Costs arising from PM_{10}

2.1.1. *PM_{10} concentrations*

The regression equations shown in Table 1.4 were used to estimate the mean annual PM_{10} concentrations for 1990 from the mean annual Black Smoke (BS) concentrations. The equation estimated from the data from all stations was used to estimate PM_{10} concentrations for those stations for which no separate equation could be estimated. The mean annual BS concentrations averaged over the regional stations in 1990 was 11.2 $\mu g/m^{-3}$ with a range from 8 to 14 $\mu g/m^{-3}$. The resulting estimated annual mean PM_{10} concentration averaged over the regional stations in 1990 was 45.5 $\mu g/m^{-3}$, and ranged from 41.1 to 55.2 $\mu g/m^{-3}$.

The spatial distribution of PM_{10} concentrations was again constructed by linear interpolation on a $5*5\ km^2$ grid surface. Concentrations per grid were used to calculate an area-weighted PM_{10} concentration for each municipality. Only the results from the regional stations were used for the interpolation. At first sight one might think that this will result in an underestimate of urban concentrations of PM_{10}. However, measurements of PM_{10} from 1992 and 1993 surprisingly show that concentrations measured at city background stations are scarcely higher than those measured at regional stations (see Table 2.1).

2.1.2. *Dose–response relationships for PM_{10}*

The common set of dose–response relationships was used. For a discussion of these functions, the reader is referred to the Synthesis Report. Since the dose–

Table 2.1. Comparison of mean PM_{10} concentrations at regional stations and city background stations in the LML (corrected for underestimation)

	1992		1993	
	Regional (10)	City (3)	Regional (9)	City (2)
Annual mean	44	44	40	41
Range	40–51	41–47	32–47	41–41

Source: RIVM (1994a, 1994b).

response relationships for mortality give the percentage change in mortality rates we have to correct for the deaths from air pollution which are implicit in the background mortality rates. Define dx_i as the percentage change in the mortality rate in municipality i (outcome of the dose–response relationships), x_i as the known mortality rate and x_b as the mortality rate at background concentrations of PM_{10} (which is unknown). If we define pop_i as the population at risk in municipality i, the number of extra deaths in this municipality is given by $(x_i * pop_i - x_b * pop_i) = (x_i - x_b) * pop_i$. Since x_i and x_b are related by $x_i = (1 + dx_i/100) * x_b$ we can write the number of extra deaths as:

$$\sum_i \left(x_i - \frac{x_i}{1 + dx_i} \right) \cdot pop_i \qquad (1)$$

The dose response relationships for morbidity impacts do not require such a correction as they are designed to give directly the extra cases of a certain impact from an increase in the PM_{10} concentration.

2.1.3. *Results for PM_{10}*

Impacts were calculated on the level of municipalities by combining the PM_{10} concentration, mortality rate and population. Data on population per municipality were obtained from the Central Bureau of Statistics (CBS, 1990). These figures give the population on January 1 1990, by age (5 year age groups) and sex. Mortality data on municipality level are only to be obtained through analyses of postal code based data. The postal code data, collected by the Central Bureau of Statistics, are not readily available because of privacy restrictions. The mortality data on province and regional public health service (GGD) level were, however, available from a Regional Public Health Profile database made by the National Institute for Public Health and Environmental Protection (RIVM, 1990).

These data are average figures for mortality due to several causes for 1985 to 1989. Later data are not available on this detailed level. Mortality due to accidents and violent causes can be subtracted from the total mortality figures. The resulting mortality rates for non-violent death on the provincial level, range from 7.60‰ to 9.49‰ for men, and from 7.26‰ to 8.09‰ for women. These mortality rates were used for the calculations. Each municipality was assigned the mortality rate of the province it belongs to.

Some of the dose–response functions for morbidity impacts have to be applied to a certain sub-population of the general population only, for instance children or asthmatics. The percentage of asthmatics is estimated at 1.7% of the population over 18 years of age (CBS, 1989). This figure is for 1986. Later figures are not available because from 1987 on only an *aggregate* estimate for asthma, chronic bronchitis and CARA (chronic a-specific respiratory diseases) is made. We have applied the adult-percentage to all age groups. This may result in an underestimate of the number of asthmatics since asthma is more prevalent in children than in adults. Dose–response functions for children were applied to all persons under 15 years of age, dose–response functions for adults were applied to all persons from 15 years of age.

The tables below (Tables 2.2 to 2.5) give the results of the calculations for the mortality and morbidity impacts from PM_{10}. Results are related to the total population of the Netherlands on the 1st of January 1990 (14,892,574 people), and the monetary value of the damages are related to the Dutch GDP in 1990 (223,287 million ECU at 1990 prices and exchange rates). Given the uncertainty concerning the background level, results are given for a background level of 5, 10 and 15 $\mu g/m^{-3}$. Also the increase in the number of cases of a certain effect when the background level is lowered by 1 $\mu g/m^{-3}$ is shown. Since the relationship for mortality is not linear (equation (1)), the increase in mortality is not constant for the whole range of background levels. Therefore,

Table 2.2. Estimated acute mortality impacts from PM_{10}

	Number of deaths (% of 1990 population)			Damage in million ECU (% of 1990 GDP)		
PM_{10} background	Low	Mid	High	Low	Mid	High
5	3072	4913	6742	7986.1	12774.4	17529.5
	(0.021)	(0.033)	(0.045)	(3.57)	(5.72)	(7.85)
10	2702	4329	5952	7024.1	11257.7	15475.4
	(0.018)	(0.029)	(0.040)	(3.15)	(5.04)	(6.93)
15	2329	3740	5151	6056.1	9724.6	13393.9
	(0.016)	(0.025)	(0.035)	(2.71)	(4.35)	(6.00)
Average increase per 1 $\mu m\ m^3$ lower background level	75	120	164	196	312	426

Table 2.3. Estimated chronic mortality impacts from PM_{10}

	Impacts, cases (% 1990 population)			Damages, million ECU (% 1990 GDP)		
PM_{10} background	Low	Mid	High	Low	Mid	High
5	12968	16425	19667	33717	42705	51135
	(0.09)	(0.11)	(0.13)	(15.1)	(19.1)	(22.9)
10	11522	14645	17594	29956	38076	45745
	(0.08)	(0.10)	(0.12)	(13.4)	(17.1)	(20.5)
15	10036	12803	15435	26093	33287	40130
	(0.07)	(0.09)	(0.10)	(11.7)	(14.9)	(18.0)
Average increase per 1 $\mu g/m^{-3}$ lower background level	293	362	423	762	942	1101

Table 2.4. Estimated acute morbidity impacts from PM_{10} at different background levels

	Back-ground level	Number of cases			Damage in 10^6 ECU		
		Low	Mid	High	Low	Mid	High
Hospital admissions for respiratory infections	5	749	1130	1517	4.946	7.459	10.012
	10	657	991	1330	4.337	6.540	8.779
Unit-value: 6600 ECU	15	565	852	1143	3.728	5.621	7.545
	per μg[1]	18.46	27.85	37.38	0.122	0.184	0.247
Hospital admissions for COPD	5	973	1372	1771	6.422	9.055	11.688
	10	853	1203	1553	6.264	7.939	10.248
Unit-value: 6600 ECU	15	733	1034	1335	6.106	6.824	8.808
	per μg[1]	24	34	44	0.158	0.223	0.288
Emergency room visits for COPD	5	3505	4352	5197	0.652	0.809	0.967
	10	3074	3815	4557	0.572	0.710	0.848
Unit-value: 186 ECU	15	2642	3279	3917	0.491	0.610	0.729
	per μg[1]	86	107	128	0.016	0.020	0.024
Emergency room visits for asthma	5	2418	3868	5198	0.450	0.719	0.967
	10	2120	3992	4557	0.394	0.631	0.848
Unit-value: 186 ECU	15	1822	2915	3917	0.339	0.542	0.729
	per μg[1]	60	95	128	0.011	0.018	0.024
Hospital visits for childhood croup	5	13176	17589	23087	2.451	3.271	4.294
	10	11552	15421	20243	2.149	2.868	3.765
Unit-value: 186 ECU	15	9929	13254	17399	1.847	2.465	3.236
	per μg[1]	325	433	569	0.060	0.081	0.106
Restricted activity days in adults	5	15715532	24660537	38695792	980.649	1538.817	2414.617
	10	13779516	21622574	33928809	859.842	1349.249	2117.158
Unit-value: 62.4 ECU	15	11843500	18584612	29161826	739.034	1159.680	1819.698
	per μg[1]	387203	607593	953397	24.161	37.914	59.492
Shortness of breath days in asthmatics	5	719216	1438431	2157647	22.511	45.023	67.534
	10	630615	1261229	1891844	19.738	39.476	59.215
Unit-value: 31.3 ECU	15	542014	1084027	1626041	16.965	33.390	50.895
	per μg[1]	17720	35440	53160	0.555	1.109	1.664
Symptom days	5	134112572	281038062	415150634	844.909	1770.540	2615.449
Unit-value: 6.3 ECU	10	117591076	143416631	364007707	740.824	1552.425	2293.249
	15	101069581	211795200	312864781	636.738	1334.310	1971.048
	per μg[1]	3304299	6924286	10228585	20.817	43.623	64.440

[1] per μg: the increase in damages when the background level of PM_{10} is lowered by 1 $\mu g/m^{-3}$.

an average increase per 1 $\mu g/m^{-3}$ over the range from 5 to 15 $\mu g/m^{-3}$ is shown. For morbidity impacts this increase is constant over the whole range of background levels.

For the monetary valuation of mortality a figure of ECU 2.6 million for the Value of Statistical Life (VOSL) was used in the initial calculations which are presented below. These can be assumed to represent 1990 prices. In the Synthesis Report all the estimates were updated to 1995 prices. Unit-values that have been used to value morbidity impacts are also shown in the tables.

These results can also be related to the number of deaths from respiratory and cardiovascular diseases in 1990, which are, respectively, 10,671 and 51,620 (totalling 62,291). Using the estimates for a background level of 10 $\mu g/m^3$ the above estimates for acute mortality represent 25.3% (low), 40.6% (mid) and

Table 2.5. Estimated chronic morbidity impacts from PM_{10} at different background levels

	Back-ground level	Number of cases Low	Number of cases Mid	Number of cases High	Damage in 10^6 ECU Low	Damage in 10^6 ECU Mid	Damage in 10^6 ECU High
	5	222390	345939	464547	30.712	47.774	64.154
Chronic bronchitis	10	194993	303322	407319	26.929	41.889	56.251
in adults	15	167596	260706	350091	23.145	36.003	48.348
Unit-value: 138.1 ECU	per μg[1]	5479	8523	11440	0.757	1.177	1.581
	5	296520	469489	637517	40.949	64.836	88.041
Respiratory illness	10	259991	411652	558980	35.905	56.849	77.195
in adults	15	223462	353815	480444	30.860	48.862	66.349
Unit-value: 138.1 ECU	per μg[1]	7306	11567	15707	1.009	1.597	2.169
	5	93656	177396	262237	12.934	24.498	36.215
Chonic bronchitis	10	82119	155542	229932	11.341	21.480	31.754
in children	15	70581	133689	197627	9.747	18.462	27.292
Unit-value: 138.1 ECU	per μg[1]	2308	4371	6461	0.319	0.604	0.892
	5	113489	228081	344875	15.672	31.498	47.627
Chronic cough	10	99508	199983	302390	13.742	27.618	41.760
in children	15	85528	171885	259904	11.811	23.737	35.893
Unit-value: 138.1 ECU[2]	per μg[1]	2796	5620	8497	0.386	0.776	1.173

[1] per μg: the increase in damages when the background level of PM_{10} is lowered by 1 $\mu g/m^{-3}$.
[2] In the absence of a better figure the value for chronic bronchitis was used.

55.8% (high) of the total deaths from respiratory diseases. Related to deaths from both respiratory and cardiovascular diseases they represent 4.34% (low), 6.95% (mid) and 9.56% (high) of these deaths.

Table 2.3 presents the estimated *chronic* mortality impacts from PM_{10}. This table is included for completeness. As discussed in the synthesis report, it is still very much unclear to what extent the estimates for acute and chronic impacts are additive.

CBS (1993) gives an estimate of 29.9 days with temporary limitation of activity per person per year for the period 1990/1991 for persons of all ages, of which 4.1 days are spent in bed (permanently bed-ridden excluded). This estimate is based on a continuously conducted survey (every two weeks another sample) which asks respondents the following question: "During the past two weeks, have you cut down on the things you usually do because of illness or injury? If yes, for how many days was this in all during these two weeks". If applicable, the number of days spent in bed was asked for.

Reported figures were converted to an annual figure by multiplying them by 26. Even though this figure includes all possible causes for restricted activity days it is interesting to compare them to the estimate for RAD's from PM_{10} made here. The figure of 29.9 days per person per year means a total of 445 million days for the total population. The number of days with limited activity for persons under 15 years of age is estimated at 12.6 days per person per year (CBS, 1993), which gives an estimated total of 34 million days for all persons under 15 years. This means a total of 411 million days with limited activity for adults (15 years and older). The mid estimate of RAD's in adults at a background level of 10 μg/m^{-3} is approximately 22 million, or 5.3% of the total of 411 million restricted-activity days.

From a health survey held in 1986, the percentage of persons aged 18 years and over that suffer from chronic bronchitis is estimated to be 5.2% (CBS, 1989). Applying this percentage to all persons aged 15 years and over in 1990 gives a total of 774,414 adults suffering from chronic bronchitis. The mid estimate for chronic bronchitis in adults at a background level of 10 μg/m^{-3} PM_{10} made here is 303,322. This is 39% of the total number of adults suffering from chronic bronchitis.

2.2. Costs arising from O_3

2.2.1. *O_3 concentrations*

The 1990 annual mean of the daily 1-hour maximum O_3 concentration at the monitoring stations ranged from 45 to 92 μg/m^{-3}, with an average of 73 μg/m^{-3}. Linear interpolation was again used to construct a spatial contribution of concentrations. Results from the regional, city background and street stations were used for this purpose. This probably results in an underestimate of damages because it allows the generally lower O_3 concentrations in urban areas and especially near busy streets to extend their influence into rural areas.

However, we felt this was to be preferred to an overestimate which was likely to result if we had used only the regional background stations. A conversion factor of 0.5 was used to transform concentrations in μg/m^{-3} to concentrations in ppb. Given the uncertainty regarding the appropriate background level, impacts were calculated relative to three background levels: 20, 25 and 30 ppb.

2.2.2. *Dose–response relationships for O_3*

The common set of dose–response functions was used. For a discussion of these, the reader is referred to the Synthesis Report.

2.2.3. *Results for O_3*

Tables 2.6 and 2.7 below show the results of the calculations. The methodology and valuations used are the same as for PM_{10}.

2.3 Costs arising from SO_2

2.3.1. *SO_2 concentrations*

A spatial distribution of annual mean SO_2 levels was constructed using the same procedure that was followed for PM_{10} and O_3 concentrations. Only

Table 2.6. Estimated acute mortality impacts from O_3 at different background levels

O_3 background	Impacts, cases			Damages, million ECU (% 1990 GDP)		
	Low	Mid	High	Low	Mid	High
20	212	317	422	550 (0.25)	824 (0.37)	1098 (0.49)
25	150	224	299	389 (0.17)	583 (0.26)	777 (0.35)
30	88	131	175	228 (0.10)	342 (0.15)	455 (0.20)
Average increase per 1 $\mu g/m^{-3}$ lower background level	12	19	25	32	48	64

Table 2.7. Estimated acute morbidity impacts from O_3 at different background levels

	Back-ground (ppb)	Impacts, cases			Damages, 10^6 ECU		
		Low	Mid	High	Low	Mid	High
Hospital admissions for respiratory infections	20	1652	2202	2626	11	15	17
	25	1167	1556	1856	8	10	12
Unit-value: 6600 ECU	30	683	910	1085	5	6	7
	per ppb[1]	97	129	154	0.6	0.9	1.0
Hospital admissions for COPD	20	1005	1602	2202	7	11	15
	25	710	1132	1556	5	7	10
Unit-value: 6600 ECU	30	415	662	910	3	4	6
	per ppb	59	94	129	0.4	0.6	0.9
Hospital admissions for asthma	20	735	1438	2228	5	10	15
	25	519	1048	1575	3	7	10
Unit value: 6600 ECU	30	304	613	921	2	4	6
	per ppb	43	87	131	0.3	0.6	0.9
Emergency room visits for asthma	20	4467	6831	9090	0.8	1.3	1.7
	25	3157	4827	6424	0.6	0.9	1.2
Unit-value: 186 ECU	30	1846	2823	3757	0.3	0.5	0.7
	per ppb	262	401	533	0.04	0.08	0.10
Minor restricted activity days in adults	20	0	3,312,071	11,103,930	0	207	693
	25	0	2,340,374	7,846,255	0	146	490
Unit-value2: 62.4 ECU	30	0	1,368,678	4,588,580	0	85	286
	per ppb	0	194,339	651,535	0	12	41
Asthma attacks in asthmatics	20	1,611,577	2,569,692	3,523,393	50	80	110
	25	1,138,835	1,815,895	2,489,836	36	57	78
Unit-value: 31.3 ECU	30	666,092	1,062,098	1,456,278	21	33	46
	per ppb	94,548	150,759	206,711	3.0	4.7	6.5
Symptom days	20	6,986,530	13,713,337	20,466,117	44	86	129
Unit-value: 6.3 ECU	25	4,937,091	9,690,647	14,462,557	31	61	91
	30	2,887,652	5,667,957	8,458,996	18	36	53
	per ppb	409,888	804,538	1,200,712	2.6	5.1	7.5

[1]per ppb: the increase in impacts when the ozone background level is lowered by 1 ppb.
[2]In the absence of a better figure, the value for Restricted Activity Days was used.

measurements from the regional stations were used, which may have resulted in an underestimate of concentrations, and hence physical impacts and damages, for urban areas. The annual mean SO_2 concentrations at the monitoring stations ranged from 5 $\mu g/m^{-3}$ to 34 $\mu g/m^{-3}$ with an average of 12 $\mu g/m^{-3}$. Impacts were all calculated relative to a background level of 1 ppb or 2.63 $\mu g/m^{-3}$ SO_2.

2.3.2. *Dose–response relationships for SO_2*

The common set of dose–response functions was used. For a discussion of these functions refer to the Synthesis Report.

2.3.3. *Results for SO_2*

Table 2.8 gives the estimated impacts on mortality and their valuation while Table 2.9 shows the estimated morbidity impacts and their valuation. The estimated number of excess deaths is 912, which corresponds to a damage of ECU 2371 mn. – 1.06% of the 1990 Dutch GDP. Note that these estimates are not additive to the estimates for mortality from PM_{10}. The value of morbidity impacts ranges from only ECU 77,000 for emergency room visits for COPD to ECU 23 mn. for hospital admissions for chronic bronchitis or emphysema. In total, damages from SO_2 are relatively small compared to those for PM_{10}.

Table 2.8. Estimated acute mortality impacts from SO_2

Impacts, cases			Damages, million ECU (% 1990 GDP)		
Low	Mid	High	Low	Mid	High
–	317	–	–	824 (1.06)	–

Table 2.9. Estimated acute morbidity impacts from SO_2

	Impacts, cases			Damages, million ECU		
	Low	Mid	High	Low	Mid	High
Hospital admissions for chronic bronchitis or emphyse Unit-value: 6600 ECU	–	3433	–	–	22.7	–
Emergency room visits for COPD Unit-value: 186 ECU	–	413	–	–	0.077	–

CHAPTER 3

3. NOISE

3.1. Introduction

A substantial percentage of the Dutch population experiences noise nuisance from one or more sources, as can be seen in Table 3.1. For instance, Table 3.1 shows that no less than 20% of the Dutch population state thay they experience serious nuisance from road traffic. This high percentage can be explained by the very high population density and the very high car density in the Netherlands; the number of cars per 100 persons in the Netherlands is the highest in Europe.

3.2. Valuation of noise by the hedonic method (road and rail)

The figures in Table 3.1 are subjective nuisance scores and can not be used directly to calculate monetary damage. For a hedonic-price type calculation, the actual noise levels are needed. Data on number of houses in certain noise bands was obtained from a survey held in the second half of 1989 (de Jong *et al.*, 1990). This survey was a follow up to a survey held in 1987 in which respondents (4061) were asked if they ever (at least once a year) heard noise from road traffic, rail traffic, air traffic or from industrial sources.

Table 3.1. Noise nuisance in the Netherlands

	Percentage of the Dutch population that experiences from this specific source	
Source	Some nuisance	Serious nuisance
Road traffic	48	20
Industry	8	3
Air traffic	28	11
Rail traffic	4	1
Domestic noise	40	15

Source: IMP-M (1985).

In the follow up survey in 1989 for those respondents who had indicated that they heard noise from one or more of these noises, the noise levels at the front side of their dwellings were established. Based on the results of the first survey, the noise levels were established for 3389 dwellings which suffered from noise from city road traffic, for 761 dwellings which suffered from highway traffic noise, and for 848 dwellings which suffered from rail traffic noise. Because of missing addresses not all dwellings were covered, but the number of dwellings missing is negligible: 13 for city road traffic, 9 for highway traffic and 3 for rail traffic.

Noise levels were measured in dB(A) (more precisely as $L_{Aeq,24}$) and were obtained from the municipalities where these dwellings were located. If this information was not available they were calculated from other information provided by the municipalities. It was not always possible to calculate a precise noise level and sometimes levels had to be estimated. Results were presented in the following noise bands: $\leq$55 dB(A), 56–60 dB(A), 61–65 dB(A), 65–70 dB(A), 71–75 dB(A) and >75 dB(A). It was felt that the dwellings in the survey were representative for the Dutch housing stock (de Jong *et al.*, 1990).

In the tables below the results of this survey will be used to calculate the damage from road and rail traffic. The results of the survey are not suitable to calculate damages from air traffic, other sources were used here. The number of houses in a certain noise band was estimated by relating the number of houses in this noise band in the survey to the total number of houses in the survey (4061) and applying these percentages to the total stock of dwellings in the Netherlands at 31-12-1989, which was 5,802,361 dwellings (CBS, 1991b). Note that by doing so we implicitly assign to dwellings for which the noise level could not be established (the category 'unknown' in the tables below) a noise level of $\leq$55 dB(A) for which category damage is not calculated. This biases the results downwards.

To calculate damages we will use the Noise Depreciation Sensitivity Index (NDSI) which gives the percentage change in house price per decibel. Nelson (1980, 1982), from a review of studies on noise levels and house price depreciation, comes to a best estimate for the NDSI of 0.40 for road traffic noise and 0.50–0.55 for air traffic noise. The average house price in the Netherlands in 1990 was D*f*l 174,500 or ECU 75,476 (ECU 1 = D*f*l 2.312).

Therefore, house price depreciation per decibel is set at D*f*l 0.004 * 174,500 = 698 (ECU 302) for road and rail traffic noise, and D*f*l 0.005 * 174,500 = 872.5 (ECU 377) for air traffic noise. To calculate the *annual* value of the damage, following Jansen (1988) it is assumed that annual costs of housing are 8% of the house price. Annual damage per decibel then becomes 0.08 * 698 = D*f*l 56 (24 ECU) per decibel for traffic noise, and 0.08 * 872.5 = D*f*l 70 (30 ECU) per decibel for air traffic.

3.3. *Valuation of noise by the hedonic method* (*aircraft*)

As noted above, the survey is not suitable for calculating the number of dwellings suffering from air traffic noise. Other sources were used here. For

Table 3.2. Calculation of annual damage from noise from city road traffic

Noise band (dB(A))	No. in sample	% of 4061	No. in total dwelling stock	Exceedence of 55 dB(A)		Damage (10^6 D*f*l)
≤55	1736					
56–60	290	7.14	414352	2		46.275
61–65	196	4.83	280045	7		109.464
66–70	208	5.12	297191	12		199.141
71–75	30	0.74	42864	17		40.690
>75	1	0.02	1429	20		1.595
Unknown	928					
Total	3389				D*f*l	397.166
					ECU	171.785

Source: Calculated from data in de Jong *et al.* (1990).

Table 3.3. Calculation of annual damage from noise from highway traffic

Noise band (dB(A))	No. in sample	% of 4061	No. in total dwelling stock	Exceedence of 55 dB(A)		Damage (10^6 D*f*l)
≤55	369					
56–60	52	1.28	74298	2		8.298
61–65	62	1.53	88586	7		34.626
66–70	20	0.49	28576	12		19.148
71–75	1	0.02	1429	17		1.356
>75	–	–	–	20		
Unknown						
Total	761				D*f*l	63.428
					ECU	27.434

Source: Calculated from data in de Jong *et al.* (1990).

Table 3.4. Calculation of annual damage from noise from rail traffic

Noise band (dB(A))	No. in sample	% of 4061	No. in total dwelling stock	Exceedence of 55 dB(A)		Damage (10^6 D*f*l)
≤55	465					
56–60	48	1.18	68582	2		7.659
61–65	45	1.11	64296	7		25.132
66–70	8	0.20	11430	12		7.659
71–75	8	0.20	11430	17		10.851
>75	5	0.12	7144	20		7.978
Unknown	269					
Total	848				D*f*l	59.280
					ECU	25.640

Source: Calculated from data in de Jong *et al.* (1990).

civil air traffic noise data was available for two airports: Schiphol Airport, which is by far the largest airport in the Netherlands and is located in a very densely populated part of the Netherlands, and for Maastricht Airport. These data relate to the situation existing in 1990. Data on noise levels around military airbases is available for 1979, so these figures may be a little outdated.

One problem with these data is that in the Netherlands noise levels for air traffic noise are usually expressed in the so-called Kosten-Unit (Kosten-Eenheid; KE). This unit was constructed by the Kosten Committee to take account of the special characteristics of air traffic noise, for instance, the peak loads caused by planes taking off. Since there is no direct relationship between the noise level expressed in dB(A) and that expressed in KE, conversion of noise levels in KE to noise levels in dB(A) remains a difficult task. Sol and Feenstra (1994) give Table 3.5 below which is based on comparing subjective nuisance scores with measured noise levels for both air traffic noise and road traffic noise, and on the work of Miedema (1988). We will use this Table for our calculations.

Fortunately, the data for the most important source of air traffic noise (Schiphol airport) are expressed in dB(A), so conversion does not have to take place there. It has to be noted that all the noise levels used in the calculation for air traffic noise are not based on measurements of noise levels but on calculated noise contours. Tables 3.6 to 3.8 present the calculations for Schiphol Airport, Maastricht Airport and military airbases separately.

3.4. Aggregation and conclusions

Adding up the damages from the different sources results in a total annual damage of 291 million ECU. Note, however, that no account has been taken of accumulation of noise nuisance from different sources. For instance, in the case of road traffic de Jong *et al.* (1990) tried to establish 3402 noise levels for dwellings suffering from city road traffic noise, and 770 noise levels for dwellings suffering from highway noise. This sums up to a total of 4172 dwellings. Since there were only 4061 dwellings in the survey, some dwellings must therefor suffer from both sources of noise. Adding up damages calculated for both

Table 3.5. Conversion of noise levels in KE to noise levels in dB(A)

KE	dB(A)
25	50
30	55
35	60
40	65
45	70

Source: Sol and Feenstra (1994).

Table 3.6. Calculation of annual damage from air traffic noise at Schiphol Airport

Noise band (dB(A))	No. of houses	Exceedence of 55 dB(A)		Damage (10^6 D*f*l)
45–50	39176			
50–55	626668			
55–60	419511	2		58.564
60–65	143626	7		70.175
65–70	13377	12		11.205
>70	1032	15		1.081
Total			D*f*l	141.024
			ECU	60.997

Source: Data on number of houses from V&W (1993).

Table 3.7. Calculation of annual damage from air traffic noise at Maastricht Airport

Noise band in KE	Noise band in dB(A)	Number of houses	Exceedence of 55 dB(A)		Damage (10^6 D*f*l)
20–25	<50	16673			
25–30	50–55	12380			
30–35	55–60	3853	2		0.538
35–40	60–65	1783	7		0.871
40–45	65–70	1623	12		1.359
45–50	>70	770	15		0.806
50–55	>70	389	15		0.407
55–60	>70	77	15		0.081
60–65	>70	5	15		0.005
>65	>70	–	15		
Total				D*f*l	3.575
				ECU	1.546

Source: Data on number of houses from V&W (1990).

Table 3.8. Calculation of annual damages from noise from military airbases

Noise band in KE	Noise band in B(A)	No. of houses	Exceedence of 55 dB(A)		Damage (10^6 D*f*l)
35–40	60–65	6502	7		3.177
40–45	65–70	2965	12		2.483
45–50	>70	2173	15		2.275
50–55	>70	1199	15		1.255
55–60	>70	297	15		0.311
60–65	>70	382	15		0.400
>65	>70	130	15		0.136
Total				D*f*l	7.935
				ECU	3.432

Source: Data on number of houses from Ministry of Defence (1981).

sources of noise separately therefore leads to overestimation of damages, because the aggregate noise level from two sources is not simply the sum of noise level from the individual sources. (For instance the aggregate noise level from two sources each with a noise level of 55 dB(A) equals 58 dB(A).)

However, since data on the individual dwellings were not available, correcting for this double counting was not possible. On the other hand damages were not calculated for the categories unknown in the tables, which leads to an underestimation of damages. It is therefore felt that, as long as accumulation of noise from different sources is not too frequent, the estimate may be considered to be reasonably reliable.

CHAPTER 4

DAMAGE TO CROPS

4.1. Introduction

Damage to agriculture from SO_2 was calculated for the following crops: winter and summer wheat, winter and summer barley, rye and peas. Also damage from O_3 to wheat was calculated but given data limitations this last result can only be considered as a very rough first estimate.

4.2. Crop yield data

Data on area cultivated of each crop in 1990 were obtained from the Dutch Central Bureau of Statistics (CBS). These data give the area cultivated (in ha) for each municipality in the Netherlands (672 in 1990). Yield per ha is estimated each year for each province (12) by special committees, and is published by the CBS (CBS, 1991a). By combining area cultivated per municipality with the estimated yield per ha of the province the municipality belongs to, the yield per municipality can be calculated. Table 4.1 shows the estimates which were thus obtained for total yield of the crops studied.

4.3. SO_2 Concentrations and background levels

The spatial distribution of SO_2 concentrations was calculated from the 1990 mean summer period concentrations. This summer period runs from April up

Table 4.1. Estimated total yields in 1990 in the Netherlands

Crop	Area cultivated in ha	Estimated yield in 10^3 kg
Winter wheat	135120	1042825
Summer wheat	5524	33390
Winter barley	9950	54785
Summer barley	30466	164133
Rye	8614	36106
Peas, fresh harvested[1]	7677	42842
Peas, dry harvested	10851	57419

[1] No specific data on yield per ha available, the yield per ha of dry harvested peas is used.

Source: Calculated from data from CBS (pers. comm.) and from CBS (1991a).

to and including September. Although this period does not exactly correspond with the period which is usually taken as the growing season it was felt that, in the absence of data for the growing season, it was more appropriate to use these concentrations than mean annual concentrations. Only the results of the regional stations were used. The average concentration at these stations was 10.3 $\mu g/m^{-3}$ (3.9 ppb), and ranged from 4 to 28 $\mu g/m^{-3}$ (1.5 to 10.6 ppb).

The spatial distribution was constructed by linear interpolation using the SPANS Geographical Information System with the nearest neighbourhood method on a $5*5\ km^2$ grid surface. The outcomes for each grid were used to calculate an area weighted concentration for each municipality. Because the dose–response relationships use concentrations in ppb, the original data which gave concentrations in $\mu g/m^{-3}$ were converted to concentrations in ppb by multiplying them by 0.38. The background concentration used is 1 ppb.

4.4. Dose–response relationships

The following dose–response relationships were used. These were taken from the ExternE study of the coal fuel cycle (CEC, 1995c):

Baker 1: $y = -0.69(SO_2) + 9.35$ (1)

Baker 2: $y = 0.74(SO_2) - 0.055(SO_2)^2\ (SO_2 \leq 13.6\ \text{ppb})$ (2)

$y = -0.69(SO_2) + 9.35\ (SO_2 > 13.6\ \text{ppb})$

Roberts 1: $y = -0.18(SO_2) + 2.75$ (3)

Roberts 2: $y = 0.20(SO_2) - 0.013(SO_2)^2\ (SO_2 \leq 15.3\ \text{ppb})$ (4)

$y = -0.18(SO_2) + 2.75\ (SO_2 > 15.3\ \text{ppb})$

Weigel 1: $y = 4.92 - 0.74(SO_2)$ (5)

Weigel 2: $y = 10.92 - 0.89(SO_2)$ (6)

Weigel 3: $y = -0.93 - 0.60(SO_2)$ (7)

where: y = % yield loss

$SO_2 = SO_2$ concentration in ppb.

Equations (1) and (2) are based on work by Baker *et al.* (1986), equations (3) and (4) on the work by Roberts (1984), and relationships (5), (6) and (7) on the work by Weigel *et al.* (1990). Equations (1) to (6) were applied to wheat, barley and rye. Equation (5) and (7) were applied to peas. To correct for the losses which are already implicit in the estimated yields and yield loss at the background concentration, yield loss was calculated using the next equation,

$$\Delta Y = \left(1 - \frac{100 + y(c_b)}{100 + y(c)}\right) \cdot Y(c) \qquad (8)$$

where:

ΔY = yield loss
$y(c_b)$ = % yield loss at background concentration
$y(c)$ = % yield loss at existing concentrations
$Y(c)$ = yield at existing concentrations.

4.5. Results

4.5.1. *Results for SO_2*

Results of the calculations are shown in Table 4.2. The international prices used to value these impacts are (cf. prices in Rotterdam), the daily average over 1990 (see Table 4.3). There are no big differences between these prices and prices in other European ports. The differences are in the order of a few ECU/t.

Table 4.2. Estimated impacts on agricultural yields from SO_2 for 1990

Crop		Baker 1	Baker 2	Roberts 1	Roberts 2	Weigel 1	Weigel 2	Weigel 3
Winter wheat	10^3 kg	***−15300***	10538	*−4166*	3042	*−17156*	*−19601*	
	%	***−1.46***	1.01	*−0.40*	0.29	*−1.65*	*−1.88*	
Summer wheat	10^3 kg	***−391***	278	*−107*	80	*−439*	*−501*	
	%	***−1.17***	0.83	*−0.32*	0.24	*−1.31*	*−1.50*	
Winter barley	10^3 kg	***−494***	389	*−135*	110	*−554*	*−632*	
	%	***−0.90***	0.71	*−0.25*	0.20	*−1.01*	*−1.15*	
Summer barley	10^3 kg	***−2315***	1577	*−630*	456	*−2596*	*−2966*	
	%	***−1.41***	0.96	*−0.38*	0.28	*−1.58*	*−1.8*	
Rye	10^3 kg	***−228***	190	*−63*	53	*−255*	*−291*	
	%	***−0.63***	0.52	*−0.17*	0.15	*0.71*	*−0.81*	
Green peas	10^3 kg					*−695*		*−595*
Fresh harvested	%					*−1.62*		*−1.39*
Green peas	10^3 kg					*−819*		*−701*
Dry harvested	%					*−1.43*		*−1.22*

A negative number refers to a yield loss, a positive number to an increase in yields. Best estimates are in italics.

Table 4.3. Prices used to value impacts on agriculture

	Domestic prices[1]		International prices[1]	
	D*f*l/100 kg (VAT incl)	ECU/t	$US/t	ECU/t
Soft wheat	38.20	165.22	190	150
Barley	40.40	174.74	192	151
Rye	36.80	159.17	146	115
Green peas	57.00	246.54		

[1]Exchange rates used: $US 1 = ECU 0.787 and D*f*l 1 = 0.433 ECU.

Source: Agricultural Economics Research Institute (The Hague), ExMis Database, based on EUROSTAT data (NIMEX Trade Statistics), and personal communication.

Note that there can be large differences in the quality of soft wheat. The average landing prices for wheat from third countries over all European ports was only 170 $US/t. Rotterdam prices were higher (190 $US/t), despite the fact that Rotterdam has (together with Hamburg) the lowest shipping rates in Europe. International trade in green peas is negligible, so here damage is not valued at international prices. Domestic prices are represented by average producer prices.

A negative number refers to a yield loss, a positive number to an increase in yields. Best estimates are in italics.

Table 4.4 below shows the monetary value of the damages to agriculture using the prices given above and the range of best estimates of physical damage presented in Table 4.2.

4.5.2. *Results for O_3*

For damage from O_3 only a very rough estimate can be calculated. The dose–response functions require the seasonal 7 or 8 hour/day mean concentration of O_3. Since we do not have these concentrations from the regional stations at our disposal we cannot calculate a spatial distribution of these concentrations. However, from RIVM (1994a) we do know the *average* national seasonal 7 hour/day mean O_3 concentration of the regional stations for 1990 which was 83 $\mu g/m^{-3}$ (measured from May upto and including September, from 10 a.m. to 5 p.m.). Since 1990 was characterised by frequent periods of summer smog this cannot be considered a representative figure for all years, for instance in 1992 this figure was 68 $\mu g/m^{-3}$ and for 1993 it was 58 $\mu g/m^{-3}$ (RIVM, 1991, 1994a, 1994b).

The EXTERNE report (EC, 1995c) gave the following dose–response relationships for ozone:

Sommerville *et al.* (1989):

$$y_r = \exp\{-(O_3/272)^{2.56}\} \tag{9}$$

Skärby *et al.* (1994):

$$y_r = 1.0 + 0.0004(O_3) - 0.00001875(O_3)^2 \tag{10}$$

Table 4.4. Range of monetary values for damage to agricultural yields from SO_2

	Domestic prices (10^6 ECU)	International prices (10^6 ECU)
Wheat	0.706–3.321	0.641–3.015
Barley	0.134–0.629	0.116–0.543
Rye	0.010–0.046	0.007–0.034
Green peas	0.319–0.373	

N.B. Based on best estimates from Table 4.2.

A Swiss report on air-borne agricultural damage gave the following dose–response relationship:

Fuhrer *et al.* (1989):

$$y_r = \exp\{-(O_3/0.144)^{2.89}\} \quad (11)$$

Finally, a Dutch study gave the following dose–response relationship:

Van der Eerden *et al.* (1987, 1988)

$$R = 0.075(O_3) - 5.0 \quad (12)$$

In the equations y_r stands for the relative yield and O_3 stands for the seasonal 7 (equation 1) or 8 (equation 2) hour/day mean O_3 concentration in $\mu g/m^{-3}$ (or in ppm in the Fuhrer equation). The R in the Van der Eerden equation is percentage yield reduction. These equations were applied only to wheat because experiments have shown that barley is not sensitive to O_3 (EC, 1994). We have applied both functions to winter wheat using the *average* national seasonal 7 hour/day mean O_3 concentration. Damages were calculated using a background level of 60 $\mu g/m^{-3}$ because according to Van der Eerden *et al.* (1987) no adverse effects are to be expected below that level.

The results are shown in Table 4.5 and indicate that impacts from O_3 on winter wheat are considerably higher than impacts from SO_2. This suggests that in future research it is recommendable to focus on damages from O_3 instead of damages from SO_2, especially given the rather low SO_2 concentrations in the Netherlands. It has, of course, to be kept in mind that the estimate given here has to be regarded as a rough first estimate.

For other crops there is scarce and sometimes contradictory evidence. For example, with the *d/r* relationship of Fuhrer *et al.* (1989) ($y_r = \exp\{-(O_3/0.179)^{3.95}\}$), maize yield reduction can be calculated as 0.2%, while the Van der Eerden *et al.* (1987, 1988) relationship ($R = 0.200(O_3) - 11.7$) gives a 4.1% yield reduction. For peas the same applies: Damage calculated using the Van der Eerden relationship ($R = 0.300(O_3) - 17.7$) is much lower than that using the Fuhrer relationship ($y_r = \exp\{-(O_3/0.287)^{1.77}\}$): the first gives a 1.3% yield loss, the latter a 6.4% loss. (NB. Data not given in Table 4.5).

This amounts to a financial loss of, respectively, about 190,000 ECU and

Table 4.5 Estimated impacts from O_3 on winter wheat yields

Dose–response function used	Yield loss in 10^3 kg (% of total yield in 1990)	Monetary value in 10^6 ECU	
		Domestic prices	International prices
Sommerville *et al.*	−26017 (2.5)	−4.299	−3.903
Skärby *et al.*	−49925 (4.8)	−8.249	−7.489
Fuhrer *et al.*	−16635 (1.6)	−2.748	−2.495
Eerden *et al.*	−17768 (1.7)	−2.936	−2.665

about 909,000 ECU. Potatoes is an important crop in the Netherlands. Unfortunately, we have only one *d–r* relationship (by van der Eerden: $R = 0.200(O_3) - 11.7$). According to this *d/r* relationship potato yield would be reduced by 4.4% given the prevalent ozone concentrations in 1990. This would cause a financial loss of about 34 mn. ECU.

For other crops (e.g. tomatoes, spinach) and pasture, there is evidence of damage by increased ozone concentrations but the quantification of that damage is not yet very reliable. Keeping this in mind, it can be noted RIVM has quoted in its at the time influential publication 'Concern for Tomorrow' a damage figure of D*fl* 400 million (ECU 173 mn.) for ozone damage to agriculture (RIVM, 1989).

CHAPTER 5

DAMAGE TO MATERIALS

5.1. Introduction

To calculate damage to materials from SO_2 a model developed by Gosseling *et al.* (1990) was used. Materials included in this model are painted steel, galvanised steel, duplex steel, galvanised and painted steel, and sheet zinc. Two types of damages are taken into account: (i) costs of increased replacement of materials, and (ii) increased maintenance. Below their methodology is outlined and results of the calculations are presented.

5.2. Dose–response relationships

The following dose–response relationships that relate SO_2 concentrations to physical damages were used:

Non-painted galvanised steel and sheet zinc

The relation between the SO_2 concentration and the decrease of the zinc layer or, in the case of sheet zinc, the thickness of the sheet, is:

$$\text{Zn dec} = 0.033 * SO_2$$

where: Zn dec = decrease zinc layer (μm/year); SO_2 = annual mean SO_2 concentration (μg/m^{-3}).

This is a very rough approximation by Jansen and Olsthoorn, (1982). The authors claim that more recent dose–response relationships, for example those of the US NAPAP research (Baedecker, 1989), include variables (deposition velocity, time-of-wetness, etc.) which may be so site specific (and unknown) that they cannot be used in a more global model. The *d–r* relationship is in the middle range of *d–r* relationships presented in the UK Building Effects Review Group Report (Everett *et al.*; fig. 7.1).

Painted steel

The relation between the SO_2 concentration and the protection period of paint on steel is:

$$B = 10 \qquad T \leq 10$$

$$B = 10.2 - 0.02 * SO_2 \qquad T > 10$$

where: B = protection period in years; SO_2 = annual mean SO_2 concentration ($\mu g/m^{-3}$).

It is assumed that paint on steel is unaffected by air pollution in the first ten years after it comes into use. This is based on empirical facts, plausible because paint can be more effectively applied to steel in factory conditions than 'on the spot'. On the physical–chemical side *d–r* relationships for paints are far more complex than those for zinc. NAPAP did not succeed in finding analytical *d–r* relationships. Baedecker (1989; 2–109) concludes that "there is no short term solution for developing models for predicting atmospheric corrosion rates".

5.3. Damage relationships

The dose–damage relationships relate the physical damage due to SO_2 to economic damage. Two types of damage have been taken into account: (i) costs of increased replacement of materials; and (ii) increased maintenance.

Damage due to increased replacement may be relevant for 'maintenance free' galvanised steel and for sheet zinc. The general equation is:

$$D = I/L(SO_2) - I/E(10) \qquad \text{for } D > 0$$

where: D = annual damage (10^6 D*f*l/year); I = replacement value of object (10^6 D*f*l/year); $L(SO_2)$ = technical life (years) as a function of annual mean SO_2 concentration; $E(10)$ = economic life (years) in reference situation, taken by Gosseling *et al.* (1990) to be the situation with an SO_2 concentration of 10 $\mu g/m^{-3}$ (3.8 ppb).

This equation states that damage will occur if the technical life of materials falls short of economic life due to increased air pollution.

'Maintenance free' galvanised steel has an initial thickness of 70 μm. It has to be replaced if thickness has reduced to 25 μm, i.e. a reduction of 45 μm. The physical *d–r* relationship is Zn dec $= 0.033 * SO_2$, therefore, $L(SO_2) = 45/(0.033 * SO_2)$. Annual damage can be expressed as:

$$D = I * (0.033 * SO_2/45) - I/E(10),$$

where: $E(10)$ is object dependent.

For sheet zinc economic life in the reference situation has been put at

30 years. The function $L(SO_2)$ is then estimated to be:

$$L(SO_2) = 31 - SO_2/10 \qquad \text{for } SO_2 > 10.$$

Damage can then be expressed as:

$$D = I*(1/(31 - 0.1*SO_2) - 0.033).$$

For maintenance damages attributable to air pollution, the *d–r* relationship for painted steel, duplex steel and heating pipes in greenhouses is:

$$D = [K*T/L]*[(L-Z)/B(SO_2) - (L - Z)/B(10)]$$

where: K = maintenance costs (Dfl/m^2); T = stock at risk (10^6 m^2); L = economic life of the object (years); Z = service period (years) of initially applied coating. This service period is *not related to* SO_2 concentrations; $B(SO_2)$ = service period of coating as a function of SO_2 concentration.

For galvanised and painted steel it is assumed that the initially applied coating can be corroded by SO_2. Therefore the formula becomes:

$$D = [K*T/L]*[(L - Z(SO_2))/B(SO_2) - (L - Z)/B(10)].$$

Together with the following assumptions:

1. Zn dec $= 0.033*SO_2$;
2. Standard thickness zinc layer = 70 μm;
3. Painting is necessary with 5% corrosion; and
4. This corrosion occurs at a reduction of the zinc layer with 35 μm.

The above formula then becomes:

$$D = [K*T/L]*[(L - 35/(0.033*SO_2))/B(SO_2)$$
$$- (L - 35/0.33)/B(10)].$$

Table 5.1 presents some materials and objects sensitive to air pollution. Per material/object the type of damage is indicated together with some economic parameters.

5.4. Stock at risk

The approach taken to estimate the surfaces of the sensitive materials is:

1. Define in which objects these materials may be present.
2. Estimate the quantities of objects in the most appropriate dimension (depends on available data).
3. Break down objects quantities into type of material and type of protection. For example, split up steel door frames into painted and galvanised ones.
4. Use coefficients to translate quantities to appropriate dimensions. For example, translate tons of steel window frames into square metres of exposed surfaces.

Table 5.1. Objects and materials sensitive to SO_2, damage

Material/object (dimension)	Type of damage	Parameters
Painted steel (surface)	Increased maintenance (painting)	– Economic life: 50 years – Costs of painting: 42 Dfl/m² – Protection period of initial paint layer: 10 years
Galvanised and painted steel (surface)	Shortening of service period of initial protection and, after that, increased maintenance	– Economic life: 50 years – Costs of painting: 25 Dfl/m² – Thickness zinc layer: 70 μm – Painting at: 35 μm
Duplex steel (surface)	Increased maintenance (painting)	– Economic life: 50 years – Costs of painting: 20 Dfl/m² – Protection period of initial paint layer: 10 years
Galvanised 'maintenance free' steel (replacement value)	Increased replacement	– Economic life: 15 years – Thickness initial layer: 70 μm – Replacement at: 25 μm
Seet zinc (replacement value)	Increased replacement	– Economic life: 30 years
Greenhouse galvanised palisades (surface)	Shortening of service period of initial protection and, after that, increased maintenance	– Economic life: 17 years – Costs of painting: 40 Dfl/m² – Thickness initial layer: 70 μm – Painting at: 35 μm
Greenhouse heating pipes (surface)	Increased maintenance	– Economic life: 17 years – Costs of painting: 42 Dfl/m² – Protection period of initial paint layer: 10 years

Table 5.2 presents the objects in which the materials are present and a breakdown of the proportions of different materials present.

The total exposed surfaces (TES) are calculated using the following equation:

$$\text{TES} = \text{annual use} * \text{specific surface} * \text{life span} * \text{exposition coefficient.}$$

Annual use was estimated on the basis of production statistics corrected for import and export of the Central Bureau of Statistics (CBS) over the period 1977–1983. The economic life span of the objects was estimated at 50 years, which is a usual rule of thumb in the construction industry. Specific surfaces of the objects were taken from Olsthoorn (1979). Exposition coefficients have been estimated on the basis of expert judgement. For residential construction an exposition coefficient of 0.10 was used.

After calculation of the national stock at risk, the objects were divided over 20 so-called COROP regions. These are regions constructed by the CBS based on certain variables like socio-economic and demographic characteristics. For the calculations some of these regions were taken together. Per object different distribution keys were used: population, employment, length of road network,

Table 5.2. Breakdown of objects by different materials

Objects	Painted steel	Galvanised steel	Duplex steel	Painted and galvanised steel	Sheet zinc
Windows, frames, doors	0.05			0.95	
Garage doors	0.05		0.15	0.8	
Facades and fronts	0.05		0.15	0.8	
Sheds and warehouses	0.05		0.15	0.8	
Skeletons	0.05		0.15	0.8	
Bridges				1	
Containers	1				
Railway coverings				1	
Railway gantries				1	
Railway semaphones				1	
Pipelines				1	
Chemical industry				1	
Steel industry				1	
Food industry				1	
Utilities				1	
Balconies and galleries				1	
Lampposts				1	
Greenhouses (pallisades)		1			
Greenhouses (heating pipes)				1	
Pylons				1	
Fences (10^6 D*f*l)		1			
Traffic posts (10^6 D*f*l)		1			
Crash barriers (10^6 D*f*l)		1			
Sheet zinc (10^6 D*f*l)					1

Source: Gosseling *et al.* (1990).

water surface, length of railway network, and surface of greenhouses. Table 5.3 gives the resultant stock at risk per region

5.5. Results

The model described above was used to calculate damages to materials using annual mean SO_2 concentrations for 1990. The results of the regional stations were used to calculate the spatial distribution of the SO_2 concentration on a $5*5\ km^2$ grid surface using linear interpolation. The results were then used to calculate an average concentration for each COROP region, which was used to calculate damages. The background level used was 1 ppb SO_2. The results are shown in Table 5.4. These results sum up to a grand total of 43.058 million ECU.

Gosseling *et al.* (1990) point out that the reliability of the model calculations is very limited, due to the many assumptions and estimates concerning the

Table 5.3. Stock at risk per region

	Painted steel 10^6 m^2	Galvanised and painted steel 10^6 m^2	Duplex steel 10^6 m^2	Galvanised steel 10^6 D*f*l	Sheet zinc 10^6 D*f*l	Greenhouse heating pipes 10^6 m^2	Greenhouse pallisades 10^6 m^2
Groningen	0.54	4.42	0.21	59	438	0.04	0.08
Friesland	0.50	5.73	0.19	66	466	0.04	0.08
Drente	0.38	2.04	0.15	53	335	0.04	0.08
Kop Overijssel	0.30	2.22	0.12	41	278	0.04	0.08
Twente	0.83	3.43	0.31	83	539	0.04	0.08
Veluwe	0.50	3.98	0.20	60	539	0.15	0.30
Achterhoek	0.41	1.85	0.15	48	278	0.11	0.23
Betuwe	0.71	3.80	0.28	74	636	0.11	0.23
Utrecht	0.70	3.58	0.28	66	729	0.11	0.23
Kop N-Holland	0.24	2.18	0.11	34	415	0.32	0.64
Zaanstreek	1.89	8.37	0.72	152	1383	0.32	0.64
Leiden e.o.	1.02	5.08	0.41	90	1191	3.42	6.84
Rijnmond	1.71	9.40	0.65	144	1261	0.29	0.59
Zeeland	0.35	11.16	0.13	51	277	0.04	0.08
Baronie	0.63	3.16	0.24	61	421	0.09	0.18
Tilburg	0.45	2.03	0.17	44	331	0.09	0.18
Oss	0.59	2.50	0.22	60	404	0.09	0.18
Eindhoven	1.08	3.85	0.39	92	489	0.09	0.18
Venray	0.57	2.54	0.21	61	357	0.44	0.89
Z-Limburg	0.70	2.76	0.26	62	488	0.09	0.18
Total	14.05	84.05	5.40	1400	11250	6.00	12.00

Source: Gosseling *et al.* (1990).

Table 5.4. Damage to materials (10^3 ECU) from SO_2 in 1990 by region and material

Region	Painted steel	Galvanised and painted steel	Duplex steel	Galvanised steel	Sheet zinc	Greenhouse heating pipes	Greenhouse pallisades
Groningen							
Friesland							
Drente							
Kop Overijssel							
Twente							
Veluwe							
Achterhoek							
Betuwe							
Utrecht							
Kop N-Holland	6	1946	1	273	1	691	
Zaanstreek	8	2372	2	319	13	14899	
Leiden e.o.	21	6198	4	431	2	1808	
Rijnmond	5	8664	1	110	0	304	
Zeeland	9	2414	2	164	1	641	
Baronie	2	519	0	67	0	215	
Tilburg							
Oss							
Eindhoven							
Venray	2.39	656	0	97	0	200	
Z-Limburg							
Total	53	22769	10	1460	17	18759	

model parameters. They argue that if it is judged important to have more accurate assessments, it might be considered to take a more 'statistical' approach, i.e. to compare actual maintenance costs in 'polluted' and 'clean' areas. So the results should be considered to give an order of magnitude estimate, rather than a very accurate estimate.

CHAPTER 6

6. DEFENSIVE EXPENDITURES

6.1. Introduction

In this chapter some figures on costs incurred in improving air quality and in noise abatement are presented. These figures are collected by the Dutch Central Bureau of Statistics (CBS) and are published in Costs and Financing of Environmental Protection (Kosten en Financiering van het Milieubeheer; In Dutch). These statistics include all costs made with the intention to protect, restore or improve the state of the environment. These costs are calculated for different sectors of the economy like the government, households, private companies etc. A difference is made between the so-called 'costs of own environmental activities' of a sector and the 'net environmental burden' of a sector. The first consist of the interest and depreciation on environmental investments and current expenses for environmental activities. The net environmental burden is obtained when the first is corrected for transfers from one sector to the other as shown in Table 6.1 below. Many environmental investments made by private companies are subsidised by the government. Correcting for these transfers gives a more accurate picture of the costs actually borne by a sector. Figures are obtained through surveys, analysis of accounts and calculations (de Boo, 1993; Dietz, 1994b). Below we will use the figures as presented in Dietz (1994a,b,c).

6.2. Air pollution

Dietz (1994b) gives a figure of D*f*l 3160 mln (1367 mln ECU) for total environmental investments in 1990. This figure is not broken down into investments for separate environmental compartments. However, for 1991 disaggregated

Table 6.1. Calculation of the net environmental burden of a sector

→	Costs of own environmental activities	+
→	Paid levies, subsidies and payments for environmental services by third parties	–
→	Received levies, subsidies and payments for environmental services performed for third parties	=
→	Net environmental burden	

figures are given. Of the total environmental investments in 1991 (D*f*l 3173 mln), 38% (D*f*l 1207 mln) was for investments in air quality. Assuming that this percentage is the same each year, a figure of D*f*l 1203 mln (520 mln ECU) for investments in air quality in 1990 can be calculated. This corresponds to 0.2% of the 1990 GDP.

Table 6.2 shows the costs of activities for abating air pollution for the period 1987–1991. As can be seen from this table there has been a steady rise in these costs over the years, from D*f*l 929 mln in 1987 to D*f*l 1589 mln in 1991. In 1990 costs were D*f*l 1420 million or 614 mln ECU. For the most part these costs are borne by the sectors private companies and traffic. Dietz (1994b) does not give data on the net environmental burden for 1990 but does give data for 1991, (see Table 6.3). As can be seen from this table, looking at the net environmental burden results in higher costs for the government sector and lower burden for all other sectors. The results for the households' sector and the traffic sector are not much influenced.

Detailed figures on environmental investments and costs for the sector traffic are calculated by Dietz (1994a). In 1990, investments in air quality by this sector totalled D*f*l 677 mln or 293 mln ECU. For the largest part (D*f*l 649 mln; 96%) these were adjustments of cars and fuels to comply with environmental standards, like the catalyst convertor. The costs made by the sector traffic for abating air pollution were D*f*l 521 mln (225 mln ECU). The most important

Table 6.2. Breakdown of environmental expenditures for air pollution abatement

Year	Private companies	Government	Households	Traffic	Total
1987	555	69	4	300	929
1988	643	60	3	340	1046
1989	728	77	3	409	1217
1990	791	106	3	521	1420
1991	849	102	3	635	1589

Source: Dietz (1994b).

Table 6.3. Environmental expenditures on air pollution abatement and net environmental burden (NEB) for 1991 (millions of guilders)

Sector	Expenditure	NEB
Government	102	186
Private companies	849	786
Households	3	−10
Traffic	635	628
Not divided		0
Total	1589	1589

Source: Dietz (1994b).

categories were adjustment of vehicles (40%), non-leaded petrol (41%) and low-sulfur content diesel (16%). A separate figure for the net environmental burden of costs made for improving air quality is not given, only for all costs. However, as can be seen from Table 6.3 for 1991, the difference between the costs of own environmental activities and the net environmental burden for abating air pollution is very small for this sector.

6.3. Noise abatement

Dietz (1994b) does not give a separate figure for investments in noise abatement in 1990. The figure of D*f*l 166 mln he gives for 1991 is 5.2% of total environmental investments in 1991. Applying this percentage to total environmental investments in 1990 a figure of D*f*l 163 mln (72 mln ECU) can be calculated for 1990. This is 0.03% of the 1990 GDP.

Table 6.4 shows the costs of noise abatement activities for the period 1987–1991. As can be seen from this table these costs have risen from D*f*l 389 mln in 1987 to D*f*l 515 mln in 1991. In 1990 costs were D*f*l 474 million or 205 mln ECU. This corresponds to 0.09% of the 1990 GDP. For the most part these costs are borne by the government sector and the sector traffic. Again, no data is given on the net environmental burden for 1990 but is given for 1991, see Table 6.5. As can be seen from this table, looking at the net environmental burden results in higher costs for the government sector and lower burden for all other sectors. The differences are quite small.

Costs for the traffic sector have been investigated in further detail in Dietz (1994a). He calculates a total of D*f*l 122 mln (53 mln ECU) for 1990 by this sector, all related to adjustments of vehicles (65%), airplanes (34%) and rail equipment (1%). Costs of noise abatement from traffic were D*f*l 315 mln (136 mln ECU; 0.06% of 1990 GDP). 61% of these costs were related to road traffic, 36% to air traffic, and 3% to rail traffic. 37% of these costs were bourne by the government, the rest was born by the sector traffic.

Government expenditures and revenues related to noise abatement have been investigated by Dietz (1994c). Total expenditures in 1990 were D*f*l 227

Table 6.4. Environmental expenditure for noise abatement

Year	Private companies	Government	Households	Traffic	Total
1987	60	184	0	144	389
1988	67	186	0	153	406
1989	70	177	0	172	420
1990	74	203	0	198	474
1991	75	214	0	226	515

Source: Dietz (1994b).

Table 6.5. Environmental expenditures on noise abatement and net environmental burden (NEB) for 1991 D*f* mn)

Sector	Expenditure	NEB
Government	214	240
Private companies	75	65
Households	0	0
Traffic	226	211
Not divided		0
Total	515	515

Source: Dietz (1994b).

million (98 mln ECU); 0.04% of 1990 GDP). See Table 6.6 for more details. Revenues were D*f*l 25 mln (11 mln ECU). These came from a levy on civil air traffic. Levies related to noise nuisance from road traffic and industry have been abolished in 1988. Total government expenditures over the period 1979–1993 (in current prices) were D*f*l 2.8 billion (1.2 billion ECU). Dietz combines this figure with data on the number of dwellings which have been remediated to derive a rough estimate of the costs per dwelling remediated. For road traffic he calculates a figure of about D*f*l 7000 (about 3000 ECU), for rail traffic a figure of about D*f*l 30,000 (about 13,000 ECU) and a figure of about D*f*l 89,000 (about 38,000 ECU).

Table 6.6. Expenditures of Central Government for noise abatement in 1990 (millions of guilders)

	Total expenditures	Of which	
Road traffic	95	Prevention and remediation	60
		New roads	35
Industry	13		
Rail traffic	9		
Air traffic	57	Civil air traffic	24
		Military air traffic	33
Subsidies for development of silent technologies	20		
Allowances to lower government	28		
Miscellaneous	6		

Source: Dietz (1994c).

7. RESULTS AND CONCLUSIONS

The Dutch case study assessed the monetary damage to health, materials, and crops due to 1990 concentrations of PM_{10}, SO_2 and O_3. It also assessed the monetary damage of noise nuisance due to road, rail and air traffic. Table 7.1 presents a brief summary of the results of the study.

Health damage stands out as the major category of environmental damage in economic terms. The estimate of health damage excludes chronic mortality impacts from PM_{10} because it is not clear to what extent this can be added to acute mortality impacts. The total health damage estimate further excludes mortality and morbidity impacts from SO_2 because these should not be considered additive to impacts from PM_{10}. The uncertainty around the health damage estimate is large; the uncertainty is the product of uncertainty on the appropriate background level of PM_{10}, the dose–effect relationships, and the valuation of health impact endpoints, especially excess mortality.

The estimate of damage to materials in this case study is restricted to damage to zinc and steel due to SO_2 pollution. It therefore underestimates 'true' damage, as it is suggested that there may also be damage to other materials, both due

Table 7.1. Summary of results for the Netherlands (ECU mn. low and high estimates)

	PM_{10} ($10\ \mu g\ m^{-3}$)	SO_2 (1 ppb)	O_3 (30 ppb)	Noise (55 dB(A))	Total
Health	8658–19970	2394	277–860		8936–20829
Materials		43			43
Crops		1–4	37–42		37–46
Forests					
Ecosystems					
Amenity				291	291
Total	8658–19970	2438–2441	314–902	291	9307–21209

Notes:
The totals in the last column may differ slightly from row totals because of rounding differences.
For health damages, the total does not add up to the row total. Health damages from SO_2 are excluded from the total since these are not additive to damages from PM_{10} and O_3.
Chronic mortality and morbidity impacts from PM_{10} and SO_2 are not included.
Health damages from PM_{10} and SO_2 should not be considered additive.
Background concentrations used are shown in parentheses.

to SO_2 and O_3. According to the UK case study, damages due to soiling, SO_2 damages to paint, and O_3 damages to rubber goods may be significant. The monetary damage of air pollution on cultural buildings has also not been assessed in this study. Hence, there seems to be scope for significant improvement in this area.

Due to a lack of reliable dose–response relationships the estimate of damage to crops is restricted to four crops only: wheat, barley, rye and green peas. The total crop damage is dominated by O_3 damage. As more reliable dose–response functions become available there is scope for a more comprehensive damage assessment. It would also be interesting to take account of economic adjustments to air pollution by farmers and the effects of these adjustments on markets.

Monetary damages from noise nuisance were assessed for road, rail, and air traffic. The main uncertainties in the damage assessment lie in the valuation part. It is uncertain whether the Noise Depreciation Sensitivity Index (NDSI), that relates noise to house prices, is an accurate estimator of economic damage in this area. There are also methodological difficulties in adding-up different sources of noise nuisance. One important source of noise nuisance, domestic noise, was not assessed.

Although there is evidence for ecosystem and forest damage in the Netherlands, no attempt was made to express this damage in monetary terms in this study. This seems to be an area where progress is much needed, but difficult to achieve.

In summary, total air and noise pollution damage in the Netherlands in 1990 was assessed to be in the order of magnitude of ECU 9.3 bn. to ECU 21.2 bn. This damage amounts to about 4.2 to 9.5% of the Dutch Gross Domestic Product (GDP) in 1990. The order of magnitude of the damage assessed in this study is to a large extent in accordance with a previous Dutch valuation study (Jansen, 1988). It is important to note that damage from increased mortality, which is probably the most controversial part of a valuation exercise like this, accounts for the greatest part of this total damage. Excluding this damage category gives a total damage of ECU 2.3 bn. to ECU 5.7 bn. (1.0 to 2.6% of 1990 GDP).

CHAPTER 8

REFERENCES

Baedecker, P.A. et al. 1989, Effects of acidic deposition on materials. State of Science and State of Technology, NAPAP-program, SOS/T-Report no. 20.

CBS, 1989, Vademecum health statistics of the Netherlands 1989. The Hague, SDU.

CBS, 1991a, Maandstatistiek van de landbouw, januari 1991. (Monthly agricultural statistics, january 1991.) In Dutch. The Hague, Central Bureau of Statistics (CBS).

CBS, 1991b, Regionaal statistisch zakboek 1991. (Regional pocket yearbook 1991). The Hague, SDU/CBS publications.

CBS, 1993, Vademecum health statistics of the Netherlands 1993. The Hague, SDU.

De Boo, A.J. (1993), Costs of integrated environmental control, Statistical Journal of the United Nations Economic Commission for Europe, Vol 10, Issue 1, pp 47–64.

de Jong, R.G., Groeneveld, Y. and Halkes, H.J.C. 1990. Geluidsniveaus in Nederland. (Noise levels in the Netherlands.) In Dutch. Leiden, Nederlands Instituut voor Praeventieve Gezondheidszorg TNO. Report no. 90.065.

CRU/ERL (1992), Development of a Framework for the Evaluation of Policy Options to deal with the Greenhouse Effect, A Report for the Commission of European Communities, Directorate General for Environment, Nuclear Safety and Civil Protection, prepared by Climatic Research Unit, University of East Anglia, and Environmental Resources Limited, London.

Dietz, E.J. 1994a, Milieukosten van het verkeer 1990–1992. (Environmental costs of traffic 1990–1992.) In Dutch. Environmental Quarterly (CBS), 94/1, pp 4–9.

Dietz, E.J. 1994b, Kosten en financiering van het milieubeheer, 1991. (Costs and financing of environmental care, 1991.) In Dutch. Environmental Quarterly (CBS), 94/2, pp 4–9.

Dietz, E.J. 1994a, Vijftien jaar Rijksuitgaven geluidhinderbestrijding. (Fifteen years of central government expenditures on noise abatement.) In Dutch. Environmental Quarterly (CBS), 94/4, pp 4–9.

Everett, L.H. et al. 1989, The effects of acid deposition on buildings and building materials in the United Kingdom. London, HMSO.

Fankhauser, S. (1994), Evaluating the Social Costs of Greenhouse Gas Emissions.
CSERGE Working Paper GEC 94-1, London.

Fuhrer, J., Lehnherr, B. and Stadelmann, F.X. 1989, Luftverschmutzung und Landwirtschaftliche Kulturpflanzen in der Schweiz. Liebefeld-Bern, Eidgenossïsche Forschungsanstalt für Agrikulturchemie und Umwelthygiene.

Gosseling, H.J., Olsthoorn, A.A. and Feenstra, J.F. 1990, Economische schade aan materialen en monumenten door verzuring. (Economic damage to materials and monuments from acidification.) In Dutch. Amsterdam, Institute for Enviromental Studies, internal report W-90/002.

IMP-M, 1985, Indicatief Meerjaren Programma Milieubeheer 1986–1990 (IMP-M). (Indicative Multi-year Program for the Environment.) In Dutch. The Hague, SDU. TK, 1985–1986, 19204, nrs. 1–2.

IPCC (1992), Climate Change 1992. The supplementary Report to the IPCC Scientific Assessment, Cambridge University Press, Cambridge.

Jansen, H.M.A. 1988, Baten van milieubeleid. (The benefits from environmental policies.) Amsterdam, Institute for Environmental Studies.

Jansen, H.M.A. and Olsthoorn, A.A. 1982, Economische waardering van de nationale schade door luchtverontreiniging. In: H. Aiking (ed.), Mozaïk van de milieuproblematiek. In Dutch. Amsterdam, VU boekhandel/uitgeverij.

Miedema, H.M.E. 1988, Coëfficiënten voor de vaststelling van de cumulatie van omgevingsgeluid. (Coefficients for the establishing of the cumulation of noise.) Leidschendam,

Ministry of Public Housing, Physical Planning and the Environment (VROM). Research programme Noise Nuisance report GA-HR-08-04.
Ministry of Defence, 1981, Structuurschema militaire terreinen. (Structure Scheme Military Sites.) In Dutch. The Hague, SDU. TK 16666, 1980–1981, nrs. 1–2.
Nelson, J.P. 1980, Airports and Property Values: A Survey of Recent Evidence. Journal of Transport Economics and Policy, Vol XIV, January 1980.
Nelson, J.P. 1982, Highway Noise and Property Values: A Survey of Recent Evidence. Journal of Transport Economics and Policy, Vol XVI, May 1982.
Pearce, D.W., Cline, W.R., Achanta, A.N., Fankhauser, S., Pachauri, R.K., Tol, R.S.J. and Vellinga, P. (1994, 4th draft), Chapter 6: The Social Costs of Climate Change: Greenhouse Damage and the Benefits of Control. IPCC Working Group II.
RIVM, 1989, Concern for tomorrow. Bilthoven, National Institute for Public Health and Environmental Protection (RIVM).
RIVM, 1990, Regionaal gezondheidsprofiel 1990. (Regional public health profile 1990.) In Dutch. Bilthoven, National Institute for Public Health and Environmental Protection (RIVM).
RIVM, 1991, Luchtkwaliteit: Jaaroverzicht 1990. (Annual Air Quality Survey 1990.) In Dutch with English summary. Bilthoven, National Institute of Public Health and Environmental Protection (RIVM). Report no. 222101015.
RIVM, 1994a, Jaaroverzicht luchtkwaliteit 1992. (Annual Air Quality Survey 1992.) In Dutch with English summary. Bilthoven, National Institute of Public Health and Environmental Protection (RIVM). Report no. 722101006.
RIVM, 1994b, Luchtkwaliteit: Jaaroverzicht 1993. (Annual Air Quality Survey 1993.) In Dutch with English summary. Bilthoven, National Institute of Public Health and Environmental Protection (RIVM). Report no. 722101014.
Skärby, L., Sellden, G., Mortensen, L., Bender, J., Jones, M., De Temmermann, L., Wenzel, A. and Fuhrer, J. (1993), Responses of cereals exposed to air pollutants in open-top chambers. In Effects of Air Pollution on Agricultural Crops in Europe (Jager, H.-J., Unsworth, M.H., De Temmermann, L. and Mathy, P. (Eds). Air Pollution Research Report 46, European Commission, Brussels, 241–259.
Sol, V.M. and Feenstra, J.F. 1994, Milieubelasting Regio Schiphol in 1989 en 2015. (Environmental Pressure in the Schiphol Region in 1989 and 2015.) In Dutch. Amsterdam, Institute for Environmental Studies, report R-94/05.
Sommerville et al. (1990), Adequacy of interval estimates of yield changes to ozone estimated from NCLAN data. Crop Science, 30, 836–844.
SPANS, 1993, SPANS – Geographic Information System, 6th edition for OS2. INTRA TYDAC Technologies Inc., Ontario, Canada.
Van der Eerden, L.J., Tonneijck, A.E.G. and Wijnands, J.H.M. 1987, Economische schade door luchtverontreiniging aan de gewasteelt in Nederland. (Economic damage from air pollution to agricultural crops in the Netherlands.) In Dutch. The Hague, Ministry of Public Housing, Physical Planning and the Environment. Report no. Lucht 65.
Van der Eerden, L.J., Tonneijck, A.E.G. and Wijnands, J.H.M. 1988, Crop loss due to air pollution in the Netherlands. Environmental pollution, Vol. 53, pp 365–376.
V&W, 1990, Uitbreiding luchthaven Maastricht: Aanvullend milieu-effectrapport mei 1990. (Expansion Maastricht Airport: Supplementary Environmental Impact Analysis report may 1990.) In Dutch. The Hague, Ministry of Transport, Public Works and Water Management (V&W).
V&W, 1993, Cumulatie van geluidshinder rond schiphol: Thematische bijlage bij het het Integraal Milieu-effectrapport Schiphol en omgeving. (Cumulation of noise nuisance around Schiphol Airport: Thematic supplement to the Integrated Environmental Impact Analysis report Schiphol Airport and surroundings.) In Dutch. The Hague, Ministry of Transport, Public Works and Water Management (V&W).

GERMAN CASE STUDY

Prepared by Ifo* Institut in collaboration with the other country teams.

* Ifo Institut für Wirtschaftforschung, Poschingerstrasse 5, D-81679 München, Germany.
Tel: +(49) 89 92 24 14 32; Fax: +(49) 89 98 53 69; E-mail: 10064.226@compuserve.com

CHAPTER 1

POLLUTION DATA COLLECTION AND MODELING

1.1. Introduction

This chapter provides a brief discussion of the issues arising in the collection of the relevant data for the German case study. The project team would like to emphasise that the current analysis should be recognised as an initial estimate of German damages. Especially in the area of health, doubts about the accuracy of data have had a limiting effect on the extent of damages covered in Germany. Further research is needed in this area and the ongoing work in the GARP II project will try to close the current gaps. In particular, progress in ozone simulation modeling will be made enabling us to quantify related health and crop damages. Also the lack of PM_{10} data on an EMEP basis and the related issue of finding a suitable background level for PM_{10} will be resolved and the range of pollutants will be widened up to include heavy metals, carcinogenics, carbon monoxide and NO_x. The follow-up project will provide us with further insights of the valuation of damage to eco-systems and forests. Moreover, more work needs to be done in order to determine the defensive expenditures for certain fields. For example, the area of historic monuments is still widely under-researched. Thus, with the completion of GARP II, the monetary valuation of environmental damage will be comparable throughout the project partner countries on a broader basis.

The Ifo Institute wants to acknowledge its great appreciation for the help of Wolfram Krewitt and Rainer Friedrich from IER who calculated the SO_2 damages in GARP I. We are also grateful to Dr Werner Schulz from the Federal Environmental Agency for his helpful advice in the initial stage of the research.

1.2. Pollution and modelling data

The ambient concentration levels of the selected air pollutant are also known as the levels of 'immission' to distinguish them from emission. Not the rate of emission, but the level of pollutant to which the receptor is actually exposed is important for the damage calculation. Precise monitoring of ambient concentration is needed to complete this first step of research.

In Germany the control of air pollution is the task of the Länder. For this reason the Länder own air quality measuring nets in certain control areas where the state and development of air pollution will be documented. The immission measuring nets have an irregular spatial distribution with mostly denser control points in urban areas and less dense control points in rural areas. To fill the information in a single grid system a special interpolation procedure is used: Within a radius of 40 km the interpolation value of the circle center is calculated from all measured values of the control points. In addition values of control points nearby the interpolation point are weighted higher than those far away from it. The obtained ambient concentration values are close values to the actual ambient concentration situation. The procedure aims at the presentation of the ambient concentration situation in Germany as a whole. Methodologically it is not permissible to derive point differences of air pollution.

In this study the EUROGRID co-ordinate system was used (grids of 100×100 km^2). However, the method of moving from point source data to grid values for pollution concentrations was not the interpolation method but simulation models of atmospheric transport. Although in Germany simulation models are available for SO_2, NO_x and PM_{10}, in the end the valuation process could be done only for SO_2. According to the results of the EXTERNE project (see European Commission 1995a–f), NO_x impacts on health can be neglected. However, NO_x becomes important as a precursor substance for ozone. The problem with PM_{10} is that there are no background levels known for Germany. In addition PM_{10} effects on health were not monetized because SO_2 and PM_{10} impacts cannot be added for reasons of double counting. For ozone simulation modelling was not possible, so it was tried to take homogenous regions, e.g. urban and rural areas, and to apply the concentration from the measuring net. The German Federal Environmental Agency recommended a comparative analysis for the calculation of ozone impacts on health by using a US study and drawing conclusions for the German situation.

1.3. Background levels

An important question is the background concentration against which current pollutant levels should be compared. For all countries the most appropriate data appears to come from North West Scotland, (see synthesis report) the most pristine environment that exists in Europe. Background levels of 1 ppb SO_2 and 2 ppb NO_x seem to apply in this area. For ozone a figure of 30 ppb for Northern Europe can be used (see synthesis report). However, ozone levels in towns are likely to be much lower than the background concentration because of the nature of NO_x emissions from car exhausts. Thus in urban areas there might be even a beneficial result due to ozone. This must be examined in further research.

For PM_{10} it was up to each country team to choose a suitable value. For Germany there is no value known. The UK values of 5–15 $\mu g/m^3$ were not regarded to apply for Germany.

CHAPTER 2

HEALTH

2.1. The approach

High levels of air pollution, especially sulphur dioxide, particulates and NO_x as well as ozone have been associated with respiratory illness in children and adults. Postulated chronic and acute health effects from air pollution usually are related to the respiratory system and include mortality, respiratory symptoms and diseases, and can have effects on the lung. However, respiratory symptoms are not only a consequence of air pollution but occur as a result of a variety of factors, e.g. smoking, allergens, etc.

The impact of air pollution on the population depends on the concentration level and the sensitivity of the people. WHO has set Air Quality Guidelines on the basis of toxocological, clinical and epidemiological studies (see Table 2.1). Significant health effects are expected if these guidelines are exceeded. But there is some debate that human health is threatened even below these threshholds (Steele/Ozdemiroglu 1994).

This study reviews the available evidence in relation to the health effects which can be attributed to the pollutants mentioned above. The review findings provide a possible range of costs which can be expected if there are no further pollution reducing measures. The four stages for monetizing the effects of air pollution for health are as follows:

Table 2.1. WHO air quality guidelines for acid rain pollutants ($\mu m/m^3$)

Pollutants	1-hour exposure	24-hour exposure	1-year exposure
SO_2	350	–	–
	–	125[a]	50[a]
TSP[b]	–	–	120
NO_x	400	150	60
O_3	150–200	100–120[c]	–

Notes:

(a) When exposure to SO_2 occurs simultaneously with exposure to TSP.

(b) TSP stands for Total Suspended Particulates.

(c) The guidelines are for an 8-hour exposure.

Source: WHO (1987).

- collection of pollution data (see chapter 1),
- determination of the size and characteristics of the population exposed,
- use of dose–response functions (as far as possible),
- assignment of a monetary value to the adverse health effects identified in the dose–response relationships.

2.2. Population at risk

This represents the 'stock at risk'. The damage caused by ambient levels of pollutants depends on the spatial distribution of the population in the vicinity of the emitting source(s) as well as on the age and vulnerability structure of that population. For example, children and the elderly may be more vulnerable to adverse health effects than those in the middle age ranges. Similarly, patients already suffering from asthma or other respiratory diseases are more likely to receive additional negative health impact than those with less sensitivity to these conditions.

2.2.1. *Population density*

The EUROGRID database did not contain data for Germany. IER in Stuttgart updated the database using data from the 'Statistik Regional' database published by the German Statistisches Bundesamt, including data for the former GDR (IER, 1994). Data on the 'Kreis'-level (smaller than EUROSTAT NUTS 3 level) was transformed to the EUROGRID format using ARC/INFO GIS software. The best way to approach the measurement of the population exposed to risks associated with the ambient concentration of pollutants is to acquire population data within a $20 \times 20\,km^2$ square around each monitoring site. Using this small scale should ensure that there is no overlapping. However, for reasons of simplicity, a $100 \times 100\,km^2$ scale was used.

2.2.2. *Risk group distribution*

Most of the damage functions used for finding the health impacts are related to a certain risk group such as children, the elderly or people with allergies.

Table 2.2. Total area and population density in Germany, as of January 1st, 1991

	Germany	Former area of the FRG	New Länder and East-Berlin
Area in km^2	356,853.95	248,635.60	108,218.35
Inhabitants per km^2	223	256	148

Source: Statistisches Bundesamt (1992).

To quantify effects within a certain risk group, the share of the risk group in the total population must be known. Mean values for the age structure were used throughout rural and urban areas (although it is likely to have varying age structure between the two places).

An estimated 5% of adults in the FRG have chronic bronchitis, Weber (1991) estimates that between 0.3% and 6.7% suffer from asthma. The latter value is used in the present analysis.

The baseline all-cause death rate for Germany (former FRG only) is 11 deaths per one thousand persons a year (IER, 1994). This implies an overestimation of mortality effects within this study because deaths from violent or external causes are added whereas the mortality exposure–response functions apply to non-violent causes only. Incremental mortality is derived by linking the baseline data, the population weighted pollution increment and the corresponding exposure–response function.

2.3. Use of dose–response functions (DRFs)

It is important to have information on the direct relationships between ambient levels of key air pollutants and both morbidity and mortality effects (dose–response-functions). *Morbidity* can be defined as a deviance from a state of well-being, namely becoming ill. *Mortality* means the effect on the death rate after removing all other effects contributing to the death rate. There is acute and chronic mortality. Given the lack of German data establishing dose–response coefficients for both morbidity and mortality effects, the functions derived in US epidemiological studies (field studies) shall be used.

2.3.1. *Dose–response functions for SO_2*

An economic assessment of the health costs of SO_2 requires the application of dose–response functions to data on levels of pollution and population at risk. This is necessary to calculate changes in levels of mortality and morbidity (adult chest discomfort, emergency room visits (ERVs) will be examined).

Mortality

In various epidemiological studies variations in the ambient air concentration of sulfur dioxide have been associated with increased acute mortality. Smog

Table 2.3. Age structure of the German population (December 31, 1990)

Age/Years	<6	6–15	15–18	18–25	25–45	45–60	60–65	>65
Share	6.7%	9.5%	3.0%	10.4%	30.0%	20.0%	5.5%	14.9%

Source: IER (1994a).

episodes with short time peak exposure over a few days might induce a significant increase of death rate within the exposed population. Mainly old people with prevalent cardio-respiratory diseases are affected. Premature deaths can shorten lifetime by a few days to several months.

Admittedly, it is not clear if the results obtained can be attributed to SO_2 alone. Combinations of pollutants certainly influence human health more than one individual pollutant. High levels of SO_2 often coincide with high levels of particulates. So any correlation between SO_2 and human health does not necessarily prove that SO_2 alone is responsible.

Most of the time this study will use the same dose–response functions developed under the EXTERNE programme (see European Commission, 1995a–f). For SO_2 and mortality the function found by Hatzakis *et al.* (1986) is used. The study related daily mortality in the greater Athens area to daily measurements of SO_2 and smoke in the years 1975–1982.

Morbidity

Again, single effects are hard to isolate. In particular it is difficult to separate out effects of SO_2 from those of particulates. These two pollutants show a high temporal and spatial correlation. In addition SO_2 can be transformed into acid sulfates which would be classified as particulate matter. Nevertheless some studies suggest some independent effects.

SO_2 especially irritates the nose, the bronchial system and the eyes. Airway constriction leads to changes of the pulmonary function. Epidemiological studies suggest that changes in daily average exposure to SO_2 effect lung function as well as the respiratory system.

- Probability of a chest discomfort day in adults: Data from daily diaries indicated a significant association between SO_2 and chest discomfort (Schwartz *et al.* 1988).
- Emergency room admission: Sunyer *et al.* (1991) showed a statistically significant relationship between emergency room visits for chronic obstructive pulmonary disease and daily average SO_2 concentrations.

2.3.2. *Dose–response functions for NO_x*

Health impacts associated with NO_x are increased infections of the lower respiratory tract in children and increased airway responsiveness in asthmatics. The primary concern about NO_x, however, is its role as a precursor substance to ambient ozone in outdoor pollution.

The dose–response function regarding new phlegm and new sore throat episodes in adults due to NO_x as well as new eye irritation episodes in adults was found by Schwartz and Zeger (1990).

2.3.3. *Dose–response functions for total suspended particulates (TSP) or PM_{10}*

Particulate matter includes ammonium sulphate, iron, lead, nitrate, carbonaceous materials and sulphates (SO_4). The size of particulates determines their health effects since smaller particles travel further in the atmosphere and are more readily absorbed in the respiratory system. Particles less than 20 micrometer can penetrate the airways with 50% efficiency, particles of 4–5 micrometer can enter the lung, and sizes below 1–2 micrometer may not be deposited at all. Thus research attention is now shifting from TSP to particulate matter of less than 10 micrometer (PM_{10}). TSP is converted to PM_{10} by multiplying TSP by either 0.45 or 0.55 (US EPA conversion in adopted by EXTERNE project).

PM_{10} pollution is an important risk factor for acute changes in lung function, respiratory function and respiratory symptoms, increased acute and chronic respiratory illness and even death among high risk groups such as asthmatics.

For the following endpoints dose–response functions were established. For further details see the Synthesis Report, Chapter 3:

- Emergency room visits (Samet *et al.*, 1981),
- Respiratory hospital admissions (Ostro, 1987),
- Childrens' bronchitis (Dockery *et al.*, 1989),
- Childrens' chronic cough (Dockery *et al.*, 1989),
- Symptom days (Krupnick *et al.*, 1990),
- Asthma attacks (Ostro *et al.*, 1991),
- Cases of childhood croup (Schwartz *et al.*, 1991)

2.3.4. *Ozone in Germany*

The method of ozone monitoring is the UV absorption technique, and the unit of measurement, ppb. Three sets of standards are used:

FRG Environmental Ministry recommendations	(half hour mean $</=$200 mg/m^3)
EC Ozone Directive Requirements	(one hour mean $>/=$100 ppb; 8 hourly mean $>/=$55 ppb; daily mean $>/=$32 ppb)
WHO Guidelines	(one hour mean $>/=$100 ppb; 8 hour mean $>/=$50 ppb; daily mean $>/=$33 ppb; Growing season $>/=$30 ppb)

Clinical studies show that no effect is likely under ambient exposure conditions for persons at rest. Irritation, however, may occur at around 300 ppb (600 microgram/m^3) after one or two hours' exercise. During severe pollution episodes these levels of concentrations are found. The irritations due to ozone often are an increase in airway resistance, a reduction in lung volume and a small increase in the responsiveness of the airways to constructor substances such as SO_2. In addition increased infection and allergy cases can be expected. There is no evidence that smokers, the elderly or asthmatics are more sensitive to ozone than others. Associations have been observed at levels under 80 ppb. However, there is no proven threshold concentration level.

In this study a comparative analysis to the study by Hall and Winer (1992) will be attempted.

2.3.5. *Discussion of the damage-function approach*

Although the damage-function approach is the logical way to evaluate health impacts, it implies some critical issues which are addressed in the following (IER, 1994b):

- At present there are no usable economic valuations of lung function changes. That is why lung function health impacts from epidemiological studies have not been considered.
- Epidemiological studies are based on immediate health effects due to day-to-day changes in air pollution. Studies of this design usually provide for the most reliable exposure–response functions. As a consequence the evaluation of health effects is also focused on acute responses.
- Some of the key relationships are linearized and annualized, assuming an independence of background levels. The linearized functions are considered to be a good approximation of the reported non-linear functions. However, the application of linearized functions without threshold to very low increments in air pollution far away from the source might lead to an overestimation of effects.
- The quantification of an impact of an air pollution mixture is difficult using the impact pathway approach. Evidence on the independence or not of the effects of the component part are hardly possible to find.
- Finally the question of transferability of exposure–response functions must be raised: Partly it depends on the health endpoint under consideration. If the health endpoint is a biological factor rather than a social event (e.g. restricted activity days) transferability is likely to be better. Restricted activity days partly depend on sociocultural factors.

2.4. Assigning a monetary value to the adverse health effects

2.4.1. *Mortality*

As discussed in the Synthesis Report (Chapter 2), the most appropriate approach is to measure willingness to pay (WTP). The Human Capital approach is rejected since it would be hardly satisfactory in this context. The measure of a life being the net present value of a person's expected stream of earnings would exclude children, retired persons, housewives/husbands and the unemployed, all persons who have no earned income. At maximum they would be given a very nominal income under the Human Capital approach.

For the WTP approach one needs data on what people are willing to pay to reduce the risk of premature death. One can express small risks as a percentage and reach unity by multiplying or aggregating up over many individuals; e.g. for 100,000 people a risk of 1/100,000 means that statistically one person will die. So one has to find the WTP to reduce this small risk and then it is multiplied by 100,000 to get the WTP for a statistical life for the chosen sample. This so-called 'value of statistical life' is statistical in the sense that not the life of an existing human being is evaluated but just the value of a statistically generated life.

There are three major ways to estimate the value placed on mortality rates:

- Hedonic Wage Risk Technique,
- Market Techniques,
- Contingent Valuation Method.

The average value of statistical life in the UK/Europe is currently measured at between $2 and $3 million (OECD 1992). If it is assumed that individuals would wish to pay not only to reduce risks for themselves, but also for others, one can add 40–50% for this altruistic component of the value of statistical life. In this study, a value of statistical life (VOSL) of ECU 3.1 mn. has been taken. However, as has been discussed in the synthesis report, this value may be misleading in cases where the number of life years lost as a result of death from air pollution are small. A valuation of the mortality impacts using a value of life years lost is given in the main report. The country case study did not deal with this issue, although it recognized the importance of it. Taking the population of Germany, the cost of mortality due to air pollution will be one of the highest items of the cost of acid rain or the benefit of abatement.

2.4.2. *Morbidity*

The measurement of morbidity effects can be categorized as following:

- response in ventilatory capacity (e.g. forced expiratory volume),
- symptom response (e.g. cough),
- treatment response (e.g. hospital admission).

The problem is that not all lung function changes lead to symptoms and not all symptoms can be monetized (Steele/Ozdemiroglu 1994). However, the chosen response measures in this study are:

- Emergency room visits for all-cause as well as respiratory diseases,
- Respiratory symptom days: daily occurrence of upper and lower respiratory symptoms as an acute effect of air pollution.

Concerning the monetary valuation one has to consider the financial cost of illness and the value placed on suffering. A sophisticated approach to estimate the morbidity effects is the 'value of statistical suffering' using the willingness-to-pay method. A complete valuation of morbidity requires an estimate of the:

- private value of an individual's WTP to avoid illness (WTP_i),

- WTP to avoid illness/injury in others (altruistic value, WTP_o),
- cost of the illness: direct cost to society of the illness (e.g. lost productivity, healthcare costs).

It is assumed that the altruistic WTP_o to avoid illness in others is around 40–50% of the individual's WTP_i to avoid one's own illness. But in case of aggregation across all health effects, double-counting must be avoided: There is likely some overlap between restricted activity days and symptom days which would need to be allowed for in the aggregation process. Data on cost of the illness was not collected so that the results presented focus only on WTP. Often WTP_i already includes loss of wages and leisure time as well as the pain and suffering incurred. In the literature it is stated that the individual WTP_i is greater than the cost of illness. In this study, however, only WTP_i values obtained from contingent valuation studies in the United States are used. Valuations were only carried out for SO_2 impacts due to problems in defining appropriate background levels for other pollutants. The values used are detailed in the table below:

Table 2.4. Monetary values of SO_2 pollution for different endpoints

Health effect	Endpoint measure	Studies	Value
Mortality (value of statistical life)	% increase in death rate (e.g. deaths/million)	Hatzakis 1986	ECU 3.1×10^6
Adult chest discomfort days	Additional days of chest discomfort per adult per year	Schwartz *et al.*, 1991	ECU 7.5
Emergency room visits (ERVs)	Additional daily ERV for chronic obstructive pulmonary disease per adult per year	Sunyer *et al.*, 1991	ECU 223

CHAPTER 3

NOISE

3.1. Introduction

Noise results from the emission of sound which is vibration of air due to perturbation by some mechanical vibration. Much noise which disturbs the human ear is anthropogenic in origin. At lower levels it can interrupt sleep or hinder verbal communication. In general, unwanted sound is recognized as a burden and affects human amenity.

The usual metric of sound is the decibel (dB) which is a logarithmic scale of the pressure generated by the vibrations of the air, and is defined as:

$$1\ \text{dB} = 20 \log_{10} (\text{Pressure}/20\ \text{mn.ikroPa}),\ \text{with}$$

$$\text{Pa} = \text{Pressure in Pascals (the SI unit of pressure, } = 1\ \text{Newton m}^{-2}).$$

The main sources of noise are road and air traffic. About 13% of the population in Western Germany live in houses which are exposed to traffic higher than 65 db(A), a range where the risk for heart disease is increased. About 1% of the population in the old Länder lives in noise protection areas near airports. For 1990 there is no exact data on noise for the new Länder. However, it can be assumed that disturbances are of a similar magnitude as in the old Länder. Although there was less traffic, this was offset by louder cars and worse roads

Table 3.1. Road traffic noise pollution of the population of West Germany in 1985

	Percentage of the population exposed to noise over ... dB(A)	
dB(A)	Day time	Night time
45	–	46
50	–	28
55	45	14
60	27	6
65	13	2
70	5	0.2
75	1	–

Note: Similar figures do not exist for East Germany.
Source: Bundesumweltministerium (1992).

Table 3.2. Railway noise pollution of the population of West Germany in 1985

	Percentage of the population exposed to noise over ... dB(A)	
dB(A)	Day time	Night time
45	–	36
50	–	20
55	13	9
60	5	3
65	2	1
70	0.5	0.3
75	0.1	0.1

Note: Similar figures do not exist for East Germany.
Source: Bundesumweltministerium (1992).

as well as a similar situation in the railway sector. Only after 1990 has the situation improved.

This chapter presents a methodology for calculating the noise disbenefits from anthropogenic sources. Both a hedonic pricing and a contingent valuation method together with an estimation of defensive expenditures are used.

3.2. Hedonic pricing method

The hedonic pricing method is relatively straightforward because it is assumed that the only significant receptors are people and that they are sensitive to noise only in their homes. In this study not house price, but *rent price differences* for different noise bands were used.

So the methodology for including noise is suggested to proceed as follows:

- Identify the affected receptors
- Identify the response of receptors to the pollutant (dose–response function)
- Identify survey data on household external noise levels (see also chapter 3.2.3)
- Use national housing data (aggregate number and rent prices of houses in the country) to calculate the numbers of houses in various noise categories, and
- Use the depreciation factor per dB(A) to calculate the aggregate national impact

Appropriate valuation for noise using hedonic pricing is as follows (Borjans, 1983 and Pommerehne, 1986):

- (No difference between continous or intermittent noise): a reduction of 0.5% vs. 1.26% in rent prices per dB(A)L_{Aeq}.

In addition, to calculate the aggregate damage it is necessary to identify some noise baseline. This baseline can be either:

- the level of noise corresponding to zero economic activity, or
- the level of noise corresponding to zero amenity loss, i.e. a threshold for the dose–response function.

The former is likely to be very low (<40 dB(A)), the latter is often supposed to be 55 dB(A) because this is the WHO 'no annoyance' standard.[1] In this study a baseline of 45 dB(A) was used.

3.3. Contingent valuation method

In order to estimate the costs of noise in Germany by using the CVM, the noise situation must be known.

3.3.1. *Noise maps*

Noise maps are available for 44 cities and communities of which 38 have more than 100,000 inhabitants (Hoffjann, 1988). These maps mainly register noise from road traffic. Noise from railways is included in 14 cities, from industry in 8 cities and from leisure only in one city. Noise from air traffic is registered in noise protection areas. During the last years the availability of noise maps in Germany has improved and is now sufficient in the field of road traffic. Nevertheless, the registration of noise by noise maps remains a problem (Weinberger *et al.*, 1990):

- The existing noise maps mostly refer only to certain parts of a city, certain noise areas and noise during daytime.
- The number of people disturbed is mostly unknown.
- Very seldom do noise maps exist for small communities, making a representative calculation impossible.
- Noise maps cannot be fully compared because measuring and calculation methods are not standardized.
- The exact site of the houses, the 'floor effect'[2] and the resulting differences in noise levels are not taken into consideration. The 'frontside view' leads to an overestimation of effects.

[1] It is not certain that this level corresponds to a zero amenity loss since amenity is a highly subjective quality. For example wind turbines can be a noise source in rural areas well below 55 dB(A). But it seems reasonable that this kind of damage is low in comparison to those at higher noise levels caused by traffic and aircraft.

[2] The 'floor effect' takes into account that the higher a buil;ding is, the more noise is reduced because the distance to most sources of noice is diminished.

Obviously, the objective noise emissions cannot be sufficiently depicted in noise maps. But one can see the individual disturbance especially for road traffic, partly also for noise from air traffic, railway and industry.

3.3.2. *The noise situation in the Battelle Model*

The Battelle Institute (1981 and 1985) calculated the noise situation for 60 representative areas in Germany and estimated a result for Germany with the help of a computer model. Thus the noise disturbance according to sources (road, rail, construction, industry) and classes of 5 dB(A) for day and night time depending on 5 different community types could be documented. The model was an essential factor in the Weinberger study. However, in order to determine the individual noise disturbance of the persons interviewed in the Weinberger study, noise maps were consulted.

Two points of caution are required with regard to the Weinburger study. First, there is the problem of the 'probability of provision bias'. People believe the noise levels will not be delivered so therefore state a high WTP. Second, the estimates probably also include people's WTP for the associated reductions in noise levels, air pollution and soiling that the reduction in road traffic noise (implicitly road traffic) would result in. Nevertheless, the results are interesting and merit further consideration.

CHAPTER 4

CROPS

4.1. Introduction

Crops can be damaged directly or indirectly by air pollutants. On the one hand direct damage such as yield loss occurs and additional liming measures must be taken; on the other hand indirect effects such as decreased resistance to pests due to air pollution can happen.

This analysis is restricted to the direct effects of SO_2 in the form of yield losses. Quality changes, other mitigating measures of the farmers or indirect effects (via soil) are left out. However, approaches to do so must be borne in mind. For yield changes due to O_3, dose–response-functions exist, but ozone dispersion calculation is not possible at the moment. An assessment of corresponding damages cannot yet be performed. Furthermore, for direct effects of NO_x and acid deposition, indirect effects via soil and synergistic effects exposure–yield-relationships are non-existent (IER, 1994a). Interactions with pathogens and with climate are not analyzed in this study, but must at least be mentioned because they are of major importance.

4.2. Crop Losses Due to SO_2

Crop species show different SO_2 sensitivity. According to several sensitivity classifications based on long-term exposure to SO_2 the following categories can be developed for plant species (VDI, 1978; Guderian, 1977; OECD, 1981):

- sensitive,
- intermediate,
- less sensitive.

Wheat, rye, barley, oats, peas and beans have been considered to show at least intermediate SO_2 sensitivity and have been chosen – with the exception of peas and beans – for the analysis.

4.2.1. *Pollution level*

Regional SO_2 levels were calculated using the Harwell Trajectory Model. Results on the EMEP grid were converted to the EUROGRID coordination system.

4.2.2. *Compilation of reference environment data*

Data on area under cultivation and yield of crops was provided by publications of the Statistisches Bundesamt (1990) at the Länder level. Data from the following crop species have been used: wheat, barley, rye, oats. Due to the former separation of Germany, East German data were gathered differently so there are inconsistencies in the data which cannot be avoided. All data have been converted to the EUROGRID coordination system.

4.3. Dose–response functions

The damage function used was discovered by Roberts (1984). He analysed 125 longtime exposure chamber studies described in the literature resulting in 16 functions for 21 crop species under SO_2 concentrations between 16 and 263 ppb (controlled conditions). The influence of the following factors was examined:

- differences in response between plant species,
- differences in pollutant flux,
- variation related to growth rate and time of year,
- variation with plant age,
- variation with plant density.

But since only few rural locations in Germany experience SO_2 levels greater than 15 ppb, the equations found were not directly applicable (as in the EXTERNE reports, European Commission 1995c/d). The approach to model the effects for low SO_2 concentrations was the following:

- Extrapolation of the linear regression. As it is assumed that low SO_2 concentrations induce beneficial effects, this procedure would overestimate the actual yield changes.
- The linear curve is adjusted to a parabel with the following characteristics: Yield change should be zero when there is no SO_2 concentration and the parabel should approximate tangentially the linear curve at the point it intersects the concentration axis.

Thus, the parabel depends on the 0%-yield change mark which itself depends on the chosen control concentrations used as reference concentrations for the 0% yield change. Thus, results vary according to the choice of the 0%-yield change mark. Table 4.1 shows the exposure–response relationships used for quantification:

The relative yield change in percent due to the measured ambient air quality is calculated in the following way (IER, 1994b):

$$\text{Relative yield change} = (y_n/y_b - 1) \times 100\%$$

Table 4.1. Exposure–response relationships used for quantification

Short name	Equation	Receptors	Reference
Roberts *a*	$y = 2.75 - 0.068\ SO_2$	Wheat, Barley, Rye, Oats	Baker (1984)
Roberts *b*	$y = 0.0068\ SO_2 - 0.0017(SO_2)^2$, $SO_2 < 40.6$ mikrog/m^3; $y = 2.75 - 0.068\ SO_2$, $SO_2 > 40.6$ mikrog/m^3	Wheat, Barley, Rye, Oats	Baker (1984)

y = yield change in percent (positive for yield increase, negative for yield loss).
SO_2: SO_2 concentration in μg/m^3 (transformation factor: 1 ppb = 2.66 μg/m^3 (20°C)).

Source: IER (1994a).

with

y_b: yield in % at background concentration C_b,

y_n: yield in % at new concentration C_n.

Multiplying the relative yield change with the specific yield Y_s (in t/ha) at background concentration levels (known from agricultural statistics) provides for the specific yield change (in t/ha) and further multiplication with the area under cultivation A yields the absolute yield change (in t):

$$\text{Absolute yield change} = A \times Y_s \times \text{relative yield change}/100\%.$$

4.4. Monetary valuation

Crop losses have been evaluated with the world market prices given in Table 4.2.

Table 4.2. Monetary unit values for crops in ECU/t

Crop species	Wheat	Barley	Oats	Rye
Crop price	110	69	65	56

Source: IER (1994a).

CHAPTER 5

FORESTS

5.1. Introduction

Air pollution through SO_2 and NO_x can damage forest ecosystems directly or indirectly via acid rain or ozone formation. Usually there are three categories of damage:

- timber production,
- leisure time and recreational activities,
- existence value.

During the first phase of the EXTERNE project, forest damage experts tried to identify the most suitable exposure–response models for the quantification of forest damage. Only one correlation–valid for German Norway spruce forests–was agreed upon. This study hopes to identify exposure–response relationships also for deciduous trees and other conifers.

5.2. The German forest damage inventories (FDI)

The Forest Damage Inventory was established in West Germany in 1984 and after the reunification extended to East Germany. The FDI groups sample trees in five damage classes according to foliage loss and yellowing:
However, the procedure of the FDI has been criticised for several reasons:

- Foliage loss is an unspecific symptom which does not allow conclusions to be drawn on the cause of damage.

Table 5.1. Damage classes of the forest damage inventory (BMELF, 1991)

Damage class	Description
0	No damage
1	Warning stage (formerly slight damage)
2	Moderate damage
3	Severe damage
4	Dead

Source: BMELF (1991).

- A clear definition of regional and species-specific standards for 'normal' foliage is missing.
- The relationship between foliage loss and growth or physiological/bio-chemical characteristics is poorly characterised.

Suggestions for a methodological improvement using damage symptoms such as nutrient deficiency, climatic stress etc. have not been put into action.

According to FDI 25% of all trees in Germany suffer from significant damage (damage classes 2–4). The worst picture is presented in the former GDR where the following studies present only a few highly damaged forest areas.

- the 'Stendaler Altmoränenland' with 64% of forest area in damage classes 2–4,
- the 'Thüringer Becken' with 58%,
- the 'Thüringer Gebirge' with 50% and
- 'Mecklenburg-Vorpommern/Küstenland' with 51%

In West Germany the most affected areas are:

- the Bavarian Forest with 40%,
- the 'Frankenwald' and 'Fichtelgebirge' with 36%,
- the Bavarian Alps with 39%,
- the 'Rhön' and the 'Fränkische Platte' both with 38% and
- the Black Forest with about 22%.

It is now accepted that pines and several deciduous trees are ozone sensitive to some extent, whereas spruce or other native conifers are less so. Since firs are extremely sensitive towards SO_2, the total disappearance of these trees in regions with high concentrations of SO_2 such as parts of Eastern Germany, the Czech and Slovak republics and Poland can be explained.

Oak and pine are the most affected of all trees (31% and 29% in damage classes 2–4). The newest FDI 1994 once again confirms the affectedness of oak trees, but also mentions that firs are in a very bad condition. The FDI stresses that the origins of forest damage are more and more difficult to assess. Not only single pollutants, but in particular the so called 'environmental stress' which are synergisms between different harmful substances are assumed to be responsible for forest damages. In Hessia, the stock of firs has been diminished so extremely that from the small number of trees left, evidence for any cause-effect relationships can no longer be seen.

5.3. The critical load concept

For Germany there is a map of critical load of acidity for forest soils. Furthermore critical loads exceedences have been estimated. Although there is no map for the critical loads of nitrogen for ecosystems, empirical values as well as exceedences of critical loads have been proposed.

5.3.1. *Excedance of critical loads for total acidity in Germany*

The UN-ECE (1990) defines critical loads of sulphur, nitrogen and acidity as 'the highest deposition of acidifying compounds that will not cause chemical changes leading to long-term harmful effects on ecosystems structure and function'.

The method mostly used to calculate critical acidity for forest soils is the steady state mass balance method (SMB). This method ignores the time to reach a new steady state after a system has been disturbed. A time-independent equilibrium between the soil solid phase and soil solution is assumed. To compute the critical load of acidity, critical values of chemical parameters are needed. Again, UN-ECE (1990) defines critical chemical values as 'the highest value of a critical chemical parameter or combination of parameters that does not cause a significant harmful response in a biological indicator'. Important chemical critical values are, e.g. aluminium concentration, aluminium/calcium ratio, pH and alkalinity.

The critical load of actual acidity is then the difference between the weathering rate of base cations and the leaching rate of alkalinity when the steady state is at the critical point, i.e. determined by the critical chemical values. Depending on the weathering rate different critical loads of acidity are obtained, e.g. soils on parent rock with a high weathering rate like limestone have a higher critical load of acidity.

Although the first German critical loads map of acidity for forest soils was published in 1991 by Hettelingh *et al.*, there are still many problems with this approach. Major difficulties occur in areas with high precipitation. Since the critical value of alkalinity for forest soils is negative, the critical load of acidity increases with high runoff. This has the somewhat misleading effect that highly damaged areas like the Black Forest show a high critical load. Similarly the Bavarian Alps, the Bavarian Forest, the 'Thhringer Gebirge' in the former GDR show high damages but high critical loads in the map. So attempts were made to calculate leaching rates in a modified way. As this map is not yet available the map already published will be used in the following.

If the critical load is exceeded, negative effects are expected for forest ecosystems. The collection of deposition data is difficult in Germany because there is no measurement network for acidic deposition in Germany. Only in Bavaria there is a network of 23 measuring stations (open field, bulk collectors) in a 64×64 km^2 grid. There are two facts restricting the quality of the data: Firstly 70% of the data is from before 1986, not representing the current situation. Secondly, only acidic deposition in the open field is measured, not below tree canopies where loads are much higher. However, total acidic deposition below the canopy is hardened by the leaching of compounds from needles, leaves or bark. With the help of 'filtering factors' used for the shares of the canopy the total acidity deposited is estimated.

It turns out that in the areas mentioned above, where high damage coincides with high critical loads, the excedance of the critical load is so significant that co-occurrence of high critical load and high forest damage can be explained.

5.4. Assessment of impacts of acidic deposition on German forests

The following impacts are considered:

- effects of acidic deposition due to SO_2 and NO_x emissions and
- effects of ozone burden
  (each on the amounts of cubic meters of standing confer and deciduous forest by region).

Dose response function(s) used.

A discussion of the dose response function appropriate to the valuation of air pollution impacts is given in the UK report. The basic methodology is one based on Kuylenstierna and Chadwick, in which areas subject to critical load excedance are estimated to have a reduced growth of timber. This loss in

Table 5.2. Forest damages in 1990 in percent of Sample Trees[1]

State	Damage class 0	Damage class 1	Damage Classes 2–4
Hamburg	48	36	16
Lower Saxony	46	37	17
North Rhine-Westfalia	58	29	13
Schleswig-Holstein	54	31	15
States in Northern Germany	52	33	15
Brandenburg	29	38	33
Mecklenburg-Vorpommern	19	32	49
Sachsen	37	36	27
Sachsen-Anhalt	28	38	34
Thuringia	19	31	50
States in Eastern Germany	27	35	38
Baden-Wuerttemberg	39	44	17
Bavaria	27	43	30
Hessia	40	41	19
Rhineland-Palatinate	50	40	10
Saarland	56	27	17
States in Southern Germany	34	42	24

Source: Bundesministerium fhr Ernährung, Landwirtschaft und Forsten (1992).

[1] The figures for Eastern Germany stand for 1991 since in 1990 data was not yet available. For Baden-Wuerttemberg, Bavaria and Saarland also 1991 data is used since there was no measurement in 1990 due to storm damage.

Table 5.3. Forest areas in Germany

Federal Republic of Germany	Forest area (ha)
Old Länder:	7360
Coniferous trees	5078
Deciduous trees	2282
New Länder and East Berlin	2468
Coniferous trees	1849
Deciduous trees	619

Source: Statistisches Bundesamt (1992).

growth is then valued. Such a valuation is admittedly only partial, but it is useful as it provides some information on losses in monetary terms. The valuation for Germany was done in the UK study.

5.5. Monetary valuation

For acid rain damage there are four stages to the valuation exercise:

(i) use of dose–response-functions based on critical loads. These relate acid deposition to crown cover/density;
(ii) the next stage relates crown density to growth;
(iii) monetary estimates need to be made of reduction in growth (based on the stumpage value of timber at international prices). An approximate value taken in the UK study was ECU 50 m^{-3}.
(iv) Valuation of damages caused to recreational activities by the various emissions. These are valued using the contingent valuation method (see Chapter 9 for details).

CHAPTER 6

ECOSYSTEMS

6.1. Introduction

Given the lack of available data, impacts of air pollution on natural ecosystems should mainly be characterized using physical indicators. The following were identified as a means of describing current threats to ecosystems:

- assessment of the area of national nature reserves under threat from excedance of critical loads of nitrogen, acidity and ozone;
- brief assessment of other threats to natural ecosystems;
- assessment of the status of species present in Germany that are on the European Red List of Species.

6.2. Assessment of the area of national nature reserves under threat from excedance of critical loads of nitrogen, acidity and ozone

In the Federal Republic of Germany (without the new Länder) 1.38% of the total area have been nature reserves in 1989. During recent years more and more areas have been converted into nature reserves, but the reserves are rather small (49% of the areas are smaller than 20 ha). Especially in the Alps huge protection areas exist. However, in nature reserves many forms of land use, including recreation are still allowed. This has the consequence that clearly disturbed areas are labelled nature reserves without restricting polluting factors. For this reason it can be assumed that some of the areas shown in Table 6.1 are under threat from the excedance of critical loads of nitrogen, acidity and ozone.

The only internationally recognized protection category are national parks. Germany has ten, five of which are in the former GDR.

- Bavarian Forest (130 km^2),
- Berchtesgaden (210 km^2),
- Wadden Sea (Hamburg) (177 km^2),
- Wadden Sea (Lower Saxony) (2,400 km^2),
- Wadden Sea (Schleswig-Holstein) (2,850 km^2),
- Hochharz (59 km^2),
- Jasmund (30 km^2),

Table 6.1. Inventory of nature reserves in Germany, status: 1/1/1989

State (land)	Number of nature reserves	Area (ha)	Percentage (%) of whole area[1]	Whole area of land (ha)
Baden-Wuerttemberg	553	36.677	1.03	3.575139
Bavaria	374	107.033	1.52	7.055286
Berlin	16	312.000	0.65	48.014
Bremen	10	1.072	2.65	40.423
Hamburg	22	2.752	3.65	75.470
Hessia	397	18.868	0.89	2.111391
Lower Saxony	540	97.371	2.05	4.743920
North Rhine-Westfalia	654	39.048	1.15	3.406794
Rhineland-Palatinate	327	21.037	1.06	1.984776
Saarland	47	1.198	0.47	24.870934
Schleswig-Holstein	123	18.457	1.17	1.572790
Federal Republic of Germany	3063	343825.000	1.38	24.870934

[1]Without areas of the North and Baltic Sea as well as the Lower Elbe and Lower Weser.

Source: Plachter (1991).

- Müritz (308 km^2),
- Sächsische Schweiz (93 km^2),
- Vorpommersche Boddenlandschaft (805 km^2).

Especially all the Wadden Sea national parks are endangered by eutrophication, i.e. high levels of nitrogen and phosphorus. Most of the anthropogenically caused pollution arrives via rivers or through atmospheric transmission into the sea.

Due to diverse land uses such as forestry, agriculture and fishery but also military, traffic or sports, the parks suffer from strong pollution in places. Indirectly, further, negative effects from soil and water pollution are added.

6.3. Brief assessment of other threats to natural ecosystems

Apart from air pollution, other activities pose a threat to natural ecosystems. The Alps suffer from a severe change in land use mainly due to mass tourism. The whole region of the Alps is inhabited by about seven million people, and every year as many as ten million tourists visit the region. In Austria, the size of skiing areas has reached the size of traffic areas. Traditional forms of mountain agriculture disappear with the consequence that the land use patterns change dramatically with negative effects on nature. Table 6.2 presents some of the important current activities in the Alps and their effects.

Table 6.2. Important current activities in the Alps and their effects

Activity	Effect
Expansion of settlements	– loss of natural ecosystems – decreasing importance of agriculture and forestry – increasing danger of avalanches and floods – change in drainage behaviour of streaming waters
Construction of roads and paths	– loss of sensitive big area ecosystems – disturbance of animals – preparation of tourism development
Water management measures	– loss of natural moist areas – loss of natural streaming waters dynamics – steeper high water drainage; increasing high water danger in the lower regions
Construction of cable railways	– optical/acoustical disturbance of ecosystems; eutrophication; soil damage caused by steps – reduction of forest area – increasing danger of erosion – steeper high water drainage
Construction of skiing pistes	– reduction of forest area – increase of erosion danger – steeper high water drainage – mechanical damage to vegetation – increase of soil density – change of soil fauna – disturbance of animals
Sports (e.g. para-gliding, climbing, mountain-bikes)	– disturbance of animals – damage of exposed vegetation – starting of erosion processes

Source: Based on Plachter (1991).

6.4. Assessment of the German species contained in the European Red List of Species

In 1970 the International Union for Conservation of Nature and Natural Resources (IUCN) for the first time summarized all worldwide threatened and extinct species in a 'Red Data Book'. This collection is continued as 'Red Lists'. The following categories have been developed (Plachter 1991):

- *extinct*: species not observed in nature during the last 50 years,
- *endangered*: species threatened by extinction and of which the chances of survival are not given if the current endangering processes are not stopped. In this category also fall species which probably are extinct, but have been seen in nature during the last 50 years.
- *vulnerable*: Species which soon will be categorized as 'endangered' if the current influences are not stopped.
- *rare*: Species with worldwide small populations which currently are not 'endangered' or 'vulnerable'.

- *indeterminate*: Species which are 'endangered', 'vulnerable' or 'rare', but data is not sufficient to name one of the categories precisely.

In Germany a different classification is used which is even stricter:

- category 0 (*extinct or lost*): Species which are extinct (reference time 1850, for mammals and birds even back to the Middle Ages) or lost, i.e. in spite of searches for these species they cannot be found anymore.
- category 1 (*threatened by extinction*): The survival of these species is improbable if the causing factors are not restricted. In this class also fall 'rare' species.
- category 2 (*strongly endangered*): Species endangered in the whole area of the Federal Republic; in some regions they have already disappeared.
- category 3 (*endangered*): Species in huge areas are endangered, in some regions they have already disappeared.
- category 4 (*possibly endangered*): Species with small populations in certain areas (insofar as they do not fall into classes 1–3 due to their current endangerment).

Table 6.3 gives the number and percentage of species in each class for the FRG.

Table 6.3. Current endangerment of flora and fauna in the FRG

Organism	Number of extinct species (percentage of all species)		Number of currently endangered species (percentage of total)	
Fern and flowering plants	60	(2%)	637	(26%)
Moss	15	(2%)	84	(8%)
Lichen	26	(1%)	380	(21%)
Mammals	7	(8%)	37	(39%)
Birds	20	(8%)	78	(30%)
Reptiles	0	(0%)	9	(75%)
Amphibians	0	(0%)	11	(58%)
Fish	4	(6%)	45	(64%)
Butterflies	27	(2%)	467	(36%)
Dragonflies	4	(5%)	39	(49%)

Source: Plachter (1991, p. 285).

CHAPTER 7

MATERIALS

7.1. Introduction

A wide range of materials is exposed to air pollution. These materials show different susceptibility to the effects of acid rain, the main 'output' of air pollution causing material damage. Using the Dutch Acidification Systems Model (DAS) the following materials can be examined (selected for Germany):

- galvanized steel
- sheet zinc
- natural stone

However, Table 7.1 shows that additional materials like concrete or brickwork are also susceptible to acid rain and are of high value for Germany. In the chosen methodology these materials will not be considered.

The damage-prone objects under consideration to estimate the pollution-induced damage are:

- transmission towers
- railway bridges and railway transmission towers
- telecommunication systems
- industrial plants (fuel tanks, sludge towers)
- residential buildings
- cultural assets

7.2. Effects of air pollutants

7.2.1. *Zinc and galvanized steel*

Steel is usually used only with a protective coating such as paint or zinc galvanizing. In moist but unpolluted environments, zinc hydroxide is formed with the electrochemical reactions taking place on the surface and along an electrochemical process producing a protective layer that inhibits further corrosion. In the presence of sulphur compounds, however, the moisture on the surface becomes acidic and causes corrosion. The amount of metal lost from the surface is strongly connected with the SO_2 air concentration.

Table 7.1. Material selection criteria

Material	Susceptibility to acid rain	Value to the FRG
Brickwork	Uncertain	Very high (high economic value due to large stock at risk)
Concrete	Uncertain	Very high (high economic value due to large stock at risk)
Stonework (limestone, calcarous sandstone, marble)	High (severely affected by SO_2)	High (small to medium stock at risk, but high cultural value)
Carbon steel	High (severely affected by SO_2)	Very low (little carbon steel exists uncoated in the environment)
Stainless steel	Very low (excellent resistance to pollutant attack)	Medium
Nickel and nickel-plated steel	High (corrode at high rates in SO_2-polluted areas)	Very low (very restricted use)
Zinc and galvanized steel	High (corrosion markedly increased in polluted environment)	Medium (stock at risk small to medium but some uses of economic importance, e.g. transmission towers)
Aluminium	Low (in general excellent resistance; pitting in very polluted environments but only initially)	Medium (various commercial uses)
Copper	Medium (increased corrosion in polluted environments, but rates decrease in long term (patina layer formed))	Low (stock at risk is low)
Lead	Very low (one of the most resistant commercial materials in the environment)	Medium/low (stock at risk low but important uses, e.g. in telephone cables)
Paint	Uncertain	Very high (very large stock at risk)
Wood	Uncertain	Low (little uncoated wood used in the environment)

Source: Weltschev, M.: Economic Losses to Society Due to Material Damages of Environmental Pollution in the Federal Republic of Germany, in: Hohmeyer, O.; Ottinger, R.L. (Eds.): External Environmental Costs of Electric Power, Berlin 1991, S. 26.

Experimental work has shown zinc corrosion to be a linear function of both SO_2 and time of wetness. The majority of the evidence is expressed in terms of zinc corrosion rates. Corrosion rates can be converted to 'time of loss of protective coatings' if an assumption about the thickness of the zinc coating on galvanized steel is made. An increase of lifetime therefore is a benefit because the protective layer is not lost so quickly and costs of repainting can be saved.

7.2.2. *Effects of air pollutants on paints*

The different types of paint formulations have different sensitivities to air pollutants, but the effects of SO_2 are minor compared to the effects of sunlight and wetness. Recent research indicates that the degeneration of paint systems (paint and paint substrate) may be accelerated by SO_2. However, there is still little evidence of how paint protects metal and the mechanisms that take place at the substrate/paint interface.

7.2.3. *Air pollution as a factor of stone decay*

Natural stone deterioration is caused by dry and wet deposition. In areas with high pollution levels dry deposition is predominant. In clean air areas, however, particularly in those with high rainfall rates, wet deposition is the prevailing mechanism. Dry SO_2-deposition leads to acid formation on the surface and within the pores under the stone surface leading to deterioration of the stone. During wet deposition the reaction product between SO_2 and carbonate – gypsum – is washed away, resulting in a weight loss. Stone degradation from wet deposition is generally lower than that caused by dry deposition.

7.3. Data and monetary valuation

The Building and Dwelling Census of 1987 provides statistics on dwellings in West Germany. No similar statistics are available for East Germany at the moment. Data on industrial buildings is not available at all. Concerning the material surfaces a study by Hoos *et al.* (as part of an ECOTEC study, 1986) identified in a field study the material surfaces from 232 buildings in Dortmund und Cologne. In the next step, the building *identikits* of Cologne and Dortmund have been directly extrapolated to West Germany. This happened on the basis of the statistics on dwellings. Thus, material surfaces for zinc and galvanized steel, sandstone, limestone, other natural stones, aluminium, concrete and rendering have been obtained (IER, 1994). However, the estimation is a result of rather crude assumptions, e.g. it is assumed that in every German town or even village the percentage of skyscrapers is as high as in Dortmund or Cologne.

Another important source of data is the study by Isecke *et al.* (1991) which quantified material damages due to air pollution. The identification of material surfaces was done by extensive interviews with German authorities.

7.3.1. *Selected industrial property groups: overhead power transmission pylons*

With the help of the prevailing practice method the air pollution related maintenance costs for transmission pylons were calculated. The transmission pylons have either a zinc layer or a coating. In case of humidity of more than

Table 7.2. Time intervals for renewal of protection in transmission pylons (years)

Type	'Unpolluted areas'	'Polluted areas'
Pylons with zinc layer:		
– first layer	7–10	1–3
– second and later layers	15–25	10–18
Pylons without zinc layer:		
– second and later layers	15–25	10–16

Source: Isecke *et al.* (1992).

70%, zinc corrosion is increased. SO_2 will accelerate this process, among other factors like temperature, and also influence the lifetime of coatings (mean value: 5–10 years).

For the renewal of transmission pylons a price of 10–15 ECU per m^2 is assumed.

7.3.2. *Selected industrial property groups: railway bridges*

For security reasons all railway bridges are regularly tested, and, in case of need, the corrosion protective layer is renewed. The maintenance interval for steel bridges in polluted areas, at 20 years, is significantly shorter than in unpolluted areas (25 years). The maintenance costs are estimated to be about ECU 58/m^2.

7.3.3. *Selected industrial property groups: railway contact line systems*

For about 5–10 years contact line systems with a zinc layer and a coating have been installed. Older systems do not necessarily have a coating. The maintenance intervals vary a lot (see Table 7.4). Maintenance costs for renewal of coating are estimated to be about 20 ECU per m^2.

7.3.4. *Selected industrial property groups: mineral oil depots*

Mineral oil depots are regularly coated. Knowing the differences in renewal cycles, external costs can be monetized. Concerning the damage area of about

Table 7.3. Time intervals of the renewal of protection for contact line systems (years)

Type of protection against corrosion	'Unpolluted' areas	Polluted areas
Duplex system (from the beginning)	45	35
Duplex system (installed afterwards)	30	20
Zinc layer	22.5	12.5
Coating	25	15

Source: Isecke *et al.* (1991).

6.5 million m^2 and the percentage of 'unpolluted' areas, additional costs of about ECU 7 million per year arise for West Germany (Isecke *et al.*, 1991).

7.3.5. *Selected industrial property groups: sludge towers*

Additional costs for sludge towers were estimated to be about ECU 1 million per year due to damage in the coatings. 90% of the sludge towers are built from concrete, the remaining 10% from steel. The sludge towers built with concrete have a special coating either of eternit or aluminium, but both are susceptible to air pollution.

7.4. Residential buildings

Isecke *et al.* (1991) remark that it is not possible to give a quantitative relationship between ambient air concentration and maintenance costs for residential buildings. Identical or similar damage can also result from weathering as well as false planning and construction of building elements. However, the authors recognize as common knowledge that:

- some types of damage in certain materials are only caused by specific air pollutants;
- many weathering processes are accelerated by air pollution;
- damage due to incorrect construction is increased by air pollution.

But the influence on materials even in the same type of atmospheric area can be widely diverse because the distribution of emission varies and is also dependent on meteorological factors. In addition the building design and the construction quality influences the intensity of damages related to air pollution.

In order to gain an approximately accurate economic valuation of air-pollution-related damage a model was used which allows the spatial disaggregation of buildings using settlement types (Roth 1980). This model was also used for calculating the air-pollution-related damage to buildings in Dortmund and

Table 7.4. Air pollution related additional costs of coating renewal in sludge towers (in ECU)

Material	Eternit	Aluminium
Lifetime in years:		
Polluted area	30	20
'Unpolluted' area	50	32.5
Costs (ECU/m^2 in 1990)	144	173
Surface in polluted areas in m^2	386,920	73,384
Additional costs in ECU per year	742,888	244,143
Total costs	987,031	

Source: Isecke *et al.* (1991).

Cologne (Hoos *et al.*, 1986). There are 9 settlement types, each having an 'ideal building' of which the number of square meters is known and divided into roof, wall and window space. The basis for estimating the surfaces damaged by air pollution is the space of the outer building ('Gebäudehüllenfläche'). Furthermore, the settlement types can be categorized according to their specific density per square km. This model for Dortmund was applied to the stock of residential buildings in Germany in the period 1987/88. All cities were listed in the following order:

A: SO_2 annual mean value 30–50 mikrog/m^3
B: SO_2 annual mean value >50 mikrog/m^3
*A**: as *A*, but with data from cities with more than 100,000 inhabitants
*B**: as *B*, but with data from cities with more than 100,000 inhabitants

The calculation of the air-pollution-related additional costs is done according to the following equation:

$$\text{Additional costs} = S \times C \times (1/L_p - 1/L_{up})$$

where:

S = Surface (m^2), length (m) or number of construction elements
C = Costs per m^2, or construction element,
L_b = lifetime of a building in a polluted area,
L_{up} = lifetime of a building in an unpolluted area.

The lifetime values are mean values of estimated values asked from building companies. The mean values include a factor of insecurity as the building companies do not take care of the maintenance measures within an equal time interval.

7.5. Cultural assets

In the Federal Republic of Germany there are about 500,000 single monuments and 1.5 million cultural assets surrounded by bigger building complexes. This is about 15% of total buildings. In the new Länder there is a number of about 45,000 registered monuments, but it is assumed that this figure may be increased by 50% (Bundesumweltministerium, 1992). At the moment, however, there is no possibility of evaluating the damages to cultural assets. A variety of reasons can be given:

- It is extremely difficult to estimate the extent to which the weathering is caused by natural or anthropogenic factors. On the one hand, natural factors such as temperature or rainfall cause weathering. But the manmade increase of carbon dioxide concentrations, for example, also influences this process. On the other hand, those pollutants in air and soil that intensify the chemical corrosion trigger severe damage effects. Current research shows that only the exact analysis of a particular object will present all damage factors.

- Dose–response functions for different natural stones are not yet accurately found. Sandstone and limestone are the most vulnerable natural stones. However, even for these stones the precise knowledge on corrosion mechanisms is missing. Only for Krensheimer Muschelkalk and Baumberger Kalksandstein are there quantitative results.
- Data on cultural assets is poor. Many Länder have just begun to collect data via questionnaires. The 'co-ordination agency for environmental damages in cultural assets' within the Federal Environmental Agency has started to collect data from the monument agencies and save them in a national archive. It must be borne in mind that the data was often for internal use in monument agencies only. Sometimes the material was collected by staff that did not always have adequate knowledge in natural science. So the data is of restricted benefit.
- As a result of the tight budget of monument agencies mostly monuments of outstanding historic importance can be conserved. The total costs of restoration borne by the monument agencies do not reflect the costs of air-pollution-related damage. As already mentioned the overall damage has a variety of causes.
- There is no willingness-to-pay analysis regarding the individual value of cultural assets in Germany.

A thorough analysis would require the following steps (Pearce, 1990):

- Inventory of cultural assets
- Classification according to:
  - type of monument (e.g. sculpture, church, building),
  - type of stone (sandstone, limestone, etc.),
  - location and ambient air concentration (SO_2, NO_x and PM_{10})
- Dose–response functions for certain types of stones and monuments under a certain ambient air concentration (accurate single analysis of objects)
- Extrapolation of the results to other cultural assets of the inventory showing the same characteristics.

Such an analysis is beyond the scope of this study. As a guide to the values associated with cultural assets, however, the defensive expenditures for cultural assets will be shown in section 8.3.5.

CHAPTER 8

RESULTS FOR GERMANY

8.1. Introduction

This chapter contains the results for the following impacts:

- Health
- Noise
- Crops
- Forests
- Materials

No quantification is offered for the impacts on ecosystems, or the impacts from global warming. The latter are covered in the Dutch report for all four countries and are also reported in the synthesis report. The chapter concludes with a discussion of the results and the implications for policy and future research. *The monetary estimates presented here incorporate all intermediate calculations and are generally based on 1990 prices. In the synthesis report all the final calculations have been updated to 1995 prices.*

8.2. Health

8.2.1. *Valuation of SO_2 health impacts*

The following Table 8.1 shows the valuation results of SO_2 pollution in Germany.

The figures are dominated by the costs of mortality; they account for over 95% of all costs. This is the result of taking the value of statistical life (VOSL); as discussed in the synthesis report, and as present there, the costs are considerably less if we take a value of life years lost approach (VLYL). As an approximation, SO_2 effects are acute and the years of life lost are estimated at about 1.25. In that event the value attached to the mortality will be only 3% of that given in Table 8.1. This issue is discussed further in the synthesis report.

As mentioned in the beginning, in co-operation with IER Stuttgart it was decided not to quantify health impacts due to NO_x and PM_{10}. The results of the EXTERNE study show that the health effects of NO_x can be neglected. As a consequence the dose–response functions developed and described for NO_x find no application.

Table 8.1. Health impacts due to SO_2 in 1990 using VOSL method

Endpoint	Size of impact	Cases	Damage in million ECU	Damage as % of 1990 GNP	Damage in ECU per capita
Acute	low	2,464	6,406	0.23	80
mortality	mid	4,928	12,813	0.45	161
	high	7,392	19,219	0.68	241
Adult chest	low	9.58 mn.	60	0.004	0.8
discomfort	mid	18.50 mn.	116	0.008	1.4
days	high	27.09 mn.	171	0.012	2.1
Emergency	low	6,944	1,292	0.095	16
room visits	mid	10,750	1,980	0.146	25
(ERVs)	high	14,349	2,670	0.197	34

Source: IER (1994).

The problem in monetizing the health effects of PM_{10} in Germany is that there is no baseline data. Transferring baseline data e.g. from the UK would give false results as differences in climate between the UK and Germany do not allow simple copying of PM_{10} baseline data.

8.2.2. *Valuation of ozone health impacts*

As already mentioned there is no dispersion model for ozone at the moment. According to calculations by the Federal Environmental Protection Agency the total damage is estimated to be about ECU 3.9 billion (1992). This figure relies on a comparative analysis based on a Californian study by Hall, Winer *et al.* (1992) which calculated ozone damage to be $20 billion at maximum using the damage function approach. The German approach results with about 20% of the US figure using homogenous regions as far as this was possible. Thus, the figure of ECU 3.9 billion is only a rough estimate which cannot be included within the overall framework of results.

Table 8.2. Health impacts due to ozone in 1992

Ozone damage category	Damage in million ECU	Damage in % of GNP	Damage per capita in ECU
Unspecified[1]	3.9	0.1373	48.9

[1]The damage cannot be divided into specific categories like shortness of breath, eye irritation etc.

Source: Personal communication to Mrs. Paulini, Federal Environmental Agency, Berlin, 18th January 1995.

8.3. Noise

8.3.1. *Hedonic pricing method*

Weinberger calculated rent price differences due to different levels of road traffic. He used depreciation values of 0.5% and 1.26% per dB(A) (Borjans 1983 and Pommerehne, 1986). These are the only values available in Europe. Table 8.3 presents the results. The baseline is 45 dB(A).

Regarding air traffic the rents are depreciated by values between 0.3 and 1.2% per db(A). The average noise level is assumed to be 68 dB(A). It was assumed that there are 2.32 persons per apartment and the average rent is 223 ECU per apartment.[1] For the results see Table 8.4.

In noisy residential areas relatively cheaper apartments are offered than in quiet neighbourhoods where expensive apartments dominate. The assumption of a constant average rent thus overestimates willingness-to-pay in noisy areas. Simultaneously, quiet neighbourhoods are usually characterised by bigger households than noisy areas (Statistisches Bundesamt, 1978). Since only the number of affected people per noise band is known and not the number of households, an underestimation of the number of affected households in noisy areas as well as an underestimation of the total willingness-to-pay is the consequence. Furthermore, higher construction costs in noise protection areas may also lead to an underestimation of the average rent in these areas. Since there is no basis for objective noise bands at the moment, differences in rents and households per noise band were foregone. Although the effects mentioned above may neutralize each other, an underestimation of the total willingness-to-pay may remain.

In addition, it is not clear whether the depreciation rate is independent of the noise level. For example Pommerehne (1986) succeeds better with a non-linear rent function. However, the survey results of Weinberger *et al.* (1991) show that a linear function allows a very good approximation of the obtained values. Another problem is the contrary effect of increasing traffic, e.g. in airport vicinities rents are often above average. This kind of compensation may also lead to an underestimation of the willingness-to-pay for less noise obtained by rent differences.

8.3.2. *Sensitivity analysis*

A sensitivity analysis shows that the influence of the chosen baseline is extreme on the results. If the baseline is increased from 45 dB(A) to 55 dB(A), the

[1] The average rent in 1985 was ECU 215 (DIW, 1989). Using the rent price index for 1987 the average rent is ECU 223 (Statistisches Bundesamt, 1988). Using the rent price index for 1990 (Statistisches Jahrbuch, 1994), the results remain unchanged.

Table 8.3. Monthly rent price differences due to road noise in West Germany in 1990

Noise level in dB	<45	45–50	50–55	55–60	60–65	65–70	70–75	75–80	Total
Affected people in %	21.1	17.1	17.0	18.3	14.2	7.5	4.0	1.1	100
Affected people in million	12.9	10.4	10.4	11.2	8.7	4.6	2.4	0.7	61
Affected households in million	5.5	4.4	4.4	4.8	3.7	1.9	1.0	0.3	26
Monthly rent in million ECU	1222	990.5	985	1060	822.5	434.5	231.5	63.5	5807
Monthly rent price difference (1.26% per dB)									
Affected people in %	0	3.2	9.5	15.8	22.1	28.4	34.7	41.0	
Total in million ECU	0	31	93	167	156.5	123	80.5	26	702
In % of GNP (in 10^{-3})	0	1.22	3.66	6.56	6.15	4.83	3.16	1.02	27.59
Per person in ECU	0	2.98	8.91	15.93	20.9	26.88	32.85	38.33	
Monthly rent price difference (0.5% per dB)									
Affected people in %	0.0	1.3	3.8	6.3	8.8	11.3	13.8	16.3	
Total in million ECU	0.0	12.5	37	66	72	49	32	10.5	278.5
In % of GNP (in 10^{-3})	0	0.49	1.45	2.59	2.83	1.93	1.26	0.41	10.95
Per person in ECU	0	1.18	3.55	5.93	8.29	10.66	13.03	15.4	

Source: Weinberger *et al.* (1991), with changes.

maximum willingness-to-pay falls from ECU 8.2 billion to ECU 3 billion per year. So the baseline has a stronger influence than the depreciation rate. However, the marginal willingness-to-pay is independent of the baseline if a linear rent function is used.

Table 8.4. Monthly rent price differences due to aircraft noise in West Germany in 1990

	0.3% per dB	1.2% per dB
Affected people (over 67 dB(A))	600,000	600,000
Monthly rent per person in ECU	98	98
Monthly rent price difference in %	6.9	27.6
Per person in ECU	6.76	27.05
Total in million ECU	4.02	16.25
Total rent price difference in million ECU per year	48.5	195
In % of GNP (in 10^{-3})	1.91	7.66

Source: Weinberger *et al.* (1991), with changes.

For road traffic the marginal monthly willingness-to-pay per person is

- 0.49 ECU per dB (using 0.5% per dB) and
- 1.20 ECU per dB (using 1.26% per dB).

For air traffic the marginal monthly willingness-to-pay per person is

- 0.3 ECU per dB (using 0.3 per dB) and
- 1.18 ECU per dB (using 1.2% per dB).

8.3.3. *Comparison with results from a contingent valuation approach*

In 1989 Weinberger *et al.* asked 6,500 households for their willingness-to-pay for less noise. The objective noise situation of the households was given by noise maps. For the state 'little noise' using the noise levels during daytime the following relationship was found:

$$\text{WTP}\ (N) = 1.67 \times N - 71.70$$

with N being the noise level.

This function was estimated for the monthly willingness-to-pay if there is street noise. It reflects the willingness-to-pay for noise levels above about 43 dB(A). The individual marginal willingness-to-pay for an reduction of the noise level for 1 dB(A) is about ECU 0.84 per month.

The yearly willingness-to-pay for a general reduction of street noise by 3 dB(A) in Germany is ECU 1.45 billion.[1] The same procedure was used for other noise sources. The willingness-to-pay for a reduction of rail traffic noise is about ECU 0.75 billion, for air traffic noise only within the noise protection areas about ECU 13 million and finally for a reduction in industrial noise about ECU 0.75 billion. In % of GNP and population weighted the following figures are obtained:

[1] The baseline level was 45 dB(A). The model involves 48.24 million people.

Table 8.5. Willingness-to-pay for noise reduction in West Germany per year

	Total WTP in ECU mn.	WTP in % of GNP (in 10^{-2})	WTP per capita
Traffic	1450	5.699	22.92
Rail	750	2.948	11.86
Air	13	0.051	0.21
Industries	750	2.948	11.86

Source: Weinberger *et al.* (1991).

8.3.4. *Total willingness-to-pay for 'almost no noise'*

Using an average information level the willingness-to-pay rises to ECU 11 billion DM per year, whereas a good information level gives a willingness-to-pay of even ECU 12 billion for 'almost no noise' due to street, railway, industrial and construction noise. The willingness-to-pay for a strong reduction of noise due to road traffic is about 60% of the total willingness-to-pay.

The figures do not include the new Länder. In 1990 the data base was not yet sufficient for a similar analysis as done for Western Germany. However, the noise situation in the new Länder may have approximated the old Länder. Under the assumption of equal preferences at least the marginal individual willingness-to-pay should be the same as in the old Länder. This is valid for the hedonic pricing method as well as the willingness-to-pay approach. But of course influences like income distribution, information etc. play a major role. For that reason an accurate number for the new Länder cannot be given at the moment.

8.3.5. *Defensive expenditures*

The damage costs of noise should also take defensive expenditures into consideration. The estimates for defensive expeditures are contained in Table 8.7.

8.4. Crops

Multiplying the identified crop loss with the world market prices yields the following economic damage (Table 8.8):

Table 8.6. Willingness-to-pay for 'almost no noise' in West Germany per year

	Total WTP in billion ECU	WTP in % of GNP	WTP per capita (ECU)
Average information level	11	0.43236	173.90
Good information level	12	0.47166	189.71

Source: Weinberger *et al.* (1991).

Table 8.7. Estimation of defensive expenditures in West Germany per year

Street noise	
Noise protection windows:	
for 75/65 dB(A) maximum	ECU 80–105 million
for 70/60 dB(A) maximum	ECU 200–270 million
Railway noise	
Noise protection windows:	
for 75/65 dB(A) maximum	ECU 46–61 million
for 70/60 dB(A) maximum	ECU 109–145 million
Air traffic noise	
Noise protection windows	ECU 49–66 million
Noise at work and industrial noise	
Investment per year	ECU 450–650 million
Noise in the neighbourhood	
Investment in new buildings per year:	
'increased noise protection'	ECU 800–1,350 million
'very good noise protection'	ECU 1,200–2,150 million
Total defensive expenditures per year	ECU 2,934–4,297 million
Total defensive expenditures in % of GNP (in 10^{-6})	1.2–1.7
Total defensive expenditures per capita	ECU 0.05–0.07

Source: Weinberger *et al.* (1991).

Table 8.8. Estimated crop damages for Germany due to SO_2 in 1990

Crop species	Size of impact	Yield loss in dt per year	Damage in million ECU per year	Damage in % of GNP (in 10^{-4})	Damage per capita
Wheat	low	1,048	10.07	3.5	0.13
	mid	2,745	26.35	9.3	0.330.54
	high	4,446	42.68	15	
Barley	low	1,408	5.62	2	0.07
	mid	2,724	14.71	5.2	0.18
	high	4,411	23.82	8.4	0.3
Rye	low	2,877	4.49	1.6	0.06
	mid	7,532	11.75	4.1	0.15
	high	1,220	19.03	6.7	0.24
Oats	low	1,367	0.77	0.3	0.01
	mid	3,579	2.01	0.7	0.03
	high	5,797	3.25	1.1	0.04

Source: IER (1994b).

The sum of agricultural damage was ECU 164.55 million in 1990. This is 4.13% of the total turnover for grain in 1990 (Statistisches Bundesamt, 1994). This figure, however, is only the minimum, since it neither contains liming measures additionally required by the farmers due to increased acid deposition nor assesses damage due to ozone. At the moment the latter can be described only qualitatively because there is no dispersion model. It can be assumed that ozone concentrations of 100 mikrog/m^3 (8-h-mean value) or 33 mikrog/m^3 (24-h-mean value) already induce yield losses. The critical load of SO_2 for sensitive crops is about 100 mikrog/m^3. This implies that not even the critical load of 140 mikrog/m^3 (German TA Luft) for the protection of human health guarantees the protection of crops. For the protection of crops, the precaution loads must not be ruled out. Furthermore, the synergistic effects of different air pollutants cannot be generalized. Depending on the type of crop, the measured parameter, and the pollutant combination, the effects partly accumulate and partly decrease (Bundesumweltministerium, 1992).

8.5. Forests

8.5.1. *Damage according to the dose–response method*

The use of the dose response method for calculating damages is demonstrated in the UK study, which also covered Germany. It values the reduced growth in timber, which comes out at ECU 49 mn. as the central estimate, with a low value of ECU 22 mn. and a high value of ECU 77 mn. This is much lower than the value obtained by Nilsson, who estimates losses at ECU 600 mn. for Germany. However, the Nilsson study is subject to much criticism, as is discussed in the UK report. As a percentage of German forest output the results of our study amount to 3.6%.

8.5.2. *Loss of recreational value/existence value*

The Dose–Response approach left no room for evaluating the loss of recreational value due to forest damages. Ewers (1986), however, considered several kinds of cost categories in the area of leisure and recreation: Costs which are paid

- by the users of forests (user costs),
- by the leisure industry,
- by people not actually visiting forests but willing to pay for the possibility of using them (non-user costs/existence value).

As a consequence of forest damages, fewer visits to forests and fewer holidays in forest areas were estimated. On the basis of these estimations a yearly damage in the field of leisure time and recreational activity of ECU 1.15 to

Table 8.9. Estimated forest damages (based on leisure time/recreational activities) in West Germany in 1986

Damage category	Absolute damage in billion ECU	In % of GNP	Loss per capita in ECU
Damage in the case of strict clean air aims	1.15–3.15	0.04520–0.12381	18.18–49.80
Damage if air pollution remains high	2.1–5.65	0.08254–0.22207	33.20–89.32

Source: Ewers (1986).

3.15 billion was calculated. This figure, however, is only valid when very strict clean air aims are realized (only West Germany). If air pollution remains at the level of the beginning of the eighties, damage in the area of leisure time and recreational activities will amount to as much as ECU 2.1 to 5.65 billion. The minimal and maximal limits depend on the discount rate the estimate is based upon. Details are given in Table 8.9.

8.5.3. *Costs of compensatory activities*

Kroth *et al.* (1989) examined the silvicultural measures which forest management applies to counteract forest damages and investigated their costs for Germany. Table 8.10 shows the total forest areas in West Germany to which these measures have been applied between 1988 and 1992. The table lists only those measures which have been agreed by experts to have mitigating effects and of which a separation from normal operation is possible. Using this method, the total costs for West Germany are in the range of ECU 41.2 and 112.9 million per year during a period of five years.

8.6. Materials

The results in this section relate only to West Germany.

8.6.1. *Industrial property*

Transmission pylons
Assuming a price of 10–15 ECU per m^2 the additional costs of maintenance in West Germany are about ECU 2.5 million per year (Isecke *et al.*, 1991).

Railway contact line system
Estimating about 20 ECU per m^2 coating the transmission related maintenance costs sum up to ECU 2 million.

Table 8.10. Costs of mitigating measures and forest areas totally affected in West Germany between 1988 and 1992, selected from Kroth *et al.*, 1989

Mitigating measure	Forest area used (1988–1992) in 1000 ha	Costs in ECU/ha	Total costs ECU mn./year (1988–1992)
Liming and supplementary fertilisation	28–230	150–500	6.85–51.3
Site mapping	344	33.5–35	2.4
Reforestation of damaged stands	0.3	0–5,000	4.95
Remodelling	1.7–4.0	2,500–10,000	14.6–32.0
Cultivation of underwood	2.4–6.0	600–5,100	5.3–12.9
Renovation of protection forests	0.16	30,000[1]	4.8
Biological protection of forests	n.a.	n.a.	1.75
Reduction of deer	n.a.	n.a.	0.5–2.75[2]
Total costs			41.2–112.9
Total costs in % of GNP (10^{-3})			1.62–4.44
Total costs per capita in ECU			0.65–1.78

[1] Mean value.
[2] Only state forests.

Source: IER (1994), according to Kroth *et al.* (1989).

Railway bridges
Applying ECU 58/m^2 on the total layer area (15 million m^2) one obtains about ECU 3.5 million additional maintenance costs (Isecke *et al.*, 1991).

Mineral oil depots
There are additional maintenance costs of about ECU 7 million per year (Isecke *et al.*, 1991).

Sludge towers
There are total estimated external costs of about ECU 1 million due to damage in the coatings (Isecke *et al.*, 1991).

The results are summarised in Table 8.11.

8.6.2. *Cultural assets*

For cultural assets only damage prevention expenditures can be shown. The following table indicates the restoration costs of monuments and other cultural assets in Germany.

Table 8.11. Maintenance costs for selected industrial property groups in West Germany in 1987

Object	Total costs per year in million ECU	Total costs in % of GNP (in 10^{-4})	Total costs per capita in ECU
Transmission pylons	2.5	1	0.04
Railway contact line system	2	0.8	0.03
Railway bridges	3.5	1.4	0.06
Mineral oil depots	7	2.8	0.11
Sludge towers	1	0.4	0.02

Source: Isecke *et al.* (1991).

Table 8.12. Damage prevention expenditures on cultural assets in West Germany

Location	Objects	Period	Costs in ECU
Federal Republic of Germany	All municipal bronze monuments and sculptures	Annual	2 million
	All metal sculptures in museums and open air	Annual	0.5 million
	All medieval stained glass	10 year cost estimate	1–1.5 million
	Artifacts in museums	During construction	15% of construction costs
Munster	Castle facade	1965–73	0.5 million
Cologne	Cathedral stained glass windows	1978	220,000
Cologne	Cathedral facade	Annually (1977–97)	1.5–30 million (estimated)
Freiburg	Cathedral stained glass windows	1978	75,000
Ulm	Cathedral stained glass windows	Total cost	1.5 million

Source: Feenstra (1984); Altshuller *et al.* (1983).

8.6.3. *Residential buildings*

The method used for calculating the air-pollution-related additional maintenance costs was the prevailing practice method also taking account of dose–response functions. This method is based on the comparison of maintenance costs in polluted areas – taking mainly SO_2 into consideration – and less polluted areas. However, the values from prevailing practice concerning

- the renewal of corrosion protection systems,
- the exchange of construction elements,

- the maintenance of damaged construction elements and
- cleaning

differ considerably, in part, from values which can be expected according to dose–response research.

In order to quantify the effect of air pollutants on materials, the atmospheric concentration is not the most important factor. Rather the effects of anthropogenically caused pollutants directly on the material play a major role. Location alone can influence the mechanisms and can lead to different exposure of materials in a region of same pollution concentration. So it is very important to consider practical values rather than dose–response functions only.

A very good corrosion protection system leads to a longer lifetime of materials which tend to reduce maintenance costs. There are no differences between urban and city regions in choosing a corrosion protection system. It is extremely difficult to give a final evaluation of air-pollution-related effects on materials because additional factors like weathering can hardly be isolated. The situation is particularly bad for residential buildings since there are no long term statistics on the effects of air pollutants. The current maintenance measures are highly dependent on the budget of owners, authorities and building companies. The data for estimating the maintenance costs of transmission towers, bridges, contact line systems, mineral depots and sludge towers are based on intensive surveys of electric companies, the German railway and various companies.

Concerning the costs in residential buildings, the data used need to be examined. Modelled data as well as the mean values for maintenance intervals need to be examined by dose–response research in the near future. Data on an industrial building identikit is totally lacking at present so that the actual damage costs may lie above those indicated in the study.

In spite of these reservations, the following estimates have been made. Table 8.13 gives the major costs of air pollution on buildings for different construction elements, classified as 'A' and 'B'. The figures are for small cities (less than 100,000 inhabitants). For larger cities, the figures are given in columns 'A*' and 'B*'.

The additional costs of about ECU 1.5 billion in smaller cities and of about ECU 0.87 million in cities with or with more than 100,000 inhabitants show only the approximate costs of air pollution. The true value might lie between those two figures, since an exact examination regarding the relationship between ambient air concentration and materials damage is missing. And the statements given by the building companies on the lifetime of the construction elements are often only rough estimates.

In addition to the costs given in Table 8.13, we have the costs of cleaning windows due to dust. A survey among cleaning companies showed that windows are cleaned every 2 months in polluted areas and only every 3 months in unpolluted areas (mean values). Taking cleaning costs of ECU 0.5/m^2 one obtains the following Table 8.14 showing the mainly dust related cleaning costs.

Table 8.13. Pollution related costs in residential buildings in West Germany (ECU mn)

Construction element	A	B	A*	B*
Coat of paint	137.165	43.09	73.81	28.14
Plaster and facades	318.95	98.665	173.485	66.025
Balconies	29.845	13.78	21.865	8.97
Gutters	61.225	19.175	32.245	12.575
Roofs	161.59	50.305	86.88	33.18
Windows	394.42	124.245	212.24	80.945
Windowsills	35.635	10.805	19.175	7.31
Chimneys	2.965	0.955	1.59	0.61
Total costs	1,141.695	361.02	621.29	237.6
Total costs in % of GNP (in 10^{-2})	4.487	1.419	2.442	0.934

Source: Isecke *et al.* (1991).

Table 8.14. Costs of cleaning windows per year in West Germany (ECU mn)

Construction element	A	B	A*	B*
Windows	179	56	96	37

Source: Isecke *et al.* (1991).

CHAPTER 9

CONCLUSIONS, IMPLICATIONS FOR POLICY AND FUTURE RESEARCH

This section reviews the results and makes a broad assessment of their credibility. The underlying methodology is re-considered in the light of the results and its limitations are noted. Each receptor category is considered in turn.

9.1. Health

In comparison to a study by Schulz (1985), which quantified the monetary value of a nationwide improvement of air quality in West Germany at ECU 14 billion, this study shows that about ECU 13 billion is needed solely to compensate for damage due to SO_2.[1] This high value describes the current state when no further SO_2 reducing measures are taken. The size of costs, however, makes clear that the benefit of additional air cleaning measures will be substantial, not only in the field of health.

The precise monetization of ozone damage is still a problem because of the lack of a simulation model. Dispersion models are the only instruments of emission-related evaluation of emission reduction. But on the one hand the emission data must be significantly improved, e.g. the VOC emissions must be examined by proper experiments. On the other hand, the research concerning the models themselves must be accomplished. For example, meteorological data must be adapted to the requirements of dispersion models. In addition, a further and more general methodological difficulty is the definition of correct baseline data. This is true for PM_{10} as well as for acid deposition. Future research should be undertaken in this area.

9.2. Noise

Road traffic accounts for over 60% of noise costs and is by far the largest contributor. Since it turned out that that the willingness-to-pay increases only

[1] In the Schulz study, the air quality improvement is described as 'air in small towns'. The willingness-to-pay shows no information deficits. It has to be mentioned that this willingness-to-pay survey is not only directed towards health impacts of air pollution, but also to damage in materials as well as forest and plant damages.

slightly as noise increases, it is recommended that politicians promote those noise protection measures that yield the greatest possible effect: general primary noise protection measures (e.g. construction of quiet machines) are more likely to produce a greater overall benefit than secondary measures (e.g. construction of sound absorbing walls) which merely serve those persons exposed to high noise impact.

9.3. Crops

The agricultural damage due to SO_2 was about 165 million ECU in 1990. Effects of other pollutants like O_3 on yield or indirect effects had to be neglected as either dispersion calculations are missing or dose–response models are not available. Research shows that the damage induced to crops depends strongly on the SO_2-sensitivity of the plants. One surprising result is that the SO_2 critical loads developed for human health are too weak to prevent yield loss. This evidence and the not negligible damage should state an incentive to further investigate critical loads.

9.4. Forests

The monetized damage costs in the field of leisure time and recreation activities were calculated to lie between 1 and 3 billion ECU if strict clean air aims are reached. Additionally the defensive expenditures amount to between 40 and 110 million ECU per year. The reduction in wood output was calculated using the dose–response approach and amounted to 22 million ECU.

For comparison with the dose–response method in the wood production area, Ewers (1986) found a minimum forest damage between 1.15 and 1.45 billion ECU per year. The forest damage relates to a reduction in wood production and compensatory activities; wood production, however, has the major percentage. Thus there is a big difference between his figures and ours, which amount to 60–120 million ECU. As we noted earlier, however, there are some much higher estimates of forest timber loss (particularly Nilsson's work). The team felt that these were on the high side.

This figure as well as the physical description of the forest state in this study prove an urgent need for forest protection measures. Also 'soft' heuristic methods such as the defensive expenditures approach should be taken into consideration.

9.5. Materials

This is perhaps the area with most methodological difficulties. While damages in form of higher maintenance costs in residential buildings and the public infrastructure are relatively easy to calculate, damage to cultural assets is still hard to quantify. If one assumes the contingent valuation method as the only correct monetizing procedure, it is evident that the costs of restoration do not reflect the 'true' damage value. It is important to know the willingness-to-pay for the 'rescue' of cultural buildings, since only then the irreversible damage, including aesthetic values, can be evaluated. However, there is an immense variety of historically valuable cultural buildings. It is recommended that analyses on every single monument must be done. In Germany, there is great need for further research in this area.

CHAPTER 10

REFERENCES

Altshuller, A.P., Lindhorst, R.A., Nadler, J.S., Niemeyer, L.E. and McFee, W.W. (1983): Acidic Deposition Phenomenon and Its Effects, Critical Assessment Review Papers. Volume 1: Atmospheric Science (Review Draft), EPA-600/8 = 16a, PB 84-171644, Raleigh, NC, USA.

Baker, C.K., Colls, J.J., Fullwood, A.E. and Seaton, G.G.R. (1986): Depression growth and yields in winter barley exposed to sulfur dioxide in the field, in: New Phytologist 104, pp. 233–241.

Barker, D.J.P. and Rose, G. (1990): Epidemiology in Medical Practice, Churchill Livingstone, London.

Battelle-Institut (1981): Erhebung über den Stand und die Entwicklung der Belastung der Bevölkerung durch Lärm, Phase 2 (Forschungsbericht Nr. 81-105 02 803/01 im Auftrag des Umweltbundesamtes, Frankfurt/Main.

Battelle-Institut (1985): Belastung der Bevölkerung durch Lärm, Phase 3 (Forschungsbericht Nr. 85-105 01 321 im Auftrag des Umweltbundesamtes), Frankfurt/Main.

BMELF (Bundesministerium für Ernährung, Landwirtschaft und Forsten) (Ed.): (1991): Waldzustandsbericht des Bundes – Ergebnisse der Waldschadenserhebung 1991. Bonn.

BMELF (Bundesministerium für Ernährung, Landwirtschaft unf Forsten) (Ed.): (1992): Waldzustandsbericht des Bundes – Ergebnisse der Waldschadenserhebung 1992. Bonn.

BMELF (Bundesministerium für Ernährung, Landwirtschaft und Forsten) (Ed.): (1994): Waldzustandsbericht des Bundes – Ergebnisse der Waldschadenserhebung 1994. Bonn.

Borjans, R. (1983): Immobilienpreise als Indikatoren der Umweltbelastungen durch den städtischen Kraftverkehr, Düsseldorf.

Bundesumweltministerium (1992): Bericht der Bundesregierung an den Deutschen Bundestag, Fünfter Imissionsschutzbericht der Bundesregierung, Bonn.

DIW (1989): Lage und Perspektiven am Wohnungsmarkt, in: DIW Wochenbe richt, 56. Jg., 1989, p. 267–274.

Dockery, D.W. et al. (1989): Effects of Inhalable Particles on Respiratory Health on Children, Am. Rev. Respir. Dis., 139: 1989, pp. 587–594.

EC DGXII (1995a–f): Science, Research and Development Programme (JOULE). Externalities of Energy (EXTERNE) Project. Report numbers: EUR16520 (Summary), EUR16521 (Methodology), EUR16522 (Coal & Lignite), EUR16523 (Oil & Gas), EUR16524 (Nuclear), EUR16525 (Wind & Hydro).

Environmental Protection Agency (EPA) (1989): The Potential Effects of Global Climate Change on the United States, Jel. B. Smith and Dennis Tirpak (Eds.), Washington D.C.

Ewers, H.-J. et al. (1986): Zur monetären Bewertung von Umweltschäden – Methodische Untersuchung am Beispiel der Waldschäden, Kemper, M., Das Umweltproblem in der Marktwirtschaft, UBA-Berichte 4, Berlin.

Feenstra, J.F. (1984): Cultural Property and Air Pollution: Damage to Monuments, Art Objects, Archives and Buildings due to Air Pollution. Leidschenham, Netherlands, Ministry of Housing, Physical Planning and Environment.

Guderian, R.(1977): Air pollution, phytotoxicity of acidic gases and its significance in air pollution control, in: Ecological studies, Vol. 22.

Hall, J.V., Winer, A.J. et al. (1992): Valuing the Health Benefit of Clean Air, in: Science, Vol. 255, p. 812–817.

Hansen, J. et al. (1989): Regional Greenhouse Climate Effects, in Topping (Ed.). Coping with Climate Change, pp. 273–281.

Hatzakis, A., Katsouyanni, K., Kalandidi, Day, N. and Trichopüoulos, D. (1986): Short-term

effects of air pollution on mortality in Athens, in: International Journal of Epidemiology, 15, pp. 73–81.
Hettelingh, J.P., Downing, R.J. and Smeet, P.A.M. (1991): Mapping critical loads for Europe, Technical Report No. 1. Coordination Center for Effects, Bilthoven, The Netherlands.
Hoffjann, Th.: Die Berücksichtigung der Umweltbelange in der Bauleitplanung. Ergebnisse aus einer Umfrage bei allen Städten mit über 100,000 Einwohnern in der Bundesrepublik Deutschland, in: Der Städtetag, 41. Jg., 1988, S. 137–141.
Hoos, D., Jansen, R., Kehl, J., Noeke, J. and Popal, K. (1986): Gebäudeschäden durch Luftverunreinigungen – Entwurf eines Erhebungsmodells und Zusammenfassung von Projektergebnissen – Forschungsbericht im Auftrag der ECOTEC Research and Consulting Ltd., Birmingham (GB), Dortmund.
IER (1994a): CEC/US Joint Study on Fuel Cycle Costs (February 1993): Assessment of the External Costs of the Coal Fuel Cycle. Prepared for DG XII of the Commission of the European Community.
IER (1994b): Calculations for this project.
Isecke, B., Weltchev, M. and Heinz, I. (1991): Volkswirtschaftliche Verluste durch umweltverschmutzungsbedingte Materialschäden in der Bundesrepublik Deutschland, Texts of the Federal Environmental Agency 36/91, Berlin.
Kane, S., Reilly, J. and Bucklin, R. (1989): Implications of the Greenhouse Effect for World Agricultural Commodity Markets, Washington: USDA.
Kroth, W., Grottker, Th. and Bartelheimer, P. (1989): Forstwirtschaftliche Massnahmen bei neuartigen Waldschäden – Nutzen-Kosten-Untersuchung forstwirtschaftlicher Massnahmen in durch Luftschadstoffe geschädigten oder gefährdeten Waldbeständen. Schriftenreihe des Bundesministers für Ernährung, Landwirtschaft und Forsten, Reihe A: Angewandte Wissenschaft, Heft 368, Münster-Hiltrup: Landwirtschaftsverlag.
Krupnick, A.J., Harrington, W. and Ostro, B. (1990): Ambient Ozone and Acute Health Effects: Evidence from Daily Data, J. Econ. Manage, 18, pp. 1–18.
Kuylenstierna, J. and Chadwick, M. (1994): Research Project on Modelling Forest Damage from Air Pollution, submitted to the EXTERNE Project.
Nilsson, S. and Pitt, D. (1991): Mountain World in Danger: Climate Change in the Forests and Mountains of Europe, Earthscan, IIA SA-Bellerive Foundation.
Oak Ridge National Laboratory and Resources for the Future (1992): Damages and Benefits of the Coal Fuel Cycle: Estimation Methods, Impacts, and Values, unpublished report for the US Department of Energy and the European Commission, Oak Ridges, Tennessee, USA.
OECD (1981): The costs and benefits of sulfur oxide control, Paris.
OECD (1992): Global Warming, The Benefits of Emission Abatement, Paris.
Ostro, B.D. (1987): Air Pollution and Morbidity Revisited: A Specification Test, J. Environ. Econ. Management; 14: 1987, pp. 7–98.
Parry, M. (1990): Climate Change and World Agriculture, London: Earthscan Publications.
Pearce, D. (1990): The Monetary Valuation of Damage Costs with Special Reference to Material Damages, Report to Bundesanstalt für Materialforschung und Prüfung Berlin, London.
Pearce, D.W., Bann, C. and Georgiu, S. (1990): The Social Costs of Fuel Cycles, HMSO: London.
Plachter, H.(1991): Naturschutz, Stuttgart.
Plagiannakos, T. and Parker, J. (1988): An Assessment of Air Pollution Effects on Human Health in Ontario, in: Ontario Hydro, March 1988.
Pommerehne, W.W. (1986): Der monetäre Wert einer Flug- und Strassenlärmreduktion: Eine empirische Analyse auf der Grundlage individueller Präferenzen, Berichte 7/86, UBA, Berlin.
Prognos, A.G. (Ed.) (1992): Die externen Kosten der Energieversorgung von Klaus, P. Masuhr/ Heimfrid Wolff/Jan Keppler unter Mitarbeit von Stefan Heltemes/Thomas Oczipka, Stuttgart.
Rind, D., Goldberg, R., Hansen, J., Rosenzweig, C. and Ruedy, R. (1990): Potential Evapotranspiration and the Likelihood of Future Drought, Journal of Geophysical Research, Vol. 95, No. D7, pp. 9983–10004.
Roberts, T.M.: Longterm Effects of Sulphur Dioxide on Crops: An Analysis of Dose-Response-Relations. Phil. Trans. Royal Society of London B 305 (1984), 299–316.
Roth, V. (1980): Wechselwirkungen zwischen der Siedlungsstruktur und Wärmeversorgungssystemen, Schriftenreihe 'Raumordnung' des Bundesministers für Raumordnung, Bauwesen und Städtebau, Bonn.
Samet, J.M., Speizer, F.E., Bishop, Y., Spengler, J.D. and Ferris, B.G. (1981): The Relationship

between Air Pollution and Emergency Room Visits in an Industrial Community. Journal of the Air Pollution Control Association; 31: 236–240.
Schwartz, J. and Zeger, S. (1990): Passive Smoking, Air Pollution and Acute Respiratory Symptoms in a Diary Study of Student Nurses. American Review of Respiratory Disease; 141: 62–67.
Schwartz, J., Wypij, D., Dockery, D., Ware, J., Zeger, S., Spengler, J. and Ferris, B. (1991): Daily Diaries of Respiratory Symptoms and Air Pollution: Methodological Issues and Results. Environmental Health Perspectives; 90: 181–187.
Schulz, W. (1985): Der monetäre Wert besserer Luft. (Frankfurt, Berne and New York).
Statistisches Bundesamt (1978): 1%-Wohnungsstichprobe 1978 Wohnumfeld – Infrastrukturversorgung und Umwelteinflüsse (= Fachserie 5; Bautätigkeit und Wohnen, Heft 6, S. 32 ff.), Wiesbaden.
Statistisches Bundesamt (1990): Land- und Forstwirtschaft, Fischerei – Ausgewählte Zahlen für die Agrarwirtschaft. Reihe 1, Fachserie 3, Wiesbaden.
Statistisches Bundesamt (1992): Statistisches Jahrbuch 1992 für die Bundesrepublik Deutschland, Wiesbaden.
Statistisches Bundesamt (1994): Statistisches Jahrbuch für die Bundesrepublik Deutschland, Wiesbaden.
Steele, P. and Ozdemiroglu, E. (1994): Acid Rain and Health, Draft, EFTEC Ltd., London.
Sunyer, J., Anto, J.M., Murillo, C. and Saez, M. (1991): Effects of Urban Air Pollution on Emergency Room Admissions for Chronic Obstructive Pulmonary Disease. With Invited Comment commentary from M. Corn and authors' response. American Journal of Epidemiology; 134: 277–289.
Triebswetter, U. (1993): Ex post-Anwendung einer Kosten-Nutzen-Analyse am Beispiel des Nationalparks 'Schleswig-Holsteinisches Wattenmeer', München, forthcoming in 'Ökosystemforschung Nationalpark' 'Schleswig-Holsteinisches Wattenmeer'.
Umweltbundesamt (1992): Daten zur Umwelt 1990/91, Berlin.
Umweltbundesamt (1994): Jahresbericht 1993, Berlin.
UN-ECE (1990): Draft Manual on Methodologies and Criteria for Mapping Critical Levels/Loads and Geographical Areas Where They Are Exceeded.
Verein deutscher Ingenieure (VDI) (1978): Maximale Immissionswerte für Schwefeldioxid, VDI-Guideline 2312 – Part 2.
Weber, I. et al. (1990): Dringliche Gesundheitsprobleme der Bevölkerung in der Bundesrepublik Deutschland, Baden-Baden.
Weinberger, M., Thomassen, H.G. and Willeke, R. (1990): Kosten des Lärms in der Bundesrepublik Deutschland: Forschungsbericht 10103110/05 im Auftrag des Umweltbundesamtes, Berlin.
Weltschev, M. (1991): Economic losses to society due to material damages of environmental pollution in the Federal Republic of Germany, in: Hohmeyer, O. and Ottinger, R.L. (Eds.): External Environmental Costs of Electric Power, Berlin.
WHO (1987): Air quality guidelines for Europe. World Health Organization, Regional Office for Europe. WHO Regional Publications, European Series No. 23.

ITALIAN CASE STUDY

Prepared by FEEM* in collaboration with the other country teams.

* Fondazione Eni Enrico Mattei (FEEM), via Santa Sofia, 27, 20122 Milano, Italy.
Tel.: +39 2 520 36934; Fax.: +39 2 520 36946; E-mail: pavan@feem.it

CHAPTER 1

INTRODUCTION

1.1. Background

The purpose of this report is to document the application of the common GARP methodology to Italy and present some preliminary estimates of national level damages. For a more detailed and scientific discussion of the general approach the reader should refer to the Synthesis Report. The study focuses on a selected range of impacts following an extensive review of the available data, applicable dose–response relationships and valuation information for Italy. Table 1.1 gives a summary of the study scope for Italy.

In contrast to the other country studies this report does not provide valuation estimates for damage caused by noise pollution on amenity values, nor of forest and materials damage. The available quantitative information on noise pollution was almost non-existent at the time of writing. The same limitations apply to the assessment of air pollution damage to building materials and to forests. However, a significant effort has been made in the collation of basic data that could provide a foundation for more advanced future research in this area.

1.2. Methodology

The analysis of damages proceeds in three parts:

- identification of pollutant levels;

Table 1.1. Pollutants and receptor variables covered in Italian assessment

	Health	Crops	Forests	Ecosystems	Materials
PM_{10}	A	NE	NE	NE	NQ
SO_2	A	A	NQ	NE	NQ
NO_x	A	NE	NQ	NQ	NQ
O_3	A	NQ	NQ	NQ	NQ

A – impact assessed.
NQ – impact not quantified.
NE – no significant impact.
SO_2 and NO_x include acid deposition impacts more generally.

- assessment of impacts;
- valuation of impacts with assessment of uncertainty and identification of key sensitivities.

Much of the basic methodology for this analysis has already been identified in the work completed within the EXTERNE project (see European Commission, 1995a–f). Indeed, one of the objectives of this study is to assess the applicability and transferability of the methodological framework established in EXTERNE to the assessment of national level damages.

The study aims at assessing the damage caused by anthropogenic emissions on selected receptors. Damage is therefore assessed on an incremental basis, with respect to background pollution levels in the absence of human activities. Damage estimates are *receptor-based* rather than *source-based*. This means that the issue of transboundary pollution is not considered in the current study, the main reason for this choice being that damage caused to other countries by anthropogenic emissions originated in Italy are not directly relevant to the welfare of Italy's residents.

The impacts on the selected receptors are assessed in physical terms using the dose–response function approach. The reader is again referred to European Commission (1995a–f) for a discussion of the applicability and limitations of such functions. In addition, the more recent literature on the subject has been carefully reviewed by the members of the project team in order to re-appraise the suitability of some of these functions, to assess their applicability to the assessment of non-marginal changes in pollution and additional impact endpoints.

The valuation stage of the assessment adopts the concept of willingness-to-pay as the underlying foundation. Where no economic valuation of the impacts was possible because of lack of physical and/or monetary data, damages have been assessed qualitatively and results reported in physical terms.

A large amount of uncertainty surrounds this sort of exercise: scientific uncertainty regarding the relationships between pollution and its physical impacts on the various receptors, uncertainty regarding baseline pollution levels in the absence of human activities, uncertainty occurring in spatial transmission mechanisms between emissions and ambient air concentrations or depositions, uncertainty regarding the economic valuation of the various impacts and its underlying assumptions.

One of the objectives of the study is to provide figures that are useful to identify real and potential problems linked to the impact of anthropogenic activities on the natural environment and that can be used in making policy decisions in the environmental area. The credibility of results from the study requires the carrying out of some sensitivity analysis both at the impact and at the valuation stages. The team agreed to identify the extent of uncertainty at the valuation stage by quoting upper and lower bounds for most figures.

Due to consideration of space and to avoid repetition, these methodological issues have not been considered here in detail. A complete and detailed discussion can be found in the Synthesis Report, Chapters 1 and 2.

1.3. Air quality data

At the outset of the project it was agreed that, where available, national data on atmospheric emissions and air quality should have been preferred to data contained in international datasets. The main reason being that, although data sources are not consistent and therefore fully compatible across countries, national data sources offer greater resolution than pan-European datasets such as EMEP and WHO. Over and above this relevant remark, in the case of Italy it has to be noticed that the availability of data from pan-European datasets is relatively limited compared to the other countries involved in the project. As far as the EMEP air quality monitoring network is concerned, for example, the Italian contribution was limited to four active stations in 1989 (Montelibretti, Stelvio, Ispra, Arabba), all located in the Northern part of the country.

Italy generates levels of SO_x, NO_x and CO_2 per unit of GDP and per capita which are well below the OECD average (ENEA, 1992; Ministry of the Environment, 1992). In 1990, emissions of SO_x amounted to 1,800,000 tonnes. From 1980 onward they have constantly declined as a result of two main factors: the gradual substitution of other fuels with natural gas and the lowering of the percentage of sulphur content in gas oil for heating and vehicles. Emissions of nitrogen oxides (NO_x) have risen since 1970 and particularly over the period 1985–1990 to reach 2,000,000 tonnes in 1990. The transport and the power sectors are the major emission sources (55% and 30% respectively). Emissions of total suspended particulates (TSP) amounted to 500,000 tonnes in 1990. Much of the 1985–1990 increase can be accounted for by the construction material sector, with decreasing contribution from the steel and diesel-engine vehicles. The transport and industrial sectors are each responsible for about 45% of total emissions.

1.3.1. *The air quality monitoring network in Italy*

The most recent study on the air quality monitoring networks in Italy dates back to 1989 (ISS, 1989) and describes the state of the art as of 1986. According to this study, in 1986 there were a total of 806 monitoring sites, over half of which measured SO_2 and a quarter measured TSP. Of the remainder there were sites measuring NO_2, NMVOC, O_3, CO, Pb, and F. The number of sites and type of pollutants monitored varied widely from region to region. Stations were not evenly distributed over the national territory. In some regions, different networks (industrial, local, regional) overlapped, while in others the geographical coverage was very poor. At the time of the report there were no monitoring sites in two regions (Molise and Basilicata, in the Central and Southern part of the country respectively). Table 1.2 shows the distribution of monitoring stations per administrative region for the year 1986 together with the type of air pollutants measured. The ISS study did not give any information on the existing monitoring networks with regard to the location of the monitoring sites (i.e. urban, industrial, versus rural areas), nor on the extension of the monitored territory, the population exposed to the pollutants, monitoring methods and procedures.

A Ministerial Decree of May 20, 1991 required an updated inventory of the air quality monitoring network in Italy to be developed, including information on the extension of the network, the number of fixed and mobile stations, the characteristics of the measurement sites, the pollutants and climate parameters measured and the measurement techniques, the efficiency of the stations as well as data handling procedures. To date, the results of this inventory are not yet available. Reporting to the Ministry of the Environment has been incomplete and, in 1991, the data received covered only 39 of the 94 administrative provinces (including public and industrial networks).

Available information show a distribution of monitoring sites which is still highly non-homogeneous over the national territory: stations are concentrated

Table 1.2. Distribution of air quality monitoring stations and pollutants measured (1986)

Region	SO_2	TSP	NO_2	Pb	CO	O_3	F	NMHC	Total
Abruzzo	2	2	2	2	0	0	0	0	8
Basilicata	0	0	0	0	0	0	0	0	0
Calabria	5	0	0	0	0	0	0	0	5
Campania	16	9	8	1	1	0	0	0	35
Emilia Romagna	39	27	6	6	4	5	2	3	92
Friuli V. Giulia	8	3	2	0	2	2	0	2	19
Lazio	12	4	1	0	0	0	0	0	17
Liguria	29	26	5	0	0	1	0	0	61
Lombardia	59	6	12	0	0	1	0	0	78
Marche	1	4	1	1	0	0	0	0	7
Molise	0	0	0	0	0	0	0	0	0
Piemonte	52	19	3	1	2	1	0	0	78
Puglia	9	3	0	0	0	0	0	0	12
Sardegna	19	14	0	0	0	0	1	0	34
Sicilia	52	14	13	0	0	1	2	8	90
Toscana	37	30	9	13	0	0	0	0	89
Trentino A. Adige	11	11	8	0	5	0	4	0	39
Umbria	4	1	2	1	0	0	0	0	8
Valle D'aosta	1		1						2
Veneto	69	40	11	0	1	5	3	3	132
Total	425	213	83	25	15	16	12	17	806

*Monitoring stations in a network or single measurement point.

Source: ISS (1989).

mostly in Northern and Central Italy and appear to be predominantly located in urban and industrial areas. Still no classification of station by urban, industrial or rural class is available.

Monitoring methods and procedures are not completely standardised and compilation and publication of results is still rather low. Data on the common pollutants are still limited in spite of their importance for policy decisions and, in particular, health protection.

A national environmental information monitoring system (SINA) was created in 1993. It is aimed at standardising and collecting information available from local and central government and other public bodies with a view to producing integrated databases. A standardised national air monitoring network is being put into place. Attention is being given to the site selection for monitors – both from the point of view of broad geographical coverage and to achieve a balance of urban, rural, industrial and transport-related sites – the localised positioning of monitors, standardisation of monitoring methods and regular quality assurance to ensure continuing reliability of data.

The air quality pollution database for the present study was constructed by combining data from various local/regional monitoring networks. Where two or more monitoring networks were overlapping (i.e. distance between sites equal or less that km 10), the average of measured concentration values was taken.

The reference year for the construction of the data base is 1990. Where data coverage for this year was poor, due to gaps in the activity of some monitoring stations, the database has been constructed by integrating data over a three year period (1989–1991). The number of 'total monitored days over the year' has been used as the selection criteria. National regulation on air quality establishes that a minimum number of observation must be available in order to estimate statistical parameters. This limit is equal to 75%, that is at least 273 days per year have to be monitored in order for meaningful statistical parameters to be estimated.

Figures 1.1, 1.2 and 1.3 show the distribution of point source data for SO_2, TSP and NO_2 obtained as a result of the above methodology.

Figure 1.3 refers to point source data availability for NO_2. This is because NO_x concentrations are not directly monitored in Italy. The national regulation on air quality (DPCM 28/3/1983 and DPR 203/1988) sets a standard for NO_2 in terms of 98° of average 1-hour concentration. Therefore data are available only for the annual 50° and 98° of average NO_2 1 hour concentration, while our analysis focuses on the impacts from NO_x concentration. In order to construct the database for NO_x the available data on NO_2 concentrations have been converted in annual average NO_x concentration via the application of statistical relationships between NO_x and NO_2 (see the following paragraph).

Ozone (O_3) concentrations are not systematically monitored in Italy and the national coverage is rather poor. According to the ISS inventory (ISS,

Fig. 1.1. Point source data for SO_2 concentrations

Fig. 1.2. Point source data for TSP concentrations

1989), only 16 stations over the whole national area measured ozone concentrations in 1986. In addition, data availability on the pollutant is rather poor due to the high inefficiency and discontinuous activity of the monitoring sites. No data on ozone concentration are reported in official statistics on air quality (ISTAT 1991 and 1993). Via different information sources, the authors were

Fig. 1.3. Point source data for NO_2 concentrations

able to collect some data for a limited number of rather non-homogeneous areas. We therefore chose to limit the assessment of damage due to ozone concentration to the Lombardia region, the only area in which the regional air monitoring network provides a sufficient data coverage for the pollutant.

Acid depositions in Italy are monitored by the National Monitoring Network for Acid Depositions. As in the case of air pollution concentrations, stations are not evenly distributed over the territory, but are concentrated mostly in Northern and Central Italy. They are mostly rural in the South and Islands, whereas those for the North are predominantly urban and industrial. Measured values include medium and mean concentrations of PH and major chemicals (ammonium, calcium, magnesium, sodium, potassium, nitrates, sulphates, chlorides, etc.). Only wet depositions are considered and measured weekly. Data are available also in an EMEP grid format (50 km × 50 km).

1.4. Data modelling

Four methodological issues will be dealt with in this section. The first issue concerns methods for moving from point source data on pollution concentrations to values for homogeneous areas. The second issue relates to the conversion of data on 50° and 98° percentile of NO_2 concentrations to data on average annual NO_x values. Thirdly, the choice of the conversion factor from TSP to PM_{10} values will be briefly addressed. Finally, the identification of background non-anthropogenic concentrations for the pollutants under consideration will be discussed.

1.4.1. *Mapping pollution concentrations*

Early in the study the working group on the project decided that the approach taken would be based on spatially disaggregated analysis of environmental damages in each country. Accordingly, the land area in each country has to be zoned following certain factors that could vary between pollutants. Two alternative approaches can be followed in order to map pollution concentrations:

- modelling the long range transport of pollutants in the atmosphere starting from emissions data;
- using measured data.

The first approach implies the use of *ad hoc* simulation models that use meteorological data and emission data to produce maps of pollution concentrations on the ground and/or of pollution depositions. Different models have been developed in the last 20 years for this purpose: box models (Fisher, 1975; Venkatram, 1985) statistical models (Rodhe, 1972; Fisher, 1978), Lagrangian and Eulerian models (OECD, 1977; Eliassen, 1978; Eliassen and Saltbones, 1983; Eliassen et al., 1988; Iversen et al, 1989). The use of simulation models is very useful when one must predict the impact on the environment of certain events (such as, for example, operation of a new industrial plant), when different emission scenarios have to be evaluated and, more generally, when a certain environmental phenomena must be related to one or more causes (i.e. identification of the specific contribution of a certain emission source to a given impact).

Our study aims at evaluating the impact of selected pollutants upon certain receptors. In order to define pollution concentrations for homogeneous zones it was decided to rely on data measured at monitoring stations.

How do we construct concentration data that apply to regions covering the whole country starting from point source concentration values? Various alternative approaches can be followed data interpolation, determination of 'areas of influence' for each monitoring station, definition of homogeneous regions, etc.

Interpolation allows one to obtain data distributed on a regular grid starting from sparse values. Different types of interpolation methods are available from the literature: nearest point, inverse of the distance, Kriging, etc. The major limit of these methodologies is that they are pure geometric algorithms and do not take into account factors that may have a significant influence in the spatial distribution of pollutants (e.g. the morphology, climatic variables, etc). In addition, interpolation methods are not appropriate to map secondary pollutants – such as ozone and acid depositions – that show a very high spatial variability.

Defining an *area of influence* for each monitoring station – an area in which the concentration is assumed to be constant and equal to the value measured at the station. Once the areas of influence have been defined for each station, the aggregation at the regional level is based on a weighted average of the concentrations measured at all the stations for which the area of influence overlaps the area of the administrative region, the weights being the ratio between the overlapping portion of the area of influence and the total area of the region. In mathematical terms:

$$C_r = P_i * C_i$$

where:

C_r average concentration over region r
C_i concentration of area of influence i
P_i a_i/a_r,
a_i portion of area of 'area of influence' of station i overlapping region r,
a_r total area of region r.

This method allows to easily and rapidly obtain data that apply to regions. However, it is purely geometric, even if some constraints could be set in order to account for geographical and climatic factors.

The third approach consists in defining a number of *homogeneous sectors* grouping those geographical areas that present similar morphological and meteorological characteristics, and have similar distributions of polluting emissions. In our study we decided to follow the third method in so far as it better allows to take into account a number of factors that may play an important role in determining the spatial distribution of pollution concentrations. Nine 'homogeneous sectors' have been identified for Italy: North-West sector, Central-East Alps, Padania region, Tirrenian sector, Adriatic sector, South-West sector, South-East sector, Sicily and Sardinia (Centro per lo Studio della teoria dei Sistemi, 1993, mimeo). They are shown in Figure 1.4.

The concentration associated with each sector is obtained by averaging the concentrations measured at the monitoring stations that fall into that sector. Once concentration values have been constructed for each of the 9 homogeneous sectors, data on pollutant concentrations at the level of administrative region have been derived. Each administrative region takes the concentration of the sector in which it falls. In the case of regions that overlap more than one sector – such as Lombardia and Veneto – the concentration value is obtained by a weighted average of the concentrations of the sectors they belong to, the weights being the portions of the territory of the region overlapping each sector. The method has been applied to map pollution concentration of SO_2, PM_{10} and NO_x.

1.4.2. *Deriving NO_x concentration levels from data on NO_2*

Official statistics on air quality in Italy do not provide any data on NO_x concentrations. Only data on 50° and 98° percentile of average 1hour NO_2 concentrations are available, according to the requirement of the national

Fig. 1.4. The 9 homogeneous sectors

legislation on air quality standards. The relations between NO_2 and NO_x were taken from a statistical study on the urban area of Milano, in Northern Italy, where time series of both NO_2 and NO_x were available (Giuliano, 1993). Unfortunately, since no other data on NO_x concentrations were available, the computed values for NO_x could not be validated. Furthermore, it is not clear to which extent the proposed relations can be extrapolated to contexts that may be different from the one in which they were derived. The reference study proposes the following statistical relations between annual mean of NO_2 and 50° and 98° percentiles:

A linear model:

$$P_{98} = 1.775 * M_{a;NO_2} + 68.04 \tag{1a}$$

$$P_{50} = 0.8953 * M_{a;NO_2} - 1.28 \tag{1b}$$

Or a log-normal model:

$$P_{98} = 2.582 * M_{a;NO_2} \tag{2a}$$

$$P_{50} = 0.8687 * M_{a;NO_2} \tag{2b}$$

In addition, the following relation between annual mean of NO_2 and annual mean of NO_x is developed:

$$M_{a;NO_2} = 308.8 * \frac{M_{a;NO_x}}{M_{a;NO_x} + 502.9} \tag{3}$$

Equations (1) and (2) showed very similar results in terms of standard error and correlation coefficient; the annual mean of SO_2 showed a higher correlation with the 50° percentile than with the 98° percentile (0.9242 vs 0.7887). Equation (3) has a linear correlation coefficient of 0.9535. We therefore used equation (1b) to compute annual mean of NO_2 from 50° percentile and equation (3) to compute the annual mean of NO_x from the annual mean of NO_2. When 50° percentile was not available, equation (1b) was used. Once the annual means of NO_x were computed, we proceeded as in the case of the other pollutants (see above).

1.4.3. *Converting TSP data in PM_{10} concentrations*

In Italy only total suspended particulates (TSP) are monitored through the national air quality monitoring networks. TSP data have be converted into PM_{10} by using a conversion factor of 0.60, which is the result of a review of other European studies (ETSU, 1993) as well as of consultation with national experts.

1.5. Identification of background concentrations

The objective of the current study dictates that damages will be estimated as marginal impacts, over and above some background concentration, defined as levels of pollution that would exist in the absence of human activity. The working group on the project considered that for some pollutants an estimate of background non-anthropogenic levels can be gauged from the concentrations measured in remote areas. In other cases, particularly for ozone, it would be necessary to rely on modelled data.

Background levels are likely to vary from region to region, even within a country. The estimated damages are likely to be sensitive to the choice of these values. In view of this, some sensitivity analysis of damage estimates to different baseline values of air pollutant concentrations will be carried out where thought to be appropriate.

As far as SO_2 and NO_x are concerned, it was suggested that all countries use background levels which come from North West Scotland, an area that can be regarded as a pristine environment in Europe. This region is remote from major industrial activities and the prevailing south westerly winds are unlikely to carry a high pollution load from other countries. UK RGAR (1990) provide values of about 1 ppb SO_2 (2.618 g/m^3) and 2 ppb NO_x for this region. In Southern Europe higher SO_2 background levels could exist because of volcanic activity. Berreheim and Jaeschke (1982) have estimated the annual non-eruptive emissions of SO_2, H_2S and SO_4 – from Mt. Etna in Sicily – the largest active volcano in Southern Europe – to be in the range of 0.05 to 0.32 Tg S y^{-1}. Non-eruptive emissions are estimated to be 14 times higher than emissions during the eruptive phase. In addition, sulphur emissions from volcanic activity are estimated to be about one third of total sulphur emissions from natural sources (Katsoulis *et al.*, 1990).

From this it follows that background level of SO_2 concentration in Southern Italy could be 30% higher than that indicated for North West Scotland (i.e. around 1.3 ppb $\approx$ 3.4 g/m^3). Some sensitivity analysis to baseline non-anthropogenic level of SO_2 will therefore be undertaken with regard to the Southern part of the country. A standard SO_2 background level will be considered for Central and Northern Italy since the wind field of the area shows that the main air flows are from west and north-west sectors. As for NO_x background level, in the absence of alternative indications, the value of 2 ppb indicated by the UK team will be used.

Background particulate concentrations are more difficult to predict. Particulate matter originate from both natural processes (volcanic activity, sea spray, etc), and human activity. The contribution of natural versus anthropogenic sources varies with the diameter of particles and with the type of area considered (i.e. coastal versus inland). In the current study a background concentration of 10 g/m^3 for PM_{10} was used, as a result of consultations with national experts. A sensitivity analysis to a lower, arbitrarily chosen background level (i.e. 7 g/m^3) has be carried out.

The definition of background concentrations for O_3 is problematic as they vary with altitude and sunlight. Ideally, modelled data should be used. The present analysis will focus on damage to receptors that are confined to lower altitudes. (Effects of O_3 on forests and uplands ecosystems will not be valued because of lack of dose–response data). Therefore, we decided to focus on the lower level ozone concentrations, where modeled baseline concentrations have been identified in 25 ppb for Southern Europe. This value is confirmed for Italy in Cirillo et al. (1988).

For the reasons explained above, damage due to O_3 concentration will be estimated only in the Lombardia region. It should be noted that this is a very industrialized and densely populated region of the country, and in some areas ozone levels are likely to be lower than the background value identified here. This is the result of the complex atmospheric chemistry of the pollutant, and its dependence on the relative levels of NO and NO_2, that together give total measures of NO_x. In the centre of urban areas, where the level of vehicle emissions is particularly high, NO dominates and consume O_3. Following the emission of NO_x into the atmosphere, sufficient NO will be converted to NO_2 for the latter to dominate, driving the reaction in favour of O_3 production. It is thus possible that the case study on Lombardia will estimate net benefits in terms of damages to human health and agriculture resulting from the effect of anthropogenic activity on O_3 levels.

CHAPTER 2

HEALTH

2.1. Introduction

Laboratory experiments and public health studies in polluted environments show that air pollutants have effects on human health. High levels of air pollution have been associated with acute as well as chronic effects related to the respiratory system including mortality, respiratory symptoms and diseases that can have effects on the lung. Given the wide range of variables which impact on human health, it is of course difficult to derive reliable relationships between exposure to air pollutants and health impacts. Moreover, even where effects of air pollution on health are unambiguously observed, it is difficult to isolate the health impacts of single pollutants, given that the main air pollutants tend to be strongly correlated – both spatially and temporally.

Existing studies are in good agreement that SO_2, NO_x, particulates and ozone are among the major pollutants responsible for health impacts and that significant interactions exist between these pollutants. For example, according to the results of epidemiological studies, synergistic effects exist between NO_2 and O_3, and between SO_2 and particulates.

The impact of air pollution on the population depends on the concentration level and the sensitivity of the people. But quantitative relationships can be difficult to establish. In the health science, a dominant concern has been the development of concentration guidelines for public health authorities, i.e. concentration levels below which there are no measurable ill effects in health subjects. Table 2.1 reports WHO air quality guidelines set on the basis of toxicological, clinical and epidemiological studies.

Significant health effects are expected if these guidelines are exceeded. There is some debate, however, as to whether human health is threatened even below these thresholds. The alternative assumption, that there is a dose–response function which has no threshold is equally plausible and consistent with many epidemiological studies. Evidence concerning health impacts at low pollution levels is less statistically reliable and the existence or otherwise of threshold concentrations remains unresolved.

The following analysis assesses the impact of SO_2 and particulates on mortality and morbidity. Both acute and chronic impacts will be considered. Attention will be given to the need to avoid double counting with respect to

Table 2.1. WHO air quality guidelines for acid rain pollutants (g/m^3)

Pollutants	1-h exposure	24-h exposure	1-year exposure
SO_2	350	125[1]	50
TSP	–	–	120
NO_X	400	150	60
O_3	150–200	100–120[2]	–

[1] When exposure to SO_2 occurs simultaneously with exposure to TSP.
[2] The guidelines are for an 8-hours exposure.

Source: WHO (1987).

the various morbidity impacts of a given pollutant as well as to the non-additivity of certain exposure–response functions linking a given outcome to different pollutants. Health impacts of NO_x on mortality and morbidity have been assessed by the authors. However, after consultation with the UK team it was decided that the results should not been included due to the high level of uncertainty. In addition, due to data limitations it was only possible to make an assessment of ozone damages in the Lombardia region.

2.2. Stock at risk

The health damage caused by ambient level of air pollutants depends first of all on the spatial distribution of the population in the country. In addition, available evidence shows that children and the very old may be more vulnerable to adverse health effects than middle age individuals. Similarly, individuals already suffering from asthma or other respiratory diseases are more prone to the negative health effects of air pollution. The age and ‘sensitivity’ structure of the population are therefore also important in determining the impacts of air pollution on human health. Most of the exposure–response relationships derived from the literature are related to certain risk groups e.g. children, old people, asthmatics, etc.

Population density data used for the assessment of the health impacts in Italy were taken from the National Population Census of 1991, disaggregated at the level of administrative regions (ISTAT, 1991). To quantify the health impacts of air pollution within a certain risk group, the share of the risk group in the total population must be known. The age structure of the Italian population is shown below (Table 2.2).

A significant difference exists between the Northern and Central part of the country on the one hand and the South and Islands on the other, with a greater percentage of young people living in the latter. We have therefore decided not to use the national mean values throughout the various administrative regions but to consider the age structure specificity of each region. Instead,

Table 2.2. Age structure of the Italian population (%) for 1991

Age	North	Central	South and Islands	Total
0–14 years	13.1	14.0	20.2	15.9
15–65	70.2	69.3	66.9	68.8
65 and above	16.7	16.7	12.9	15.3

Source: ISTAT (1991).

no distinction was made between urban and rural areas, although the age structure is likely to vary between the two. Exposure–response functions for children were based on those aged less than 14 years. The proportion of *asthmatics* in the population was estimated as 7% regardless of age (ISTAT, 1991).

Mortality data at the regional level are published annually by the National Statistical Office (ISTAT). The baseline *all-cause death rates* for each region have been used to assess the mortality impact of the selected pollutants. This includes deaths from violent or external causes whereas the selected mortality exposure–response functions apply to non-violent causes only. An overestimation of the mortality effects is therefore implied. Average all-cause death rates were applied to the population at risk in each administrative region to give expected deaths in the at-risk population in the absence of anthropogenic emissions. Incremental mortality is then derived by linking the baseline data, the incremental pollution generated by anthropogenic emissions and the corresponding exposure–response functions. The all-cause death rate for Italy is equal to 0.0095 deaths per person per year, equivalent to 539,391 deaths per year.

2.3. Exposure–response functions and data sets

The common set of exposure–response functions were used, as presented in the Synthesis Report (Tables 3.1 to 3.5). All the selected exposure–response relationships are linearized without thresholds, in accordance with the advice given by the epidemiologists advising the GARP team.

The *mortality* analysis includes both SO_2 and PM_{10} effects on a national basis and O_3 effects for the region of Lombardia. The occurrence of SO_2 and particulate matter shows a strong spatial and temporal correlation and it is rather difficult to separate out the specific effects of particulates from those of SO_2. A commonly used approach is to choose one index of air pollution as primary and to treat the relationship of mortality – as well as morbidity with that index – representing the more general mixture to which people had been exposed. Consequently, the results obtained from the exposure–response functions for SO_2 and particulate matter have to be considered as alternative to one another rather than additive. Given the need to choose, the weight of evidence strongly support the use of PM_{10} mortality as the primary result. The results from acute and chronic mortality are also not additive.

The analysis of *morbidity* effects included most of the endpoints presented in the Synthesis Report (Chapter 3, Table 3.9).

The use of exposure–response relationships requires spatially resolved data on ambient air concentration of the relevant pollutants. It is assumed that the annual average concentrations are a good measure of exposure. The estimated impacts are based on a dataset of mean annual concentrations of SO_2, NO_x and PM_{10} monitored throughout the entire national territory and adjusted for homogeneous areas as described in Chapter 1. Damage has been assessed on a marginal basis with reference to the chosen background concentrations.

Given that the selected exposure–response functions are independent from background levels, the link up of population and incremental pollution data can be summarised as a population weighted annual average of incremental pollution. This is obtained by multiplying the estimated incremental pollution in each region by the estimated population at risk in that region, summing the products over regions and dividing by the total population at risk.

2.4. Estimated impacts and valuation

Results of the implementation are given in a series of tables organised by pollutants and type of health impact. Sensitivity analysis to different background levels (i.e. 7 $\mu g/m^3$ vs. 10 $\mu g/m^3$ for PM_{10}; 1.3 ppb vs. 1 ppb in the Southern part of the country for SO_2) was carried out in order to assess the distribution of the results.

For the monetary valuation of mortality a figure of ECU 2.6 million for the Value of a Statistical Life (VOSL) was used in the initial calculations which are presented below. These can be assumed to represent 1990 prices. In the Synthesis Report all the estimates were updated to 1995 prices. The various endpoint valuations for morbidity impacts were also originally calculated with 1990-based values. Again, the final values and updated estimates can be found in the Synthesis Report.

The estimated impacts for deaths, particularly deaths due to chronic effects appear rather high (see Table 2.3). There was a consensus among the GARP team that the function used to estimate chronic deaths should be regarded as an upper estimate of damages. For validation we have compared the results against independent data, i.e. mortality statistics. In 1991, in Italy there were 539,391 deaths, 240,738 of which were from diseases of the circulatory system and 32,902 from diseases of the respiratory system. The figure for acute mortality due to PM_{10} are equivalent to 41.7% of those who die from diseases of the respiratory system and 5.1% of those who die from either respiratory or

Table 2.3. Acute and chronic mortality impacts from incremental PM_{10} and SO_2

Endpoint	Background	No. of cases Low	No. of cases Mid	No. of cases High	Damages (ECU million) Low	Damages (ECU million) Mid	Damages (ECU million) High
PM_{10}	10 μm^{-3}	8,653	14,061	19,605	22,498	36,560	50,974
Acute effects	7 μm^{-3}	9,689	15,744	21,951	25,191	40,936	57,074
PM_{10}	10 μm^{-3}	39,886	52,191	64,495	103,706	135,696	167,687
Chronic effects	7 μm^{-3}	44,660	58,437	72,213	116,117	151,936	187,756
SO_2	10 μm^{-3}		6,835			17,773	
Acute	7 μm^{-3}		6,711			17,450	

circulatory system diseases. The figures for SO_2 are 20.7% and 2.5% respectively. The figures for chronic mortality are an incredible 158% of those who die from diseases of the respiratory system, and 19% of those who die from either respiratory or circulatory system diseases! Further investigation of these exposure–response functions is clearly warranted.

Italian GDP in 1990 amounted to about ECU 862 billion (where 1 ECU = LIT 1522). The total economic damage from the selected air pollutants range from 2.56% to 4.74% of GDP depending on whether the acute mortality figure

Table 2.4. Acute morbidity impacts from incremental PM_{10}

Endpoint	Background	No. of cases (000s) Low	No. of cases (000s) Mid	No. of cases (000s) High	Damages (ECU million) Low	Damages (ECU million) Mid	Damages (ECU million) High
Respiratory	10 μm^{-3}	1.4	2.1	2.8	9.1	13.8	18.4
HAs	7 μm^{-3}	1.6	2.4	3.2	10.5	15.9	21.3
HAs for	10 μm^{-3}	1.8	2.5	3.3	11.9	16.7	21.6
COPD	7 μm^{-3}	2.0	2.9	3.8	13.6	19.2	24.9
ERVs for	10 μm^{-3}	6.5	8.0	9.6	1.2	1.5	1.8
COPD	7 μm^{-3}	7.4	9.3	11.0	1.4	1.7	2.0
ERVs for	10 μm^{-3}	4.5	7.1	9.6	0.8	1.3	1.8
asthma	7 μm^{-3}	5.1	8.2	11.0	0.9	1.5	2.0
Visits for	10 μm^{-3}	24.3	32.4	42.6	4.5	6.0	7.9
child group	7 μm^{-3}	28.0	37.4	49.1	5.2	6.9	9.1
RADs	10 μm^{-3}	35,500	55,600	87,300	2,212	3,472	5,448
	7 μm^{-3}	40,900	64,100	100,600	2,550	4,002	6,280
Asthma	10 μm^{-3}	546	10,900	16,400	17.0	342	513
attacks	7 μm^{-3}	630	12,600	18,900	19.7	394	591
Symptom	10 μm^{-3}	245,300	591,300	766,000	1,545	3,725	4,826
days	7 μm^{-3}	282,800	725,400	883,000	1,781	3,231	5,563

Notes:

HA Hospital Admissions.

COPD Chronic Obstructive Pulmonary Disease.

ERV Emergency Room Visits.

RAD Restrictive Activity Days.

Table 2.5. Estimated chronic morbidity impacts from incremental PM_{10}

Endpoint	Background	No. of cases (000s) Low	Mid	High	Damages (ECU million) Low	Mid	High
Chronic bronchitis in adults	10 μm^{-3}	421.4	655.6	880.3	58.2	90.5	121.6
	7 μm^{-3}	485.0	754.4	1,013.0	67.0	104.1	139.9
Respiratory illness in adults	10 μm^{-3}	561.9	889.7	12.081	77.6	122.9	166.8
	7 μm^{-3}	646.7	1023.8	1,390	89.3	141.3	192.0
Chronic bronchitis in children	10 μm^{-3}	151.8	287.6	425.2	21.0	39.7	58.7
	7 μm^{-3}	176.6	334.5	494.5	24.4	46.2	68.3
Chronic cough in children	10 μm^{-3}	184.0	369.8	559.1	25.4	51.0	77.2
	7 μm^{-3}	214.0	430.1	650.3	29.5	59.4	89.8

Table 2.6. Estimated chronic morbidity impacts from incremental SO_2 pollution

Endpoint	Background	No. of cases Low	Mid	High	Damages (ECU million) Low	Mid	High
HAs for chronic bronchitis	10 μm^{-3}	479	484	488	3.76	3.81	3.84
ERVs for COPD	10 μm^{-3}	753000	767000	779000	168	171	174

from SO_2 pollution is considered as the primary result as opposed to PM_{10}. The latter is equivalent to about 4.2% of the GDP while the former represents around 2%. As far as morbidity impacts are concerned, the dominant factor appears to be the short term impact of PM_{10}. However one should note that these results are influenced also by the greater availability of exposure–response studies on PM_{10} as opposed to the other pollutant.

2.5. Conclusions

The above estimates of health damages caused by anthropogenic emissions show a considerable range of variation according to the exposure–response functions that are applied. Given the lack of Italian studies establishing exposure–response coefficients for both mortality and morbidity effects, the analysis has relied upon functions derived in the US and in other European countries. The extent to which it is feasible and correct to transfer health dose–response relationships developed in other contexts is debatable.

In general, it is argued that transferability depends on the endpoint being considered: if the endpoint represents a pure biological event as opposed to a health event that involves social and cultural decisions, then transferability is likely to be more justifiable. This is because in the case of biological events the health impact only depends on the health status of the individual and its identification and reporting is not influenced by cultural factors and/or by the social system governing health care (availability of health care services and their costs). Hence, for example, transferability is more correct in the case of exposure–response functions relating air pollution concentrations to asthma attacks, while greater care should be paid when considering the impacts of air pollutants on restricted activity days (RAD), hospital admissions (HA) and emergency room visits (ERV). The same remarks apply to mortality impacts.

However, even in the case of biological events, extrapolation is subject to other limitations. Mortality and morbidity impacts of air pollution vary with the age as well as health profile of the population and the available functions are not sufficiently refined to take these differences into account. Therefore, the extrapolation of relationships derived in other countries is likely to result in some biases in the final estimates of health impacts.

Finally, the issue of transferability also emerges when considering the specific pollution mixture (context) to which the relevant population is exposed. At the beginning of the chapter we have underlined that a significant spatial as well as temporal correlation exists between various air pollutants and that, consequently, the health impact on the population is the result of the specific pollution mixture to which the population is exposed. In transferring the exposure–response functions available in the literature we have assumed that the Italian population is exposed, throughout the entire national territory, to the same air pollution mixtures as the population of the original studies. This may of course not be true, particularly when considering the urban versus rural context.

In our study, we have tried to extend the analysis to cover both acute *and* chronic health effects in order to take account of the full health impacts of the pollutants under consideration. In addition, a number of different morbidity endpoints have been considered, both for acute and chronic impacts.

In some cases these endpoints cannot be added together This is true both in the case of one single endpoint linked to two or *more air pollutants* and in the case of different endpoints relating to the same pollutant. Hence, for example, in the case of SO_2 and PM_{10} the mortality outcomes are not additive and results have to be considered separately. However, we have seen that the choice of one pollution index as opposed to the other greatly influences the results of the analysis. Moreover, it is not clear to what extent the relationship with one of the two pollutants acts as a surrogate with the relationship with the other, and with the pollution mixture as a whole. This, in turn, leads us back to the transferability issue in the presence of different types of pollution contexts.

On the other hand, the issue of the additivity of various exposure–response functions exists also when considering different endpoints relating to the *same pollutant*. This is the case, for example, for the various morbidity impacts of particulates. The range of available exposure–response functions linking morbidity with particulates is quite extensive and care must be taken when implementing those functions to allow for the fact that some double-counting is implied in merely adding up all the various estimated impacts. In our estimates we have considered all the functions as additive with the exception of the dose–response relationship for symptom days. The issue is less pressing in the case of SO_2 and NO_x, given that the range of endpoints for which exposure–response functions have been derived in the literature up to now is more limited.

We have already underlined that the reliability of exposure–response estimates from cross-sectional air pollution studies is often questionable because of difficulties in taking account of confounding variables, which are likely to result in an overestimation of the air pollution impact. The estimates obtained above for chronic mortality seem to confirm these difficulties. The numbers obtained are excessively high and comparisons with independent data (i.e. mortality statistics) confirm that these functions need to be carefully revised. On the other hand, focusing the evaluation exclusively on short-term or acute responses has drawbacks in terms of full coverage of health effects, in so far as short-term time series studies do not take account of any later or delayed impact of air pollution on mortality.

Some recent studies suggest that longer term impacts might be mediated via development of chronic respiratory diseases. If this is the case – i.e. if longer term mortality is preceded by respiratory symptoms – one could, in principle, assume that chronic mortality effects are indirectly taken into account by including the development of chronic respiratory diseases in the valuation of air pollution impacts on human health.

However, at present there is still scientific uncertainty on this point. In addition to this, even if we overcome the limited scientific knowledge and assume that chronic respiratory symptoms can be an adequate 'surrogate' for chronic mortality impacts, one further issue has to be raised. In order to adequately account for chronic mortality impacts through the use of chronic morbidity valuations, the economic value measures used to monetize the impacts of air pollution on chronic respiratory symptoms should take into account the risk of any adverse health outcomes in the *longer term* and the willingness to pay of individuals to avoid that risk. On the contrary, the economic valuation of respiratory symptoms used in the present analysis refers 'only' to willingness to pay to avoid *current* symptoms and does not take into account the increased risks of mortality which might be associated with these symptoms.

The essence of the above discussion is that the estimates obtained in this study should be regarded as first round assessment of the orders of magnitude for health damages. It is very much hoped that the second phase of the work will improve the confidence in both the underlying scientific aspects and the economic valuations.

2.6. Appendix – Ozone concentrations in Lombardia

Due to insufficient monitoring stations at the national level it was only possible to undertake an assessment of health related ozone damages in the Lombardia region. A regional ozone air quality monitoring network exists which holds data on O_3 concentrations. These monitoring stations are located mainly if not exclusively in urban areas. Given the particular dynamic of this pollutant (see Chapter 1) this is likely to result in a downward biased estimate of average concentration in rural zones. In Lombardia more than 75% of the population lives in communities with a population of more than 10,000.

Incremental ozone pollution due to anthropogenic polluting emissions has been calculated with reference to a background level of 25 ppb. The total population is around 8.85 million, with 15% of age 65 or over. Total registered asthmatics equate to 7% of the population. The exposure–response functions used to estimate mortality and morbidity effects are presented in the Synthesis Report (Tables 3.1 and 3.3). The application of those functions to the risk group populations in the region gave the following results.

In Lombardia there were 84,132 deaths in 1991, 40.4% of which from diseases of the circulatory system and 6.3% from respiratory diseases. The above figure for acute mortality due to ozone represent 1.73% of those who died from diseases of the respiratory system and 0.23% of those who died from either respiratory or circulatory system diseases.

Table 2.7. Acute mortality impacts of incremental ozone concentrations in Lombardia

	No. of cases			Damages (ECU million)		
Endpoint	Low	Mid	High	Low	Mid	High
Acute mortality	61	92	122	159	239	319

Table 2.8. Acute morbidity impacts of incremental ozone concentrations in Lombardia

	No. of cases			Damages (ECU million)		
Endpoint	Low	Mid	High	Low	Mid	High
HAs for respiratory infections	411	548	653	2.71	3.62	4.31
HAs for COPD	250	399	548	1.65	2.63	3.62
HAs for asthma	182	369	554	5.7	11.6	17.4
ERVs for asthma	1,111	1,700	2,262	0.2	0.3	0.44
MRADs (000s)	0	1,009	3,381	0	127	259
Asthma attacks (000s)	1,651	2,633	3,611	51.7	82.4	113
Symptom days (000s)	1,476	2,898	4,325	9.3	18.3	27.2

CHAPTER 3

CROPS

3.1. Introduction

The impacts of atmospheric pollution on crops have been extensively studied both in the laboratory and in the field. Crops can be damaged *directly* or *indirectly* by pollutants. Direct damage is in the form of yield loss and additional liming measures to be taken by farmers. Indirect effects include reduced resistance of crops to pests due to air pollutants, decreased resistance to stress caused by physical and chemical agents, etc. Moreover, significant interactions between pollutants and other stresses on agricultural crops may exist. These include the natural effects of climate, pests and pathogens (Warrington 1989; Houlden *et al.*, 1990; Sommerville *et al.*, 1990).

Most experimental research in this area has investigated the direct effects of individual pollutants on agricultural crops. Relatively little attention has been given to synergistic and indirect effects. As a consequence, published dose–response functions typically incorporate only direct effects of pollutants on plants. Existing studies are in good agreement that the most important impacts are due to sulphur dioxide (SO_2) and ozone (O_3). Any contributions from acid and nitrogen depositions are thought to be largely masked by addition of fertiliser and lime in traditional agricultural practice (European Commission, 1995b).

The following analysis assesses the direct effects of SO_2 concentrations on selected crop species in the form of yield changes. Quality changes, mitigating measures of the farmers, indirect effects and other interactions have not been considered. Yield changes due to O_3 have not been assessed at the national level because of the very limited availability of data on ozone concentrations and a lack of applicable dose–response relations for Europe.

3.2. Stock at risk

The selection of crops with which to measure yield changes due to air pollution has been carried out in such a way as to give an order of magnitude of the economic damages occurring in the agricultural sector. This effort was hampered by both the limited scientific knowledge on the sensitivity to air pollutants of some major national crops and the limited availability of applicable dose–response functions for some of these crops.

Crops species show different sensitivity to air pollutants. Several sensitivity classifications based on acute as well as long term exposure to SO_2 and O_3 are available in the literature (Taylor, 1986; Guderian, 1977; OECD, 1980). According to these studies crop species have been classified into: 'sensitive', 'intermediate', 'less sensitive'. Based on these classifications a first selection of crops for valuing air pollution damage was made. The second criteria for the selection of crops was their relative importance in terms of percentage coverage of total cultivated agricultural land. In other words, from the above sensitivity list only those crops that are grown in large quantities – i.e. more that 1% of total farmland – in the country have been considered for the analysis. A further criteria considered is the value of produce as a proportion of total agricultural production.

Table 3.1 classifies crops according to their sensitivity class, percentage cover of total agricultural land and contribution to the total value of agricultural production.

3.3. Dose–response functions and datasets

Given the above list of selected agricultural crops, only those crops for which applicable dose–response functions were available have been considered for the analysis. Such functions were mostly derived from the EXTERNE project. In addition, a comprehensive review of the Italian literature on the subject has

Table 3.1. Crop classifications

Crop	Sensitivity SO_2	Sensitivity O_3	Farmland (%)	Total agricultural production (%)
Vine	–	vs	4.0	1.44
Lettuce	s	ls	0.1	1.39
Onion	ls	s	0.06	0.59
Pea	s	s	0.1	–
Spinach	vs	s	0.02	0.23
Strawberry	s	ls	0.02	0.97
Winter wheat	s	s	5.5	8.47
Spring wheat	s	s	4.0	(*)
Rice	–	s	0.8	1.93
Corn	–	s	2.9	4.04
Soyabean	–	s	1.9	n.a.
Potato	ls	–	4.6	1.89
Tomato	s	vs	4.5	2.83
Apple	s	–	0.3	2.35

Note: vs = very sensitive; s = sensitive; ls = low sensitive.
(*)Value included in the figure for winter wheat.

been carried out in order to identify applicable dose–response relationships for specific national crops. In particular, a series of dose–response functions obtained through regression analysis of open field experimental data have been reviewed (Camposano *et al.*, 1986; Cirillo *et al.*, 1988; Triolo 1994). However, for some of the major crops known to be sensitive to SO_2 and O_3 it has not been possible to identify applicable *d/r* functions. Olive trees, for example, cover 4.4% of total agricultural land, vines account for 4% and citrus trees for 0.7% which together constituted a values of LIT 1871 Bn. in 1991.

In the following we present the dose–response functions that have been applied to each of the selected crops: winter and spring wheat, potato, tomato and apple. Where possible, two sets of functions have been considered: one set including *d/r* relationships from the EXTERNE study and a further set incorporating some more *ad hoc* functions developed from regression analysis on open field experimental data (Camposano *et al.*, 1986). The aim was to test the sensitivity of the estimates to the use of transferred *d/r* relationships as opposed to more location specific functions. The use of experimental results implies a higher degree of uncertainty in the estimates.

3.3.1. *Dose–response functions for SO_2*

The direct impacts of SO_2 on winter and spring wheat were estimated using the Baker *et al.* relationship as modified in EC (1995b) to enable the impacts of small SO_2 concentrations to be dealt with, together with the Weigel *et al.* model.

Baker *et al.* (1986):

$$y = 0.74(SO_2) - 0.055(SO_2) \qquad \text{(from 0 to 13.6 ppb)}$$

$$y = -0.69(SO_2) + 9.35 \qquad \text{(above 13.6 ppb)}$$

Given that SO_2 concentration levels never exceed 13.6 ppb in the various areas in which we have divided the national territory, only the first equation has been applied in the analysis.

Weigel *et al.* (1990):

$$y = 4.92 - 0.74(SO_2) \qquad (SO_2 \text{ in ppb})$$

Where y is the percentage yield variation and SO_2 is the mean annual concentration from anthropogenic sources.

These results were then compared with those from the application of more location-specific dose–response functions, derived from regressions analysis of open field experimental data (Cirillo *et al.* 1988):

$$y = 1.48 - 0.33 \log_{10}(SO_2) \qquad (SO_2 \text{ in } \mu g/m^3)$$

For potato, tomato and apple, no dose–response relationships were available

Table 3.2. Summary of dose–response functions used

Crops	Dose–response functions for SO_2
Winter wheat	Baker *et al.* (1986), Weigel *et al.* (1990), Cirillo *et al.* (1988)
Spring wheat	Baker *et al.* (1986), Weigel *et al.* (1990), Cirillo *et al.* (1988)
Potato	Camposano *et al.* (1986)
Tomato	Camposano *et al.* (1986)
Apple	Camposano *et al.* (1986)

from the EXTERNE work. Direct effects of SO_2 concentrations on these crops were therefore estimated using dose–response functions from Camposano *et al.* (1986).

potato:

$$y = 1 - 0.0020(SO_2) \qquad (\mu g/m^3)$$

tomato:

$$y = 1.06 - 0.00042(SO_2) \qquad (\mu g/m^3)$$

apple:

$$y = 1.12 - 0.0042(SO) \qquad (\mu g/m^3)$$

3.3.2. *Pollution concentrations*

The use of the above listed dose–response relationships requires spatially resolved data on ambient air concentration of the relevant pollutants as well as on the areas under cultivation and yields. In the field, crops are subject to doses of pollutants that vary over time. Peak concentration may be much higher that average, particularly for ozone. Changes in yield may be a function of either peak or average pollution levels, depending on whether the damage mechanism is acute or chronic. However, there is currently a general consensus that long term mean levels of pollution rather than peak value are the dominant factor. This is the parameter used for estimating the crop yield losses via the selected dose–response models.

The estimated impacts from SO_2 are based on a dataset of mean annual concentrations monitored throughout the entire national territory and adjusted for homogeneous areas as described earlier (see Chapter 1). As the measurement stations of the network are mainly located in highly polluted areas (in the center of a city, roads with high traffic density, etc.) the concentrations per zone are likely to be above the real ambient concentrations in rural areas.

Damage has been assessed on an incremental basis with reference to SO_2 concentrations in the absence of anthropogenic emissions. The chosen background concentration for SO_2 is 1 ppb for the whole of Italy with the exception of the Southern part of the country, where a baseline level of 1.3 ppb has been used to take into account volcanic activity. This baseline concentrations are subtracted from the current mean annual concentrations derived above to give the pollution resulting from anthropogenic emissions. In Italy, the background value was subtracted from the calculated mean annual concentration for each region. Because of lack of pollution concentration data, agricultural damage due to SO_2 concentration has not been estimated for three regions: Abruzzo, Marche and Molise.

3.3.3. *Spatial distribution of selected crops*

A spatially resolved dataset of yield production per crop within the country has been constructed by multiplying data on areas assigned to each crop by the mean annual yield per unit area (ha) for the relevant crop. Data on land coverage for crop species and on mean annual yield per hectare in each zone have been taken from national sources (ISTAT 1991c). The results are presented at the level of administrative regions.

3.4. Quantification of impacts

Yield changes from the dose–response functions are expressed as a % yield variation compared to a baseline of no yield change at concentration of zero (Y_0). However, crops in the field are clearly exposed to some level of SO_2 and agricultural data used already incorporates some SO_2 effect. Direct application of the dose–response functions will thus introduce some degree of error. To account for this, we compute the absolute change in specific yield (Y) with respect to the background SO_2 concentration as in the following. From the *d/r* functions we compute:

$$y_b = [(Y_b - Y_0)/Y_0] * 100 \tag{1}$$

$$y_n = [(Y_n - Y_0)/Y_0] * 100 \tag{2}$$

where:

y_b = yield variation at background concentration as % of yield at 0 SO_2 ppb (eq. (1))
y_n = yield variation at actual concentration as % of yield at 0 SO_2 ppb (eq. (2))
Y_0 = yield at 0 SO_2 ppb (unknown)
Y_b = yield at background SO_2 concentration (unknown)
Y_n = yield at actual SO_2 concentration (known from statistics)

From equations (1) and (2) we derive:

$$Y_0 = 100 * Y_n/(001 + y_n) \quad \text{and} \quad Y_b = Y_n * (100 + y_b)/(100 + y_n)$$

Table 3.3. Estimated yield losses due to SO_2 (%)

Dose–response study	Winter wheat	Spring wheat	Potato	Tomato	Apple
Baker *et al.* (1986)	−2.1	−3.2	–	–	–
Weigel *et al.* (1990)	−2.36	−3.6	–	–	–
Cirillo *et al.* (1988)	−1.08	−1.15	–	–	–
Camposano *et al.* (1986)	–	–	−0.87	−1.01	−1.08

Table 3.4. Estimated impact of direct effects of SO_2 on selected crops (tonnes)

Dose–response relationship	Winter wheat	Spring wheat	Potato	Tomato	Apple
Baker *et al.* (1986)	−108,328	−137,955	–	–	–
Weigel *et al.* (1990)	−121,721	−155,100	–	–	–
Cirillo *et al.* (1988)	−55,793	−49,231	–	–	–
Camposano *et al.* (1986)	–	–	−19,480	−54,930	−19,468

The absolute change in specific yield for Y is what is needed to make a valuation assessment. It is given as:

$$Y = Y_n - Y_b = Y_n * (y_n - y_b)/(100 + y_n)$$

The application of the above dose–response relationships and datasets produced the final results presented in Tables 3.3 and 3.4. To test the dependence of the results on the background concentration, the baseline concentration has been varied as in the case for health damage assessment. Variation in results was small.

3.5. Estimation of economic impacts

In the case of agricultural crops, air pollution affects products that are directly marketable. The valuation of these impacts is therefore straightforward. Yield changes have been valued at international prices for internationally traded goods, while wholesale prices taken from the Commodity Exchange and local markets have been used to estimate the economic impact from yield changes of non-traded crops. Of course, prices on world markets reflect costs (or benefits) more accurately as they should be free of national subsidies. Crops were valued at the following prices (ECU/tonne):

Winter wheat	107
Spring wheat	160
Potato	210
Tomato	200
Apple	406

Sources: FAO (July 1995), Bologna Board of Trade, ISTAT (1993).

Table 3.4 presents estimates of the physical losses attributable to SO_2 and Table 3.5 coverts those losses into valuations. The figures represent 1992 prices; see Synthesis report for updated values.

3.6. Conclusions

Results show significant differences in the estimates obtained with the applications of dose–response relationships transferred from the international literature as opposed to those derived from regression analysis of open field experimental data. For example, in the case of winter and spring wheat, yield changes due to SO_2 pollution are estimated to be between 1% and 2.4% (for winter wheat) and between 1.1% and 3.6% (for spring wheat) depending on which dose–response model is considered. In the case of potato, tomato and apple, only one dose–response model was used. Estimated yield variations are equal to 0.9%, 1% and 1.1% of the total national yield respectively.

To test the dependence of the results on the background concentrations, the baseline SO_2 concentration has been varied in the Central and Southern areas of the country according to the methodology described in Chapter 1. Variation of estimated impacts appears to be small. On the whole, the estimated economic damage from SO_2 pollution on the selected agricultural crops ranges from 0.0004% to 0.003% of 1990 GDP. The above estimates are subject to a number of caveats, some of which have already been mentioned. They can be summarised as follows.

i. The use of spatially resolved data on ambient air concentrations of SO_2 that were derived from monitoring stations mainly located in highly polluted areas. This implies that the calculated spatially resolved concentrations are likely to be above the real ambient concentrations in rural areas. This is likely to result in an *overestimation* of SO_2 impacts on the selected agricultural crops.
ii. Incomplete geographical coverage. Due to the lack of reliable data on pollution concentrations in some Central and Southern areas of the country, SO_2 damage to agriculture has not been estimated for three administrative regions: Abruzzo, Marche and Molise. This results in an *underestimation* of the impact on most selected crops at the national level (except

Table 3.5. Estimated crop damages for Italy due to SO_2 (ECU million)

Dose–response relationship	Winter wheat	Spring wheat	Potato	Tomato	Apple
Baker *et al.* (1986)	11.8	22.3	–	–	–
Weigel *et al.* (1990)	13.3	25.1	–	–	–
Cirillo *et al.* (1988)	6.1	7.9	–	–	–
Camposano *et al.* (1986)	–	–	4.0	10.9	8.0

winter wheat of which 16.7% of total production is from these areas).

iii. The use of dose–response functions which have been either transferred from the international literature or derived through regression analysis of open field experimental data, which implies a higher degree of *uncertainty* in the estimates.

iv. The use of market prices for the economic valuation of the impacts. The agricultural sector is subsidised and agricultural prices cannot be looked upon as efficiency prices. Both prices and production levels are artificially high and crop losses calculated with these prices are probably an *overestimation* of real losses. It is therefore questionable whether the crop losses incurred by the farmer are entirely external costs. Future calculations should be based on shadow prices. World market prices may be regarded as first approximation for shadow prices.

The analysis of SO_2 impacts has been selective and general conclusions on damages can not be drawn from these figures. Quality changes, mitigating measures of the farmers, indirect effects and other interactions have also been left out. The damage data given above can therefore only be seen as the *underestimate* of the total SO_2 damage to the agricultural sector in Italy.

CHAPTER 4

ECOSYSTEMS

4.1. Introduction

The environmental degradation of terrestrial ecosystems deriving from air pollution can take several forms and follow different impact pathways that have been described in the literature. The dose–response mechanism for ecological impacts is particularly complex with varying responses to pollutants and considerable uncertainty over critical loads. Valuation endpoints are, therefore, difficult to establish and even if conducted would be somewhat questionable. The present study has focused on highlighting the land areas in which habitats and ecosystems are considered to have a particular ecological value. Therefore, physical indicators on the state of national natural ecosystems are reported.

An assessment of major impact categories on these pristine areas is also attempted, with particular reference to threats to the alpine ecosystem, acidification of lakes and other natural reserves, the status of endangered species, and threats related to land use impacts. The assessment of ecosystems impacts from global climate change is beyond the scope of the present study and, therefore, particular potential threats of this nature are only mentioned and appropriately referenced.

4.2. Preservation of natural ecosystems

The share of national territory under environmental preservation has increased in recent years from a level – well below European standards – of 4.3% in 1988 to the reported 8.2% in 1991 (Ministry of Environment, 1993). These protected areas, which total approximately 2.6 million ha, are distributed among 17 national parks, 147 national natural reserves, 247 regional natural parks and reserves, 34 wetlands, and 5 marine parks (Ministry of Environment, 1992; WWF, 1994). In addition, there are a number of wilderness reserves which are privately owned and preserved. Most of them are the WWF-Italy owned and/or managed reserves (63 areas) totaling 25,000 Ha, and 15 areas totaling 2,500 Ha managed by LIPU (Italian League for Bird Protection). The regional distribution of protected natural areas is detailed in Table 4.1.

Table 4.1. Regional distribution of protected natural areas (1992)

Region	Areas (Ha)	% of land	Ha/100 Inhabitants
Piemonte	163921	6.3	3.8
Valle D'aosta	41255	12.6	35.6
Lombardia	503485	21.1	5.7
Trentino A.a.	439522	21.3	34.4
Veneto	77173	4.2	1.8
Friuli V. g.	45172	5.8	3.8
Liguria	62379	11.5	3.7
Emilia-Romagna	131125	5.9	3.4
Toscana	123075	5.4	3.5
Umbria	17957	2.1	2.2
Marche	61459	6.3	4.3
Lazio	97981	5.7	1.9
Abruzzo	106153	9.8	8.5
Molise	5606	1.3	1.7
Campania	2222	0.2	–
Puglia	9788	0.5	0.2
Basilicata	95432	9.6	15.6
Calabria	127090	8.4	6.1
Sicilia	330516	12.9	6.7
Sardegna	14147	0.6	0.9

Source: Ministry of Environment (1993).

4.2.1. *The national law on protected areas No 394/1991*

The implementation of the new Law on Protected Areas No 394/1991 will raise to 10% the percentage of national territory under environmental protection and gives birth to a profound change toward a thorough land planning of natural ecosystems on the national territory. According to this law, the territory within the national parks and the other preserved natural areas is submitted to a series of gradual constraints, thereby harmonising the conservation of natural ecosystems with the economic activities and anthropogenic effects. This reorientation of economic activities toward more environmentally compatible activities is regulated and managed with the introduction of a zoning system.

The territory of a protected area is subdivided into four categories according to varying degrees of ecosystem conservation: Class A areas are Integral Reserves, in which ecosystems preservation is complete; Class B areas are Oriented General Reserves, where the territory cannot be transformed (i.e., no building activity) but some human management of the natural environment is permitted; Class C areas are Protection areas, in which many traditional and low-impact economic activities are stimulated; and Class D areas, Areas of Social and Economic Promotion, are mostly subject to anthropogenic modification (i.e. tourist infrastructure).

4.2.2. *National financing for protected areas*

A Triennial Programme for Protected Natural Areas has been instituted which defines the entire public expenditure for the environment. Criteria exist for the creation, enlargement and management of natural ecosystem areas and for the distribution of financial resources. The first Programme (1991–93) was assigned approximately ECU 91 million to be divided between direct transfers to the regional authorities (ECU 43 million) and funds directed to national parks (ECU 48 million). The financial resources assigned to regionally protected areas are distributed among the regions according to a set of criteria that take account of population, land area and the number of protected areas in a jurisdiction. The public financing for National Parks has been allocated among 12 parks: Arcipelago toscano, Aspromonte, Cilento, Dolomiti bellunesi, Foreste casentinesi, Gargano, Gran Sasso-Monti della Laga, Maiella, Monti Sibillini, Pollino, Val Grande, Vesuvio.

It should be noted that there is a substantial discrepancy between the public financial resources available for natural ecosystem management and the limited level of public money actually being spent. Procedural delays and difficulties in the implementation of the law (many national parks still have to form a managing entity) reduce the capacity of investing the majority of these funds (Renzi, 1994). For those national parks that have an established organisational structure, paradoxically, the funding is still modest: both the Abruzzi National Park and the Gran Paradiso National Park receive only around LIT 5.0 billion annually as ordinary expenses contribution.

It is possible to view the public funds set aside for natural ecosystems areas as a very rough proxy for the aggregated *willingness to pay* in Italy for ecosystems protection, as expressed by the national level public authorities.

4.2.3. *Other expenses for natural areas management*

In 1990 the Province of Trento commenced a project to promote 'green' employment and more effective natural resources preservation. Since then, the equivalent of ECU 15 million has been spent annually with an investment of about ECU 25,400 per new employment opportunity (WWF, 1994b). If these figures are considered in the context of the 82,500 Ha of preserved land in the Province (as of 1991), it translates into ECU 183 ECU per Ha of preserved land.

Another example of funds spent for natural ecosystem preservation is given by the 62 WWF-Italy owned and/or managed natural reserves. For the fiscal year 1993, the balance sheet of these WWF reserves shows a total cost of ECU 1.65 million of which about 78% are operating costs and 22% are investment costs (WWF, 1994b). If the WWF management of ecosystems had to be taken as an example of natural resources preservation cost (for areas of limited extension), this is equivalent to ECU 66 per Ha of preserved land.

4.3. Environmental impact

4.3.1. *The alpine ecosystem*

The environment of the Alps has been under major pressure due to the increasing anthropogenic presence and activities. One of the major ecosystem elements under stress is the forest vegetation of these mountains. In addition to impacts related to tourism and logging, air pollution has affected alpine forests. While the scientific debate is still open on the determination of clear dose–response damage correlations, the decline is clearly related to acid depositions and is manifested in the foliar decay, growth rate reduction and habitat area reduction. The effects of acid depositions seem to be especially powerful in high mountain forests, because the vegetation is immersed in the clouds during part of the growth season. Table 4.2 shows the existing prevailing vegetation species for a range of Alpine locations and the concentrations of pollutants in the wet depositions that can contribute to damage from acidification. There usually is a level of natural acidity in rainfall – with the parameter pH at a level of 5.6 – but the table shows how in all locations this benchmark is passed (lower levels mean more acidity on a logarithmic scale) possibly due to the influence of SO_x and NO_x of anthropogenic origin.

During the 1980s, some of the Regions of the Alpine area (Friuli V.G., Trentino A.A., Veneto) started monitoring the damage of acid rain on forests. The Trentino-Alto Adige Region, for instance, has found that, since the 80s, the percentage of plants damaged is between 5 and 10% for the medium/high class of damages (Conte, Melandri, 1994).

Eutrophication processes are creating environmental pressure on another important natural resource of the Alpine region, namely the great lakes of Como, Iseo and Maggiore. The Lake of Garda presents oligotrophic conditions

Table 4.2. Vegetation and concentrations in the wet depositions in the Alpine area – year 1990 (SO_4 and NO_3 are expressed in meql^{-1})

Alpine location	Prevailing vegetation	pH	SO_4	NO_3
Oulx (TO)	*Vaccinio-Piceetalia, Seslerietalia variae, Elynetalia*	5.22	30	19
Sestriere (TO)	*Vaccinio-Piceetalia, Seslerietalia variae, Elynetalia*	4.98	57	30
Domodossola (NO)	*Arrhenathere-talia*	4.33	68	46
Lunecco (NO)	*Caricetalia curvulae, Vaccinio-Piceetalia*	4.29	55	58
Alpe Gera (SO)	*Caricetalia curvulae*	5.29	42	24
Madesimo (SO)	*Caricetalia curvulae*	4.51	54	33
Lago D'arno (BR)	Fagetalia sylvaticae, Eu-Fagion	5.04	54	29
Malga Gallina (BZ)	*Vaccinio-Piceetalia, Seslerietalia variae*	5.00	47	35
Renon (BZ)	*Pinus sylvestris*	4.81	26	26
Pian Cansiglio	*Fagetalia sylvaticae, Eu-Fagion*	5.04	62	32
Tarvisio (UD)	*Fagetalia sylvaticae, Eu-Fagion, Vaccinio-Piceetalia*	4.67	47	26

Source: Ministry of Environment (CNR 1993).

and it does not seem to show an increasing trend of nutrient concentrations. Especially for the lakes of Iseo and Como, the data show a trend of increasing concentrations of nutrients. Lake Maggiore presents similar trophic conditions due to the high level of phosphates, but the trend of nitrates concentration shows that they decline to oligotrophic conditions (< 10).

The Alpine ecosystem shows sufferance in its rich and unique flora and fauna. These biodiversity impacts have been denounced by many authors. A recent study by the Alp Action and IIASA (Nilsson and Pitt, 1991) reports that several Alpine wildlife species are at risk – among them, the edelweiss flower, the yellow saxifrage, Apollo butterfly, the Alpine ptarmigan, the viparous lizard, the black salamander, the show campagnol, the marmot and the Siberian cricket – not only due to pressure for human incursions but also due to air pollution and global warming threats. A survey conducted in the area of two Alpine lakes, Caldonazzo and Levico (Trentino Alto Adige), concluded that 34 plant species out of 153 disappeared, with a 22.2% loss of biodiversity (Pedrotti, 1990).

4.3.2. *Acidification of lakes*

The Thermal and Nuclear Research Centre of the Italian Electricity Board (ENEL) has been involved, in collaboration with the Italian Hydrobiological Institute of the C.N.R., in a study concerning both the sensitivity and the acidification of lakes, with particular reference to alpine and subalpine lakes. The research effort is still going on, but some preliminary results were released in 1988. The database regards 648 lakes which are mainly alpine and subalpine lakes (634). The remaining 14 are distinguished in 7 Appennino mountains lakes, 3 volcanic lakes, 2 coastal lakes and 2 artificial lakes (in Sardinia). The results presented show that freshwater acidification does not appear to be a serious problem for most of the Italian lakes. The alpine lakes present a higher sensitivity to acid deposition than the subalpine and other lakes due to both their limited dimension and the particular geological conditions. But only 0.8% of the monitored alpine lakes have a pH level below 5.0. Similar conclusions on the generally good state of Italian lakes as far as acid deposition is concerned, have been drawn by more recent studies (Mosello, 1993).

4.3.3. *The endangered species*

The endangered flora

Italy is a country with a very rich and particular flora, 5599 species according to a recent census and 732 endemic species that are exclusive of the Italian territory (WWF-Italy, 1992). This particularly high biodiversity can be explained by the geographic position and the variety of environmental conditions. Following the IUCN plant red data book, there are four endemic species particularly at risk: the *Androsace brevis*, the *Cytisus emeriflorus*, the *Saxifraga florulenta* and the *Primula palinuri*. More recently, the status of the Italian flora has been studied and summarised by a joint scientific effort of WWF-Italy and the Italian Botanic Society (WWF-Italy, 1992). This effort produced a red list of plants containing 458 species, equal to 8.2% of the entire Italian flora. The breakdown was as follows:

Extinct species	15
Highly vulnerable	82
Vulnerable	179
Rare species	178
Other species	4

The endangered fauna

On the basis of a recent regulation (Law 157/1992), the National Institute for Wild Fauna has to perform a census of the wind animals population and study its status and evolution. The existing data gathered by the Ministry of Environment from the above cited and other institutions are presented in Table 4.3 below. Some estimates are available (Conte, Melandri, 1993; Ministry of Environment, 1992) on the residual population of few wild species highly impacted by both pollution and human pressure on their habitat.

These mammals (Table 4.4) are species that can be considered good indicators of the environmental conditions of natural ecosystems. They suffer from anthropogenic pressure on their habitat and they need large pristine natural areas to feed and reproduce. The Italian wolf (*Canis Lupus*) numbers only 300–400 with a distribution along the entire Appennino mountain chain while the brown bear that once populated the Appennino mountains (*Ursus arctos marsicanus*) is today only present in the Abruzzo National Park with numbers of around 80–90.

Finally, a recent effort by the WWF-Italy published as Red Book of Italian Butterflies (Prola, 1990), has highlighted the status of these Lepidoptera which can be an indicator of ecosystems health. The study indicates that 13 species

Table 4.3. Census of endangered animal species

	Threatened species	Vulnerable species	Rare species	Undetermined species
Mammals	4	7	5	27
Birds	18	25	33	50
Reptiles	6	2	2	–
Amphibious	1	3	3	18
Freshwater fish	10	8	3	–
Total	39	45	46	95

Source: ISTAT, 1993.

Table 4.4. Estimated population of some rare mammals

Species	Numbers
Bear (*Ursus Marsicanus*)	80–90
Abruzzo Chamois (*Rupicapra rupicapra*)	400
Alpine Chamois (*Rupicapra pyrenaica ornata*)	340000
Sardinian Deer (*Cervus elaphus corsicanus*)	150
Steinbock (*Capra ibex*)	20000–25000
Wolf (*canis lupus*)	300–400

Source: Conte, Melandri (1993), Ministry of Environment (1992).

of diurnal butterflies are close to extinction and 32 species are endangered, while among the nocturnal butterflies 6 species are in close danger of extinction.

4.4. Land use impacts

The Ministry of Environment has produced a mapping of the areas "at high risk of environmental crisis," with the goal of planning a series of land reclamation and preservation programs. Table 4.5 presents the major information about these areas, their population and the major sources of environmental risk. They represent about 5.7% of the national territory, with 11 million population (about 20% of the national population).

Some of these areas at risk are close to important and vulnerable ecosystems and therefore contribute to their degradation. Other areas, such as the Po

Table 4.5. Areas at high risk of environmental crisis

Land areas	km^2	Population	Environmental risk source
Lambro-olona-seveso	3336	4857963	Industrial and urban pollution
Val Bormida	1562	191937	Chemical and paint industry
Po Polesine	1962	264020	Small industries and zootechny
Burana-po Di Volano	2596	386953	Ceramic industry and zootechny
Conoidi	3556	1146453	Ceramic industry and zootechny
Massa Carrara	n.a.	131232	Area Law 195/1991
Manfredonia	n.a.	58157	Area Law 195/1991
Sarno	378	418818	Industrial and urban pollution
Province of Naples	1141	2968082	Industrial and urban pollution
Brindisi	548	126949	Energy plants and chemicals, oil
Taranto	505	279977	Energy and cement, refineries
Crotone	167	55635	Industrial pollution
Priolo-augusta	569	209517	Petrochemical industry and shipping
Gela	676	104783	Petrochemical industry
Portoscuso	383	62835	Aluminum, lead and zinc industries

Source: Ministry of Environment, 1992.

valley areas with high intensity of zootechnic production and industry have a potentially high impact on soils, groundwater and ultimately on the ecosystems of the Po river estuary and on the sea environment. In several of the industrial sites, the risk includes the accident event with effects on human health and on the surrounding land. The accident at Seveso in 1976 where substantial dioxins were released is estimated to have resulted in ECU 128 million for third-party damage and off-site soil contamination damages (OECD, 1994).

4.5. Forest ecosystems damage

At the national level, the Ministry of Agriculture & Forestry has launched a program for a wide collection of survey data on forest decay in the recent years (Indefo, 1988). The database is build every year on a cluster of about 160,000 trees and focuses on forest damage related to the following causes: (a) Fires; (b) Parasites; (c) Climatic phenomena; (d) Other causes (which include air pollution impacts). This source signalled the damage correlated to air pollution and acid rain to be present on 10% of the Italian forest patrimony. Since 1989, though, the program started to survey forest damage by class of decay and not by cause, due to the high level of approximation in the attribution of impacts. The total percentage of trees damaged (foliar decay > of 10%), summing the different classes, is equal to 41.8%. The coniferous species seem to have suffered slightly more than the other species as far as medium and high class damages.

Estimates from a different source, the European Monitoring Evaluation Programme 1990, give a range of percent of forest damage between 20 and 51% (foliar decay > of 10%), but also in this case no attribution of damage was attempted.

CHAPTER 5

MATERIALS

5.1. Introduction

The effects of air pollution on building materials are qualitatively well documented. Damage related to air pollution includes: discoloration, failure of protective coatings, material loss and structural failure. The rates at which these types of damages occur have increased significantly over the period since the Industrial Revolution and pollution is the only possible cause. Major impacts are due to acid deposition – both in the form of direct effects of sulphur dioxide and the effects of acidity resulting from both SO_2 and NO_x emissions – and to the effect of particulates in the form of soiling (i.e. discoloration of stone and brickwork.). The direct effects of NO_x and the impact of ozone are believed to be of secondary importance.

The literature on pollution related materials damage relies both on field studies on real buildings and upon studies on idealised test material either in the field or laboratory experiments. In general, experiments on laboratory samples allow better control for individual pollutants in a multi-pollutant environment. On the other hand building materials rarely correspond to laboratory samples in composition and are subject to ambient conditions which differ from those in controlled laboratory studies. Therefore, a combination of laboratory experiments and field studies is required to give reliable information on the relationship between pollutants and impacts.

5.2. Methodological approaches

In common with the impact assessment made for other receptors, the reference point has been the methodological approach designed and implemented under the EXTERNE Project. In addition, an alternative approach developed by IES in the framework of the Dutch Acidification Systems Model (DAS) has been considered.

5.2.1. *The EXTERNE approach*

The EXTERNE study concentrated on acid corrosion damage and on a limited assessment of cleaning costs due to particulate emissions. The analysis is confined to the components of buildings, in so far as they comprise the largest quantity of materials exposed directly to the atmosphere. Since the surfaces of different materials have very different susceptibilities to acid attacks, it is necessary to consider each material separately. Stone, mortar, concrete, paint, steel, zinc and aluminium cover the major building materials in Europe. Hence, the damage assessment has been undertaken starting from these components. Other common building materials such as brick and slate are believed to be relatively insensitive to acid pollution and have not been considered.

The methodological approach includes the following steps:

(a) identification and quantification of the stock at risk
(b) identification of reliable dose–response functions for the building materials under analysis
(c) estimation of the impacts
(d) estimation of the damages

Stock at risk

The materials at risk are quantified in terms of exposed area. In order to calculate the aggregate area, data is required on the *types* of buildings in the country, their *number*, average size and *structure* (the so-called 'building identikits'). Separate estimates should be made for each of the major building categories considered in the analysis. Spatially disaggregated data on the number of buildings in each category are required in order to know the distribution of building and construction materials in the country. Once these data have been collected, a *materials inventory* for the country can be calculated by multiplying the number of each type of buildings in each area by the average area of each material per individual building, as provided by the building identikits. Respective areas for the various building types have then to be added together in order to obtain the total surface area of each type of material for each zone.

Dose–response functions and datasets

The dose–response functions applied were identified by literature review. These are derived through a combination of controlled studies and field measurements. Most of them assume that responses are linear with respect to both time and to the concentration of individual atmospheric pollutants. Table 5.1 gives an overview of the functions used in the study per type of building materials considered: calcareous stone, mortar, concrete, (carbonate based) paint, steel, zinc and aluminium.

where:

ER = erosion rate (in m per year)
P = precipitation rate in m/year
SO_2 = sulphur dioxide atmospheric concentration in g/m^3
H^+ = acidity in meq/m^2/year

Table 5.1. Dose–response functions for damage due to acid pollutants

Material	Dose–response function	Source
Calcareous stone	$ER = 18.8 \cdot P + 0.052 \cdot SO_2 + 0.016 \cdot H^+$	Lipfert, 1989
	$ER = 2.56 + 5.1 \cdot P + 0.32SO_2 + 0.083 \cdot H^+$	Butlin, 1992a
Mortar/concrete	$ER = 56.4 \cdot P + 0.162 \cdot SO_2 + 0.048 \cdot H^+$	Lipfert, 1989
	$ER = 7.68 + 15.3 \cdot P + 0.96 \cdot SO_2 + 0.249 \cdot H^+$	Butlin, 1992a
Carbonate-based paint	$ER = 15.8 \cdot f + 0.119 \cdot f \cdot SO_2 + 0.017 \cdot H^+$	Haynie,1986
Steel	$M = \exp(5.74 - 0.15P) \cdot (f \cdot t)^{0.54} \cdot (f \cdot SO_2)^{0.26} \cdot (H^+)^{0.018} \cdot (Cl^-)^{0.13} \cdot D^{0.022}$	Lipfert, 1987
Zinc	$M = (t^{0.78} + 0.46 \log_e(H^+)) \cdot (4.24 + 0.55 \cdot f \cdot SO_2 + 0.029 \cdot Cl^- + 0.029 \cdot H^+)$	Lipfert, 1987
Aluminium	$M = 0.2 \cdot t^{0.99} \cdot X^{0.88}$	Lipfert, 1987

f = the 'time of wetness' defined by $f = 1 - \exp(-0.121R_H * (100 - R_H))$
R_H = percentage average relative humidity
M = corrosion rate in g/m^2
f = fraction of time relative humidity of 90%, 85% or 80% is exceeded respectively in the case of steel, zinc and aluminium
t = time in years
Cl^- = chloride deposition rate in $mg/m^2/day$
D = dust concentration in $mg/m^2/day$

Estimated impacts

Most of the above relationships give the response in terms of weight or thickness loss as a function of time. In order to calculate measured impacts on buildings, information on the level of material loss at which repair or replacement is required has to be collected (i.e. repair/replacement frequencies for each types of materials). This was done using expert judgements as well as common experience on building maintenance practice.

Valuation method and estimation of economic damage

Ideally, the valuation of the impact should be made on the basis of the willingness-to-pay of people to avoid the damage due to anthropogenic emissions. However, no assessments of this type are available. Instead, the study used repair and replacement costs of building components as a proxy for economic damage due to acid deposition (assuming rational behaviour). In the case of soiling damage due to the deposition of particulates on external surfaces, the cost of cleaning soiled buildings is used as a lower bound for the estimate of the economic impact. The assumption of homogeneous repair/replacement time distribution for the building stock at risk has to be made. This is equivalent to assuming that the building stock has a homogeneous age distribution. The frequency of repair/replacement is then an adequate basis for valuation with costs assumed to occur in the year of emission. Specific unit cost factors for each of the materials were calculated.

Multiplying the frequency of repair/replacement and the total area of each material in each zone for these unit cost factors, one obtains the repair/replacement costs. The total economic impact in each zone is then given by the difference between the repair/replacement costs with and without anthropogenic emissions.

5.2.2. *The DAS model approach*

An alternative approach has been developed in the framework of the Dutch Acidification Systems Model in order to assess materials damage due to SO_2 air pollution in the Netherlands. A brief description of the methodology is given hereafter in order to identify the information requirements for its implementation in Italy. For a more complete discussion of the method the reader is referred to Gosseling, Olsthoorn and Feenstra (1990) and Kuik (1994). The model includes impacts on painted steel, galvanised steel, duplex steel, galvanised and painted steel, sheet zinc and natural stone in monuments. The damage assessment is performed based on physical and economic dose–response functions. The physical dose–response functions relate physical impacts on materials (expressed in terms of erosion rate (ER), weight loss (L) or loss in thickness (V)) to SO_2 pollution levels and are of the form:

$$ER = f\,(SO_2 \text{ concentration})$$
$$L = g\,(SO_2 \text{ dry deposition})$$
$$V = h\,(SO_2 \text{ dry deposition})$$

The economic dose–response functions relate economic damage (in the form of increased replacement or maintenance) to SO_2 concentration levels. Damage due to increased *replacement* occurs if the technical life of the material under analysis falls short of its economic life. The economic valuation of the damage is based on the replacement value of the material. Damage due to increased *maintenance* is expressed as a function of the economic life of the material, its service period and the SO_2 concentration. The physical damage is valued using the maintenance costs per unit.

The DAS Model provides information on the type of air pollution impact as well as some economic parameters for selected types of materials and building objects (i.e. economic life, critical loss of thickness/weight, costs). The above physical and economic dose–response functions are then applied to the stock of materials exposed. In the DAS Model, the stock at risk is derived thorough analysis of the material structure of each of the selected building objects, their economic life span and the application of exposition coefficients (Kuik, 1994).

5.2.3. *Feasibility of the damage analysis for Italy*

The applicability of the above methodologies crucially depends on the availability of data on the stock of building material affected in the geographic area under analysis: types and number of building per type, structure of buildings, costs of repair/replacement/maintenance. In Italy no building identikits of the type available for the UK, Germany or the Netherlands have been developed so far. In addition, there are no suitable ready-to-use sources of data on buildings distribution covering the whole country and information on construction activities are only available on an incremental basis (i.e. construction rates for very broad categories of buildings).

In the absence of such data one option could have been to extrapolate available information for other countries to Italy. The possibility to apply available identikits to Italy has been considered by the authors and discussed with national experts (architects, engineers). The general conclusion was that the assumptions necessary to overcome the lack of quantitative information for Italy would, on the whole, have been too restrictive and would have introduced considerable uncertainties in the estimates. If, on the one hand, it should be reasonable to assume the distribution of buildings to be similar in countries at relatively similar levels of economic development (adjusted for population density), on the other hand, it was argued that building materials vary quite significantly across countries, regions and areas (i.e. urban versus rural).

Consequently, it was considered that identikits developed for Northern European cities cannot be taken to be representative of Italy. Given this, and since acceptable identikits for various types of buildings in Italy could not be generated within the time frame and financial resources of the project, no work has been done on the economic damage caused by acid pollutants to building materials in the country through the application of dose–response functions.

Instead, the few existing national studies on the impact of air pollution on materials have been reviewed. These studies date back to the 1970s; an extrapolation of their results to the year 1988 has been carried out. The result is clearly only indicative; estimates obtained have to be considered as order of magnitude figures, subject to considerable uncertainty. They are reported only for completeness.

5.3. Air pollution and building materials in Italy

The only existing study attempting an estimate of air pollution damage to building materials in Italy was carried out by ISVET (1970). Damage to residential, commercial and industrial buildings was assessed, based expert judgement and extrapolation of findings for other countries. Based on the above work, Cullino (1992) estimated defensive expenditures of the household sector aiming at reducing material decay in buildings due to air pollution in the period 1986–1988. The definition of defensive expenditure adopted in the study is derived from the UN classification (UNSO, 1990). Estimates obtained are regarded as an approximation of the environmental damage (Maler, 1990). They are based on previous findings duly adjusted for population density in polluted areas. This figure is estimated to be equal to about 25.8% on the basis of the distribution of emissions of SO_x and TSP per administrative provinces (Ministry of the Environment, 1992). Cleaning and re-painting costs were the main items considered. The amount of money spent on cleaning and repairing was estimated at LIT 162 billion in 1986, LIT 397 billion in 1987, and LIT 440 billion in 1988. These are equivalent to around 0.02% to 0.05% of GDP respectively. We would like to stress once again that, because of the methodologies for evaluation adopted in the study, the values obtained are merely indicative. They provide a rough first evaluation of the phenomenon and constitutes a stimulus for further research, based on more recent data and methodological approaches.

5.3.1. *Air pollution and cultural heritage*

Italy has an immense cultural heritage to safeguard, much of which is potentially affected by air pollution. Many buildings and monuments were constructed with materials such as natural stones, marble and bronze which are sensitive to acid pollution attack. Museum and art galleries face different problems of indoor air pollution. Action is being taken on a number of fronts, including research into the effects of air pollution and into possible remedies and restoration work. The great part of the ongoing work focuses on the study of the *physical* impacts of both outdoor and indoor pollution on monuments and buildings as well as painting, frescoes, tapestries etc. A survey of the whole of Italy is planned to provide a map at commune level of the distribution of SO_2 and total suspended particulates. These data will then be integrated with collected information about buildings and monuments at risk to help set priorities for future action. Such datasets would be a valuable source of information for future research effort on the subject.

The valuation of air pollution damage to cultural buildings and art works is far more problematic than for non-historic buildings because it is highly dependent on the cultural significance of the object in question. Ideally willingness-to-pay data for cultural heritage should be collected. Since this kind of data are not available, a valuation of the damages on this basis is not possible. Alternatively, replacement and maintenance costs could be used. These are probably the easiest to evaluate but are subject to two major limits. In the first instance, restoration expenditures are dependent on the amount of money which is available for these activities (i.e. the budget for restoration). Secondly, they do not adequately reflect the loss of aesthetic and cultural values. For historic buildings and cultural monuments the aesthetic value may exceed the use value of the buildings by a large factor and may be more severely affected by air pollution damage. Estimation of the effects of air pollution on cultural values is however problematic because the associated costs are extremely site-specific, both in terms of the merit of the item under consideration, and in the way in which it can be treated.

For these reasons, together with the impossibility of collecting suitable and reliable data on restoration expenditures for the whole country within the time frame and resources available of the project, no assessment of damage to historic monuments and buildings has been completed.

5.3.2. *Future research work*

The above discussion highlights the data and information gaps that need to be filled in order to be able to produce reliable assessments of damage to building materials. With regard to physical data on the stock at risk, a materials inventory detailing the type and structure of buildings in the country, together with information on the number and distribution of the various types of building over the national area needs to be collected.

As far as data on specific unit repair/replacement costs for each material, the work done in the framework of this study has allowed us to locate a number of suitable sources of information. Finally, an in-depth survey of available information on expenditures on cleaning and repair both at the public as well as at the private level would allow a lower bound figure to be obtained. Suitable sources of information exist at a spatially disaggregated level although they are rather disparate. Ongoing studies on historic buildings and monuments at risk due to air pollution impacts will produce valuable information on buildings and monument at risk over the national territory which will constitute an important basis for future work in the area.

CHAPTER 6

CONCLUSIONS

The present study is part of a wider project on Green Accounting. The overall objectives of the project are to see whether there is a consistent method for preparing monetary valuation of environmental damages caused by economic activities that can be applied across the community and to use such a methodology to make some preliminary estimates of national level damages. The purpose of this report has been to discuss the issues arising and results obtained for Italy.

While the quantitative results of the research have been presented and discussed in detail in the previous chapters, the aim of this final chapter is to summarise the major difficulties and obstacles in carrying out environmental damage assessment and valuation in Italy and to derive some implications for future research work. The implementation in Italy of the methodological framework developed within the project has been hampered by several obstacles.

A primary issue is the lack of country specific studies on the relationships between air pollution and impacts on the various receptors. The national literature on this subject is very poor compared to other developed countries. This implies the need to rely on the transfer of dose–response functions that have been developed in other national contexts. The extent to which this is feasible and correct is debatable. Significant spatial as well as temporal correlations exist between various air pollutants. The impact on the various receptors is the result of the specific pollution mixture to which it is exposed.

This is particularly true in the case of health and crop damages. In addition, other interacting variables are important: climatic variables, cultural factors, the social system governing health care, etc. In transferring some of the dose–response functions we have taken this issue into account to the extent possible, thereby selecting the dose–response models developed in contexts more 'similar' to the Italian case. However, we have assumed that the Italian population and agricultural crops are exposed, throughout the entire national territory, to the same air pollution mixtures, climatic and other conditions as in the original studies. This may of course not be true, particularly when considering the urban versus rural context, and such extrapolation is likely to result in some biases in the final estimates of the impacts.

Data on the relevant pollution and receptor variables have also proved difficult to obtain. The national coverage for ozone is very poor and no monitoring stations exist in some Central and Southern regions of the country for SO_2, NO_x and TSP. In addition, where monitoring sites do exist no classification of their location and no precise definition of the monitoring procedures are available. As far as the various receptors are concerned, there is the need to develop comprehensive databases describing both their state and their distribution over the national territory, at a sufficiently spatially disaggregated level. There is also a lack of information and data on willingness to pay for pollution reductions, which again makes it necessary to rely on the transfer of monetary values estimated in other contexts.

These country-specific obstacles add to the limitations and to the scientific and economic uncertainties surrounding the valuation of environmental goods and services and the application of the dose–response approach. In particular, the basic soundness of the original studies data, possible model mis-specification, use of inappropriate statistical methods, multiple comparison problems, use of linearized health functions without thresholds, additivity of different health outcomes etc. are all problems that need to be documented.

In order to follow the evolution of environmental policy-making at the EU and international level and to enhance the knowledge on environmental damage assessment in Italy it is therefore important to rapidly cover the existing data and information gaps, and to develop the scientific knowledge on the relationships between the various pollution phenomena and the different receptors at the country-specific level. The estimates obtained in this study have to be regarded as order of magnitude figures, whose reliability can be greatly improved throughout further research efforts both on the scientific aspects as well as on the evaluation side.

CHAPTER 7

REFERENCES

Baker, C.K. and Fullwood, A.E. (1986), Leaf damage following crop spraying in winter barley exposed to sulphur dioxide. Crop Protection, 5, 365–367.

Bates, D.V., Baker-Anderson, M. and Sizto, R. (1990). Asthma attack periodicity: a study of Hospital emergency visits in Vancouver. Environmental Research; 51: 51–70.

Berresheim, H. and Jaeschke, W., (1982). Sulfur emissions from volcanoes. In Chemistry of the Unpolluted and Polluted Troposphere (edited by H.W. Gergii and W. Jaeschke), pp. 325–337. reidel, Dortrecht.

Bocola, W., Cirillo, M.C. et al. Progetto CORINAIR. Inventario delle emissioni di inquinanti dell'aria in Italia nell'anno 1990, mimeo.

Bouscaren, R., Veldt, C., Zierock, K.H. and Lubert, B. (1987). CORINAIR Project, Programme CORINE, National Experts Meeting in Brussels, 7–8 May 1987.

Burnett, R.T., Dales, R.E., Razienne, M.E., Krewski, D., Summers, P.W., Roberts, G.R., Raad-Young, M., Dann, T. and Brook, J. (1994). Effect of low ambient levels of ozone and sulfates on the frequency of respiratory admissions to Ontario hospitals. Environmental Research; 65: 172–194.

Butlin, R.N. (1992) Preliminary Results form the Analysis of Stone Tablets from the National Materials Exposure Programme (NEP), Atmospheric Environment, 26B, 189.

Camposano, G., Cirillo, M. and Triolo, L., (1986), ENEA, Direzione Centrale Studi.

CEC (1993): Assessment of the External Costs of the Coal Fuel Cycle, Final Draft report. Commission of the European Communities, DG XII.

Centro per lo Studio della Teoria dei Sistemi – CNR Politecnico di Milano, (1993). Applicazione di un modello di trasporto a lunga distanza per la valutazione della correlazione tra acidita' delle piogge e magigori fonti di inquinamento. Convenzione ENEL-CNR, Sottoprogetto 3, mimeo.

Cirillo, M., Bocola, W., Triolo, L. (1988). Environmental impact of air pollutant with enphasis on crops. The Piacenza case study. In Marani, A. Advances in environmental modeling, ELSEVIER Amsterdam.

Cody, R.P., Weisel, C.P., Birnbaum, G. and Lloy, P.J. (1992). The effect of ozone associated with summertime photochemical smog on the frequency of asthma visits to hospital emergency departments. Environmental Research; 58: 184–194.

Conte, G. and Melandri, G. (Eds.) (1994), Ambiente Italia 1994, Legambiente, Koinè Edizioni, Rome.

COST 613/2 Committee on Air Pollution Epidemiology. (1992). Exposure assessment. Luxembourg: Commission of the European Communities. (COST Committee Report No. 1) (EUR 14345).

Cullino, R. (1992). Una stima delle spese difensive ed ambientali. FEEM Working Paper, n. 19.92.

Dockery, D.W. et al. (1989). Effects of Inhalable Particles on Respiratory Health on Children, Am. Rev. Respir. Dis., 139: 1989, pp. 587–594.

EC DGXII (1995a–f). Science, Research and Development Programme (JOULE). Externalities of Energy (EXTERNE) Project. Report numbers: EUR16520 (Summary), EUR16521 (Methodology), EUR16522 (Coal & Lignite), EUR16523 (Oil & Gas), EUR16524 (Nuclear), EUR16525 (Wind & Hydro).

Eliassen, A. and Saltbones, J. (1983). Modeling of long-range transport of sulphur over Europe: a two-year model run and some model experiments. Atmospheric Environment 19, 727–737.

Eliassen, A. (1978). The OECD study of long-range transport of air pollutants: long-range transport modelling. Atmospheric Environment 12, 479–487.

Eliassen, A., Hov, O., Iversen, J., Saltbones, J. and Simpson, D., (1988). Estimate of airborne transboundary transport of sulphur and nitrogen over Europe. EMEP/MSC-W Report 1/88, Norwegian Meteorological institute, Oslo.
ENEA, (1992). Dati di qualita' dell'aria in Italia, mimeo.
ENEL, (1993). Rete nazionale ENELMAF per lo studio delle deposizioni atmosferiche in Italia. Risultati dettagliati dell'attivita' svolta nel 1992, mimeo.
FAO (July 1995). FAO Quarterly Bulletin of Statistics. FAO, Rome.
Fisher, B.E.A. (1975). The long-range transport and deposition of sulphur dioxide. Atmospheric Environment 9, 1063–1070.
Fisher, B.E.A. (1978). The long-range transport and deposition of sulphur oxides. In J.O. Nriagu (Ed.) Sulphur in the environment, John Wiley and Sons, New York, pp. 243–296.
Freeman, A. Myrick (1993) The measurement of environmental and resource values: theory and methods, Resources for the Future, Washington D.C.
Gosseling, H.J., Olsthoorn, A.A. and Feenstra, J.F. (1990). Economische schade aan materialen en monumenten door verzuring. IVM/W-90/002, Vrije Universiteit, Amsterdam.
Guderian, R. (1977). Air pollution, phytotoxicity of acidic gases and its significance in air pollution control, in Ecological Studies, Vol 22.
Hatzakis, A., Katsouyanni, K., Kalandidi, A., Day, N. and Trichopoulos, D. (1986). Short-term effects of air pollution on mortality in Athens. International Journal of Epidemiology; 15: 73–81.
Haynie, F.H. et al. (1986). Effects of Gaseous Pollutants on Materials – A chamber Study. US Environmental Protection Agency Report EPA-600/3-76-015.
Holland, M. (1994). Assessment of the external costs of the natural gas fuel cycle, Draft report for the ExternE study, April 1994.
Houlden, G., McNeill, S., Aminu-Kano, M. and Bell, J.N.B. (1990). Air pollution and agricultural aphid pests. I. Fumigation experiments with SO_2 and NO_2. Environmental Pollution 67, 305–314.
Istituto Nazionale di Statistica (1991). 13° Censimento generale della popolazione, ISTAT – Istituto Centrale di Statistica, Roma.
Istituto Nazionale di Statistica, (1991). Statistiche ambientali, volume 2, ISTAT, Roma.
Istituto Nazionale di Statistica (1991b). Indagine statistica sulle condizioni di salute della popolazione e sul ricorso ai servizi sanitari, ISTAT – Istituto Centrale di Statistica, Roma.
Istituto Nazionale di Statistica (1991c). Statistiche dell'agricoltura, zootecnia e mezzi di produzione, ISTAT – Istituto Centrale di Statistica, Roma.
Istituto Nazionale di Statistica (1993). Statistiche ambientali, ISTAT, Roma.
Istituto Superiore di Sanità, (1989). Reti di rilevamento degli inquinanti atmosferici operanti in Italia al luglio 1986. Rapporti Istisan, edited by Dott. M.A. Bertolaccini.
ISVET (1970). Inquinamento e patrimonio dei beni culturali. Stima dei danni economici provocati dall'inquinamento al patrimonio nazionale dei beni culturali, documenti ISVET n. 29.
Iversen, J., Saltbones, J., Sandnes, H., Eliassen, A. and How, O. (1989). Airborne transboundary transport of sulphur and nitrogen over Europe – Model descriptions and calculations. EMEP/MSC-W – Report 2/89, Norwegian Meteorological institute, Oslo.
Katsoulis, B.D. and Whelpdale, D.M., (1990). Atmospheric sulfur and nitrogen budgets for South East Europe. Atmospheric Environment, Vol. 24A, No. 12, pp. 2959–2970.
Katsouyanni, K., Hatzakis, A., Kalandidi, A. et al. (1990a). Short-term effects of atmospheric pollution on mortality in Athens. Arch Hellen Med; 7: 126–132.
Katsouyanni, K., Karakhatsani, A., Messari, I., Touloumi, G., Hatzakis, A. and Kalandidi, A. (1990b). Air pollution and cause specific mortality in Athens. Journal of Epidemiology and Community Health; 44: 321–324.
Kinney, P.L., Ito, K., Thurston, G.D. (1994). Daily mortality and air pollution in Los Angeles: an update. (Abstract only). American Review of Respiratory and Critical Care Medicine; Supplement: p. A660. (Reference incomplete).
Krewitt (1994). Assessment of the External Costs of the Oil Fuels Cycle, Draft report for the ExternE study, January 1994.
Krupnick, A.J., Harrington, W., Ostro, B. (1990). Ambient Ozone and Acute Health Effects: Evidence from Daily Data, J. Econ. Manage, 18: 1990, pp. 1–18.
Kuik Onno (1994). Material Damage in the Dutch Acidification System Model, mimeo, prepared for the EC/US Joint study on External Costs of Fuel Cycles.
Lipfert, F.W. (1987). Effects of Acid Depositions on the Atmospheric Deterioration of Mateirlas, Material Performance 12, 12–18.

Lipfert, F.W. (1989). Atmospheric Damage to Calcareous Stones: Comparison and Reconciliation of Recent Experimental Findings. Atmospheric Environment 23, 415–429.

Mäler, K.-G. (1991). National Accounts and Environmental Resources. Environmental and Resource Economics Vol. 1, No. 1, pp. 1–16.

Ministero dell'Ambiente – Servizio Valutazione Impatto Ambientake, (1991). Rete Italiana per lo Studio delle Deposizioni Atmosferiche (RIDEP), Documenta dell'Istituto Italiano di Idrobiologia, vol. 43 pubblicazione n. 12, Consiglio Nazionale delle Ricerche – Istituto Italiano di Idrobiologia, Verbania Pallanza.

Ministero dell'Ambiente (1993). Relazione sullo Stato dell'Ambiente in Italia, Istituto Poligrafico e Zecca dello Stato, Roma.

Mosello, R. Il ruolo ambientale delle deposizioni atmosferiche, in Acqua- Aria, Vol. 6, 1993, pp. 636–643.

Mosello, R., Serrini, G., Muntau, H., Morselli, L. (1992). La rete italiana per lo studio della chimica delle deposizioni umide (RIDEP). I controlli di qualità. Proceedings of Convegno ANIPLA: 'La qualità dell'aria – Il controllo delle emissioni', Milano, 21–22 October 1992.

Nilsson, S., Pitt, D. (1991). Mountain World in Danger: Climate Change in the Forests and Mountains of Europe, Earthscan, IIASA-Bellerive Foundation.

OECD (1977). The OECD programme on long-range transport of air pollutants – Mesurements and findings. OECD, Paris.

OECD (1980). Agricultural crop damage. Aus: The costs and benefits of sulphur oxide control, Ch. 8; OECD, Paris (1980).

OECD (1994). OECD Environmental performance review. Italy. OECD, Paris.

Ostro, B.D., Rothschild, S. (1989). Air pollution and acute respiratory morbidity: an observational study of multiple pollutants. Environmental Research; 48: 238–247.

Ostro, B.D. (1987). Air Pollution and Morbidity Revisited: a Specification Test, J. Environ. Econ. Manage. 14: 1987, pp. 7–98.

Pedrotti, F. (1990). Stato dell'ambiente dei laghi di Caldonazzo e di Levico attraverso l'analisi della flora e della vegetazione, in Nuovo Giornale Botanico Italiano, n. 124 (1), 1990.

Pope, C.A., Schwartz, J. and Ransom, M.R. (1992). Daily mortality and PM10 pollution in Utah valley. Archives of Environmental Health; 47: 211–217.

Prola, G. and C. (1990). Libro Rosso delle Farfalle Italiane, Quaderni WWF, n. 13, WWF-Italy, Rome.

Renzi, F. (1995). League for the Environment, Personal Communication, Rome.

RGAR (1990). Acid Deposition in the United Kingdom. First Report of the Quality of Urban Air Review Group. Department of the Environment. London.

Rodhe, H. (1972). A study of the sulfur budget for the atmosphere over Northern Europe. Tellus 24, 128–138.

Samet, J.M., Speizer, F.E., Bishop, Y., Spengler, J.D. and Ferris, B.G. (1981). The relationship between air pollution and emergency room visits in an industrial community. Journal of Air Pollution Control Association; 31: 236–240.

Samet, J.M. (1991). Nitrogen dioxide. In: Samet, J.M., Spengler, J.D. (Eds.), Indoor air pollution A Health perspective. Baltimore: Johns Hopkins University Press: 170–186.

Schwartz, J. and Dockery, D.W. (1992a). Increased mortality in Philadelphia associated with daily air pollution concentrations. American Review of Respiratory Disease; 145: 600–604.

Schwartz, J. and Dockery, D.W. (1992b). Particulate air pollution and daily mortality in Steubenville. American Journal of Epidemiology; 135: 12–19. (Discussion and response: 135: 20–25).

Schwartz, J., Hasselblad, V. and Pitcher, H. (1988). Air pollution and morbidity: a further analysis of the Los Angeles student nurses data. Journal of the Air Pollution Control Association; 38: 158–162.

Schwartz, J., Slater, D., Larson, T.V., Pierson, W.E. and Koenig, J.Q. (1993). Particulate air pollution and hospital emergency room visits for asthma in Seattle. American Review of Respiratory Disease; 147: 826–831.

Schwartz, J., Spix, C., Wichmann, H.E., Malin, E. (1991). Air pollution and acute respiratory illness in five German communities. Environmental Research; 56(1): 1–14.

Schwartz, J., Wypij, D., Dockery, D.W., Ware, J., Zeger, S., Spengler, J.D. and Ferris, B.G. (1991). Daily diaries of respiratory symptoms and air pollution: methodological issues and results. Environmental Health Perspectives; 90: 181–187.

Schwartz, J. and Zeger, S. (1990). Passive smoking, air pollution, and acute respiratory symptoms in a diary study of student nurses. American Review of Respiratory Disease; 141: 62–67.
Schwartz, J. (1992). Air pollution and the duration of acute respiratory symptoms. Archives of Environmental Health; 47: 116–122.
Schwartz, J. (1993a). Air pollution and daily mortality in Birmingham, Alabhama. American Journal of Epidemiology; 137: 1136–1147.
Schwartz, J. (1993b). Particulate air pollution and chronic respiratory disease. Environmental Research; 62: 7–13.
Schwartz, J. (1994). Air pollution and hospital admissions for the elderly in Detroit, Michigan. American Review of Respiratory and Critical Care Medicine; 150: 648–655.
Sommerville, et al. (1990). Adequacy of interval estimates of yield changes to ozone estimated from NCLAN data. Crop Science, 30, 836–844.
Sunyer, J., Anto, J.M., Murillo, C. and Saez, M. (1991). Effects of urban air pollution on emergency room admissions for chronic obstructive pulmonary disease. With invited commenatry from M. Corn and authors' response. American Journal of Epidemiology; 134: 277–289.
Sunyer, J., Saez, M,, Murillo, C., Castellsague, J., Martínez, F. and Antó, J.M. (1993). Air pollution and emergency room admissions for chronic obstructive pulmonary disease: a 5-year study. American Journal of Epidemiology; 137: 701–705.
Taylor, H.J., Ashmore, M.R. and Bell, J.N.B. (1986). Air Pollution Injury to Vegetation (a guidance manual commissioned by H.M. Industrial Air Pollution Inspectorate of the Health and Safety Executive). IEHO, London.
Thurston, G.D., Kazuhiko, I., Hayes, C.G., Bates, D.V. and Lippmann, M. (1994). Respiratory Hospital Admissions and Summertime Haze Air Pollution in Toronto, Ontario: Consideration of the Role of Acid Aerosols. Environmental Research; 65: 271–290.
US-DOE (1992): Residual Damages and Benefits of the Coal Fuel Cycle: Estimation of Methods, Impacts and Values, Draft Report prepared for the US Department of Energy, May 1992.
Venkatram, A. (1985). Air quality model over long distances. In D.D. Hougthton (Ed.) Handbook of applied meteorology. John Wiley & Sons, New York, pp. 744–753.
Warrington, S. (1989). Ozone enhances the growth rate of cereal aphids. Agriculture Ecosystems and Environment, 26, 65–68.
Weigel, H.J. et al. (1990). Yield responses of different crop species to long-term fumigation with sulphur dioxide in open-top chambers. Env. Poll., 67, 15–28.
World Health Organization, Regional Office for Europe. (1987). Air quality guidelines for Europe. Copenhagen: WHO. (WHO Regional Publications, European Series No. 23).
WWF-International (1992). Can Nature Survive Global Warming?, WWF International Discussion Paper, Gland, February.
WWF-Italy (1994b). Dossier Economia & Parchi, Rome.
WWF-Italy (1994a). La Risorsa Parco, Rome.
WWF-Italy (1992). Libro Rosso delle Piante d'Italia, in collaboration with the Italian Botanic Society, Rome.

UK CASE STUDY

Prepared by AEA Technology* in association with Eyre Energy Environment, IOM.ITE and Metroeconomica, and in collaboration with the other country teams.

* AEA Technology PLC, Culham, Abingdon, OX14 3DB, UK. Tel.: +(44) 0 1235 463687; Fax.: +(44) 0 1235 463574; E-mail: mike.holland@aeat.co.uk

CHAPTER 1

INTRODUCTION

1.1. The need for the present study

Requirements for sustainable growth and environmental protection are included in the Treaty on the European Union, and hence are now a guiding principle for future development in Europe. National legislation reflecting these concerns has been passed in most European countries. The European Commission's White Paper on Growth, Competitiveness and Employment (COM 93, 700) highlights the importance of an integrated economic approach. Another recent paper produced by the Commission (COM 94, 670) states that there is a need for "*a harmonised European system of integrated economic and environmental indicators and accounts which addresses the problems of the various economic sectors and policy fields at various scales and which will allow for comparison between Member States.*" However, the same paper concluded that no instruments for policy guidance and public information are available.

The present report concerns the use of the methodology developed by the ExternE Project of the EC Directorate General XII JOULE II Programme in just such an application, to quantify pollution damages at the national level. The ExternE Project is a major multi-disciplinary study which has developed a detailed 'bottom-up' approach for the assessment of the social costs of fuel cycles (European Commission, 1995). By following this methodology we have been able to use state-of-the-art models and data which have been the subject of international peer review.

The methodology was originally developed for assessment of the impacts of the major fuel cycles for electricity generation, including fossil, nuclear and renewable technologies, and has previously been applied almost solely to quantify the effects associated with individual power plants. In applying this methodology to quantify national level damages we thus seek to find out whether there are major barriers to its application for a purpose different to that for which it was developed.

1.2. Scope of the present study

A consequence of the application of the ExternE methodology is that the present report is concerned mainly with effects arising from energy use. No attempt has yet been made to compare the magnitude of damages that we have monetized against those that have not been investigated in this report.

The present phase of the work has concentrated on quantifying damage associated with several of the major air pollutants (Table 1.1). The same group of impacts has been studied by each of the four national teams involved in the present study (Germany, Italy, the Netherlands and the UK, which is the subject of the present report). At the present time no attempt has been made to attribute this damage to individual sectors, such as manufacturing industry, chemicals or agriculture. However, this can be done, at least partially, using available models of atmospheric transport and chemistry and could form part of a follow-up study. We have also taken no account of the transboundary nature of air pollution. Some of the damage that we predict here thus arises from pollutants 'imported' into the UK, whilst other damages, that we have not quantified, arise from UK pollution 'exports'. Although we have not accounted for transboundary effects, models are available which would allow damage in any country to be attributed to the country in which the pollutants responsible originate. An example is the EMEP model (EMEP, 1992). Again, use of these models could form part of a follow-up study.

Table 1.1 Scope of the UK case study

	Health	Materials	Crops	Forests	Ecosystems[2]	Amenity[3]
PM_{10}	A	A	NE	NE	NE	NQ
SO_2[1]	A	A	A	A	A	NQ
NO_x[1]	A	A	A	A	A	NQ
O_3	A	A	A	NQ	NQ	NE
Noise	NQ	NE	NE	NE	NE	A

A impact assessed
NE no impact expected
NQ impact likely, but not quantified
1. SO_2 and NO_x include acid deposition more generally.
2. Ecosystems include freshwater and terrestrial systems.
3. Effects of PM_{10}, SO_2 and NO_x on amenity arise through effects on visual range. This is not generally considered to be a serious environmental issue in Europe.

In addition, a preliminary assessment of global warming impacts has been carried out by the Dutch team for all four countries involved in this study (see Synthesis Report).

1.3. Methodological issues

Alternative approaches to providing an integrated system of environmental indicators are available, using Delphi and other decision analysis techniques (e.g. Jesinghaus, 1993). Comparison of different types of impact using these approaches is performed on the basis of qualitative judgements. The techniques for extracting these judgements are being greatly refined at the present time. However, the final results lack the transparency that detailed impact assessment followed by monetary valuation is able to offer. Meanwhile, the use of economic valuation of environmental and health damages remains controversial. In part this arises from a misunderstanding of what valuation attempts to do. Rather than seeking to quantify the intrinsic value of, for example, a human life, an ecosystem or a building of great cultural significance, valuation seeks instead to quantify society's preference for the allocation of resources to these and other causes. Whilst some of the implications of this notion may be disturbing it is necessary to realise that decision makers deal routinely with this type of issue, when deciding on the allocation of budgets to education, health care, defence, overseas aid, environmental protection, etc.

The main advantage of using economic valuation on top of a detailed assessment of impacts is that the basis for comparison of different effects is made much more explicit than when one relies on the output of a debate conducted on a more qualitative basis. This is especially the case when using the impact pathway approach because of the way in which physical, chemical, biological or social impacts, and their associated economic values, are quantified in detail.

It will be seen that several types of impact still cannot be valued. One of the reasons for the lack of valuation data is that the type of impact assessment used here is a recent innovation, and valuation studies which properly link in with the impact assessment have yet to be carried out. Of more concern is the fact that valuation of some impacts is extremely complicated (e.g. for natural ecosystems) and indeed may never be successfully achieved. However, some information is available to at least partially describe impacts in most cases. For the pollutants of interest to this project we have sought to demonstrate how far it is possible to take assessment even for cases where no valuation of damages has been possible.

Given the purpose of GARP, to provide analysis for the development of satellite accounts within an extended national accounting system, we have attempted to quantify damages arising from human activity as well as total damages associated with the pollutants of interest. A difference between the two arises because of generation of SO_2, PM_{10} and O_3 from natural sources.

We have attempted to attribute these damages to human activity by subtracting out background (natural) levels of pollution, where applicable. This allows partial verification of the methodology by comparison between damage estimates and national statistics (for example of mortality rates or manufacturing production) which are of course derived quite independently of this work. The results provided should be regarded as first estimates of damage, which can and should be further refined.

1.4. Air quality data

1.4.1. *Sources of air quality data*

Within the UK two sources of air quality data have been used. The first of these provided data at a resolution of 5 × 5 km (Stedman, 1995; Campbell, personal communication, 1995). These maps were derived from data collected at monitoring sites and surrogate statistics. The second source was results generated by the Harwell Trajectory Model (Lee, personal communication, 1995). This provided data at a coarser resolution (20 × 20 km, which was then aggregated to 100 × 100 km). By using two sources of air quality data we have been able to demonstrate the use of the alternatives that are available. These methods were used in preference to the pan-European datasets of EMEP and WHO because of the greater resolution of data that they contain for the UK.

In the present study no account has been taken of air quality impacts in Northern Ireland. This is a potentially significant omission because high levels of SO_2 and particulates have been reported in this part of the UK, caused largely by domestic coal burning. On the other hand the population in Northern Ireland is only a small fraction of that elsewhere in the UK (1.6 million against 57 million). The integration of Northern Ireland into this assessment is identified as a requirement of a future study of this type. In addition to the description of the data sources, background (natural) levels of each pollutant have also been estimated.

1.4.2. *Estimated high resolution maps of air quality in the UK*

Oxides of nitrogen (NO_x)

The derivation of the 5 × 5 km map of NO_x is described by Stedman (1995). The first stage was the production of a map for NO_2 based on measurements made in 1991 at over 300 sites in the UK using diffusion tubes. Data was extrapolated on the assumption that NO_2 levels are determined by vehicle density which is in turn related to population. The relationship between population and the difference between the urban NO_2 measurements and a 5 × 5 km grid square average rural NO_2 concentration was then derived;

$$\text{urban\_NO}_2\,(\text{ppb}) = \text{rural\_NO}_2 + 0.317(\text{population\_5 km})^{0.35} + 0.56 \quad (1)$$

where population_5 km is the 5 × 5 km grid square population and rural_NO_2 is a 5 × 5 km grid square map of NO_2 concentrations interpolated by kriging from measurements using diffusion tubes at 39 rural sites for the period July to December 1991. For each grid cell in the map used here NO_2 levels were weighted by the percentage of each cell defined as urban from the satellite land cover map published by Fuller *et al.* (1994). Good agreement was found when comparing the estimated map against data collected at monitoring stations.

The next stage was to quantify the relationship between NO_2 and NO_x. Using data from urban monitoring sites, urban NO_x (ppb) was taken as 2.5 times urban NO_2 levels ($r^2 = 0.5$). Rural NO_x was estimated as 1.2 times rural NO_2, based on measurements at the Lullington Heath monitoring site on the south coast of England. This site was selected as being representative of rural areas.

Agreement between estimates and monitored data was not as good for NO_x as for NO_2, presumably as a result of site to site variation in the ratio between levels of the two pollutants through factors such as distance from roads.

Fine particulate matter (PM_{10})

NO_x is a reasonably conservative species and the map so produced was used to calculate the distribution of PM_{10} levels. The following equation was used, based on a regression analysis of hourly PM_{10} and NO_x data collected at the London Bloomsbury monitoring site between January 1992 and December 1993 ($r^2 = 0.35$);

$$PM_{10}\,(\mu g/m^{-3}) = 0.18(NO_x, \text{ppb}) + 15.4 \qquad (2)$$

The large constant in this equation reflects the presence of significant rural levels of PM_{10} which are not associated with urban vehicle pollution. This equation seems likely to overestimate rural levels, though perhaps not significantly (see Stedman, 1995). Agreement between estimated and monitored PM_{10} data was generally good, though the range of concentrations included was small.

The map of annual UK mean PM_{10} levels is reproduced in Figure 1.1.

Ozone (O_3)

O_3 maps were generated from an analysis of annual O_3 statistics from the Photochemical Oxidant Review Group database (PORG, 1993) and for EUN sites for 1992 and 1993. Two maps were produced, the first showing annual mean O_3 levels and the second showing the annual mean of the hourly maximum concentration for each day. The relationship between measured urban annual mean O_3 and rural levels was assessed ($r^2 = 0.47$);

$$\text{urban\_O}_3\,(\text{ppb}) = \text{rural\_O}_3 \times (0.007(\text{percentage\_rural}) + 0.3) \qquad (3)$$

A map of rural mean O_3 levels for 1990 was calculated by bilinear extrapolation from the network of rural monitoring sites showing no significant urban influence. Again, a land cover correction was applied based on the percentage of land classified as urban and rural in each grid cell in the land cover map of Fuller *et al.* (1994). Agreement between estimated and monitored data was again good, though levels at some rural sites seemed to be overestimated. This seemed to be consistent with their being more heavily influenced by road traffic than the percentage rural land cover in their grid cell would indicate. The map of the annual mean of hourly maximum levels for each day (used in the health assessment) was then produced using the equation 4, which was derived from

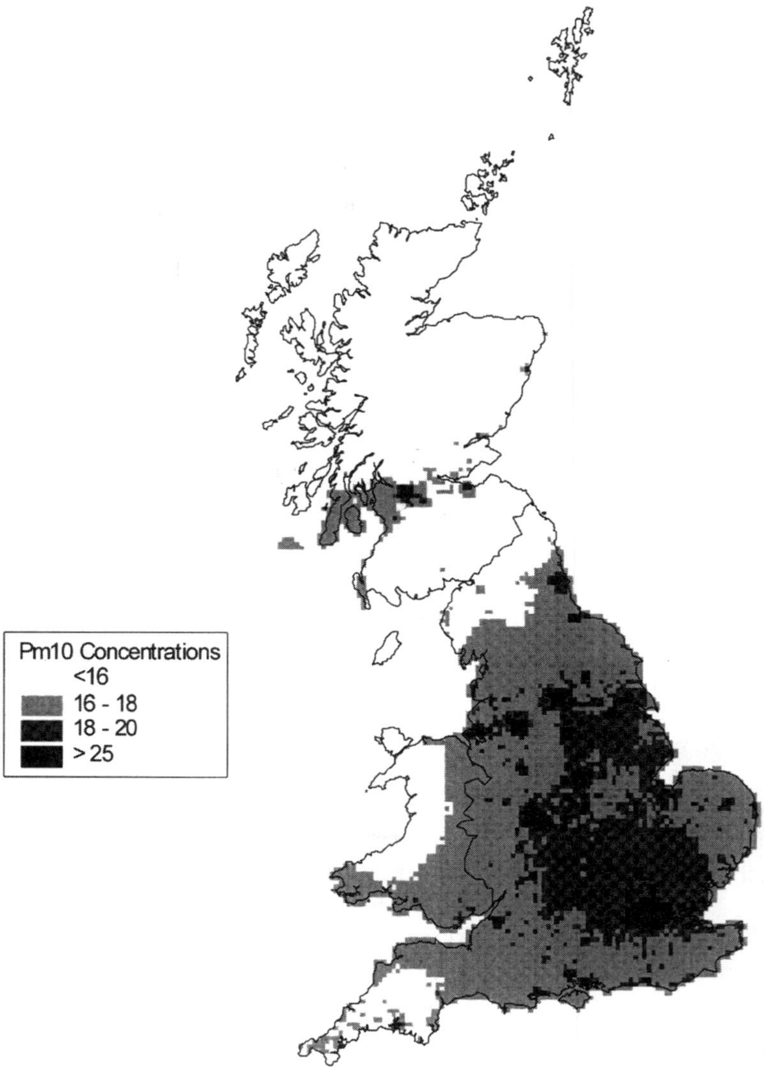

Fig. 1.1. Map of annual mean PM_{10} levels ($\mu g/m^{-3}$) for the UK. This map has been plotted at a resolution of 5 × 5 km.

1993 data. The range over which this relationship is applicable is clearly limited, but it is sufficient for UK ozone levels. Data are mapped in Figure 1.2.

$$\text{mean peak hourly } O_3\,(\text{ppb}) = 0.75(\text{mean } O_3) + 15 \quad (4)$$

Sulphur dioxide (SO_2)

Urban estimates of SO_2 were derived from the mean of urban measurements in each grid square. Data from only one network (diffusion tubes) were used.

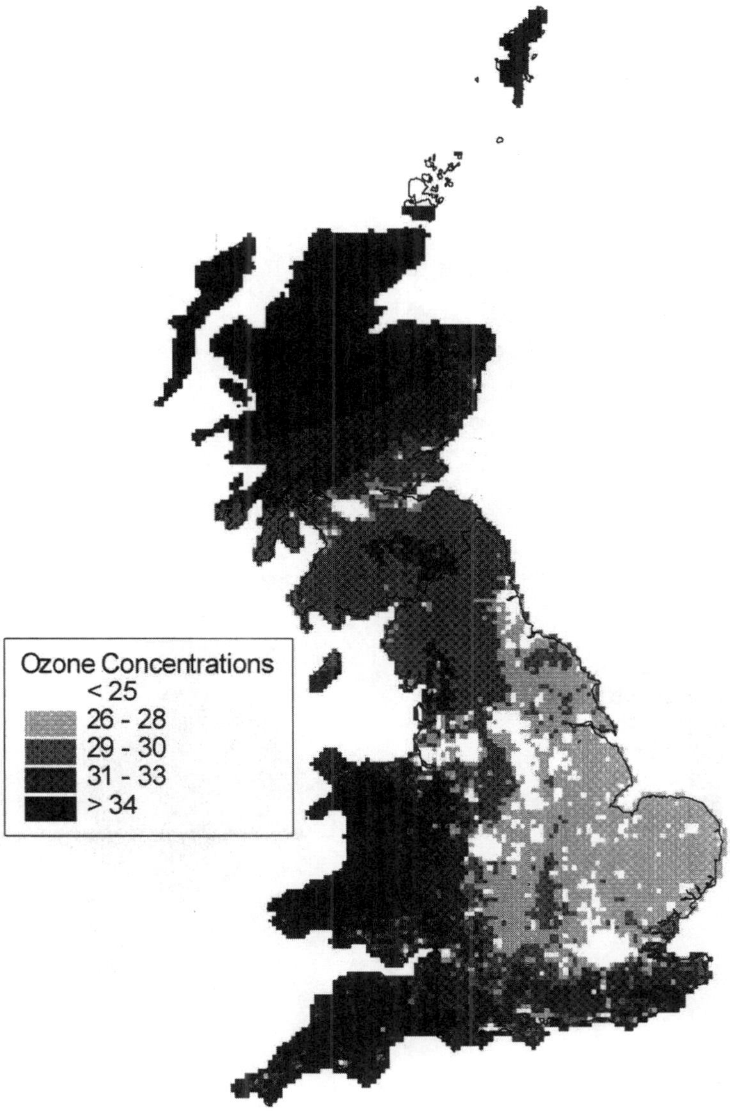

Fig. 1.2. Map of annual mean of hourly maximum O_3 levels for each day for the UK. This map has been plotted at a resolution of 5 × 5 km.

Rural estimates were derived from interpolation of rural measurements. The map was then derived by taking a weighted mean of estimates of urban and rural concentration in each grid square, again using data from Fuller *et al.* (1994), and is shown in Figure 1.3.

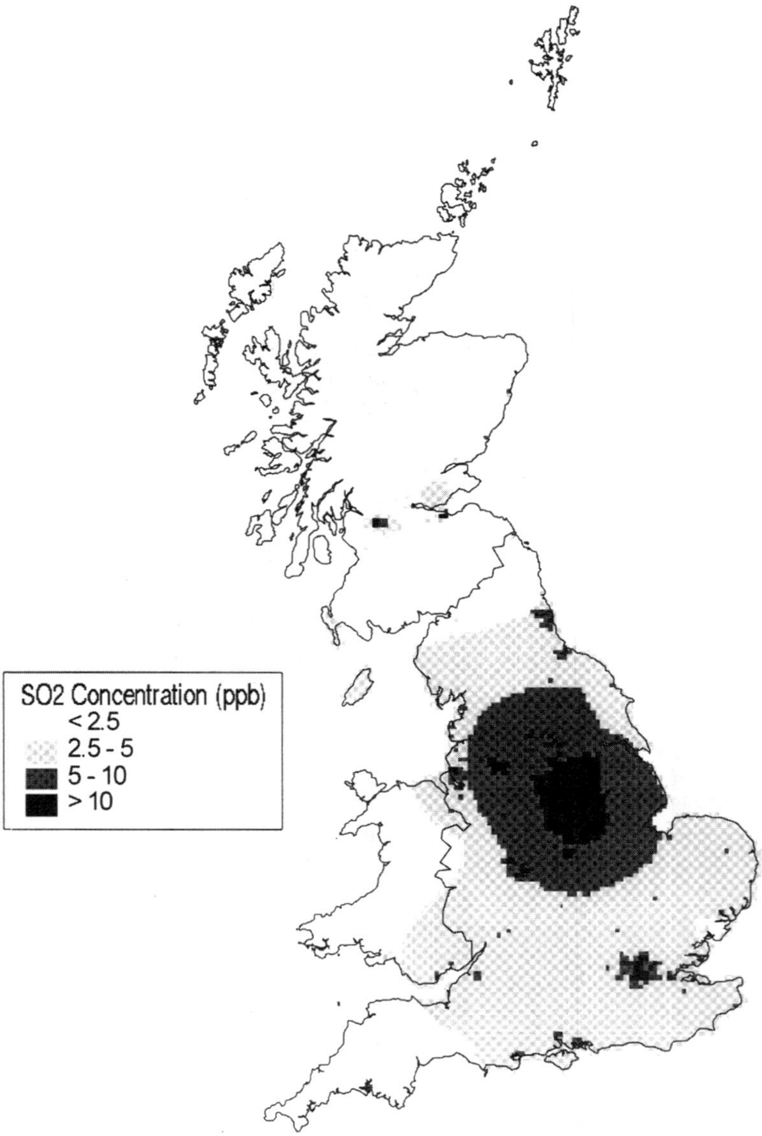

Fig. 1.3. Map of annual mean SO_2 levels for the UK. This map has been plotted at a resolution of 5 × 5 km.

Acidity (H^+)

A map of acid (H^+) deposition has been generated for Great Britain at a scale of 20 × 20 km. The map was derived from interpolation of measured data combined with information on rainfall, and is shown in Figure 1.4.

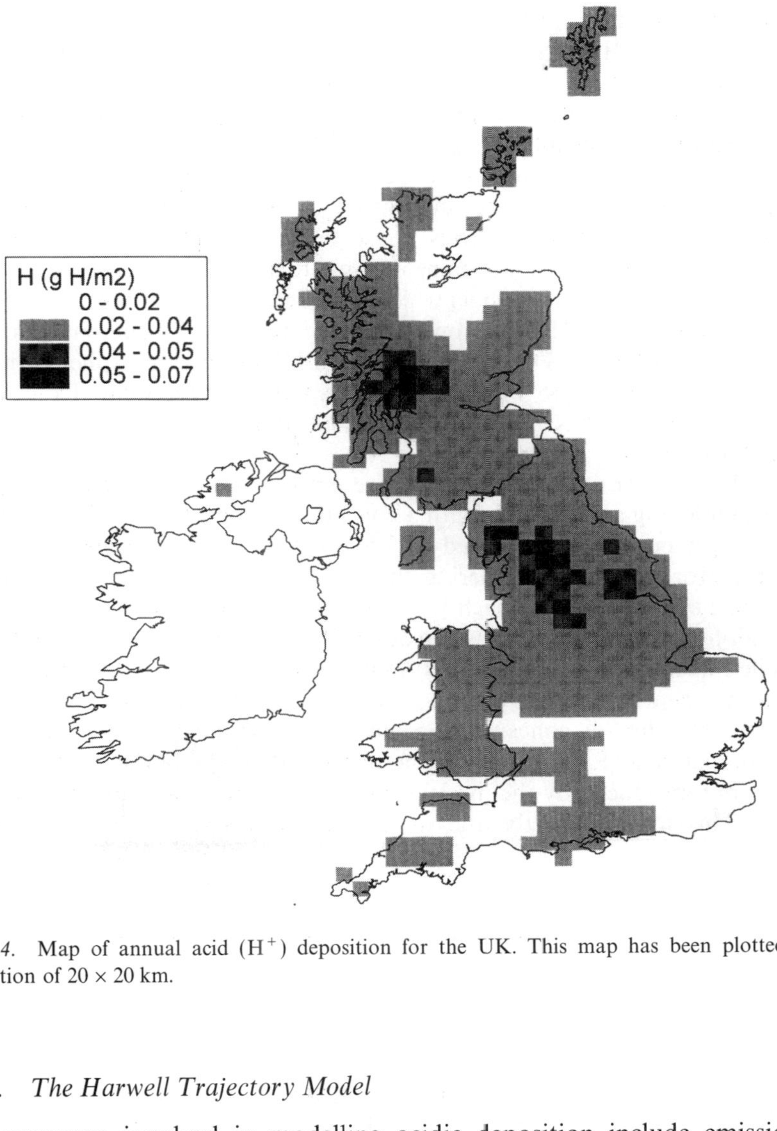

Fig. 1.4. Map of annual acid (H^+) deposition for the UK. This map has been plotted at a resolution of 20 × 20 km.

1.4.3. *The Harwell Trajectory Model*

The processes involved in modelling acidic deposition include emission of pollutants, dispersion and atmospheric transport over regional scales, chemical transformations and dry and wet deposition processes. Because of the large scale of the acid deposition phenomenon, the emissions database required is extensive. The principal emissions of interest are SO_2 and NO_x. Ammonia (NH_3) emissions from animal wastes and other sources play an important regulatory role in the formation of nitrate and sulphate aerosols. Meteorological processes are important to several aspects of acid deposition. These include initial dispersion and long range transport processes, reviewed by Eliassen

(1980; 1984), Johnson (1983) and Schwartz (1989), together with those processes involved in the formation of cloud droplets and rainfall. These processes are important because of the role of aqueous phase chemistry in the oxidation of SO_2 to sulphate. Chemical processes include gas phase reactions including the formation of tropospheric oxidants and the oxidation of SO_2 and NO_x to H_2SO_4 and HNO_3. SO_2 is also rapidly oxidised to sulphate in cloud water. The role of ammonia is important because of its influence in determining the oxidation rate of SO_2 in aqueous phase reactions (Behra *et al.*, 1989), and because of its role in the formation of sulphate and nitrate aerosols.

The Harwell Trajectory Model which was used to estimate the concentration and deposition of acid species on a national scale in the present study was originally described by Derwent and Nodop (1986) for atmospheric nitrogen species, and extended to include sulphur species by Derwent, Dollard and Metcalfe (1988). It has since been modified and updated by Lee (personal communication).

The Harwell Trajectory Model is a receptor-orientated Lagrangian plume model employing an air parcel with a constant mixing height of 800 m, moving with a representative wind speed of 7.5 m s^{-1}. The results were obtained at each receptor point by considering the arrival of 24 trajectories weighted by the frequency of the wind in each 15° sector. The trajectory paths were assumed to be along straight lines. Emissions of NO_x, SO_2 and NH_3 in the UK were compiled from data presented in RGAR (1990). The model chemical scheme is shown schematically in Figure 1.5. The distribution of modelled and measured sulphur dioxide concentrations and wet deposited sulphur over the UK were compared by RGAR (1990). The broad features of both sets of measurements are reproduced by the model, but the measurements show greater variability. This may be partly due to the relatively large scale of the model grid used.

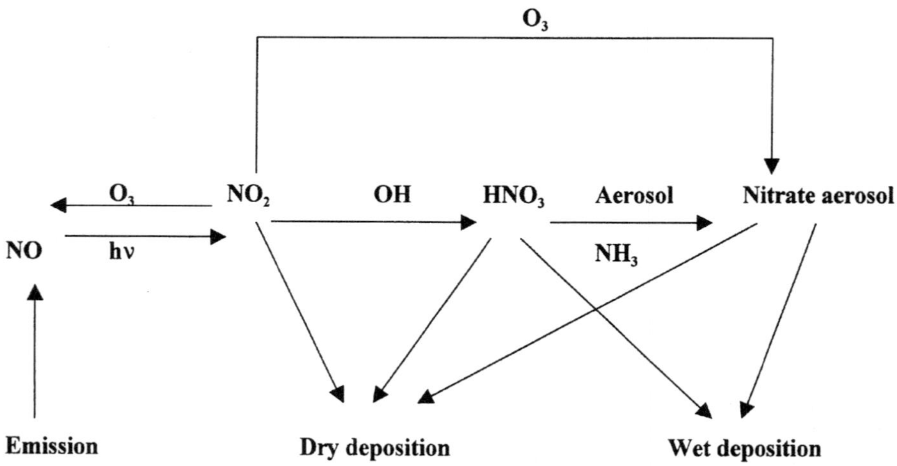

Fig. 5. The chemical scheme used by the Harwell Trajectory Model.

1.4.4. *Identification of background concentrations*

The objective of the current study, to relate environmental impacts to human activity, dictates that the damages resulting from 1990 emissions of pollution are compared to the level that would exist in the absence of human activity. For some pollutants an estimate of these levels can be gauged from the concentrations measured in remote areas.

The most appropriate data for SO_2 and NO_x levels in Northern Europe seems likely to come from North West Scotland. This region is remote from major industrial activity and the prevailing south westerly winds are unlikely to carry a high pollution load from other countries. The Review Group on Acid Rain (RGAR, 1990) provide values of about 1 ppb SO_2 and 2 ppb NO_2 for this region. Data modelled for the UK using the Harwell Trajectory Model agree well with these figures. Background levels of acid deposition (H^+) were estimated from a 20 × 20 km rainfall map for the UK and the assumption of pH 5 rain in pristine areas (RGAR, 1990).

Background PM_{10} concentrations are more difficult to predict, because of diverse sources of particulates of natural origin (Table 1.2).

The sources of fine particulate matter (the range thought to cause most health damages) includes natural emissions and agricultural emissions in addition to energy related emissions. The list given in Table 1.2 suggests that about 70% of the total arises from human activity, assuming that 'natural' NH_3 emissions arise mainly from agriculture. The extent to which some of these sources (particularly sea salt) might influence the map that we have used in this study seems limited, given that it was derived principally from data collected in central London. In line with these uncertainties and the comments made by Stedman (1995) we have used three different estimates of background PM_{10}, 5, 10 and 15 $\mu g/m^{-3}$ in the assessment of health damages given in the next chapter.

O_3 is also problematic as background concentrations vary with altitude, sunlight and other climatic conditions. The annual mean natural concentrations assumed here were between 10 and 20 ppb. Using equation 4 these translate

Table 1.2. Composition of atmospheric particles <2.5 μm in diameter (QUARG, 1993)

Component	Major source	% of total
Ammonium	Natural ammonia emissions	6
Nitrate	Energy sector	5
Sulphate	Energy sector	18
Chloride	Sea salt, coal burning	6
Base cations	Sea salt, dusts	6
Carbonaceous matter	Smoke emissions	37
Insoluble minerals	Wind-blown dust	22

to a range of between 20 and 30 ppb for the annual mean of hourly maximum levels for each day.

1.4.5. *Uncertainties*

There is clearly some uncertainty associated with the pollution maps used in this study because of the need to extrapolate data from a relatively small number of monitoring sites to a large number of grid cells. Further work is necessary to establish how important the uncertainties introduced through this sequence of calculations are. Errors are likely to be most significant at the level of individual 5 × 5 km grid cells. These are likely to be averaged out to some extent when values are combined.

A systematic error may be present in the subsequent analysis. This could arise through differences in the rationale underlying selection of sites for measuring pollutants for derivation of exposure–response functions, and the sites used to derive our pollution maps. The problem may be accentuated in the present study by the fact that the exposure–response functions used here are taken from studies performed in several different countries, where different regulations regarding air quality monitoring may be in force. Further uncertainty arises because of other forms of variation in the pollution climate between sites. One important issue is the existence of synergistic relationships between different pollutants acting on a wide variety of receptors. Another important issue applies to particulate matter, which is not chemically specific and which may vary significantly in composition from place to place.

Selection of background (natural) conditions is most satisfactory for SO_2, NO_x, and H^+. For PM_{10} and O_3, however, background levels tend to contribute more significantly, and are more variable, both in time and space. For these reasons a range of background levels have been proposed in each case.

These uncertainties are identified as issues that require further research. Given current concern over health effects in particular, some of these points are already being addressed.

CHAPTER 2

HEALTH

2.1. Introduction

Effects of air pollution on public health are the cause of considerable concern in the UK at the moment, following several episodes where pollution levels exceeded recommended levels. However, at the time of writing this report, the only published attempt to quantify UK damages is that performed by Schwartz on behalf of the New Scientist magazine (Brown, 1994), who estimated that particulate air pollution is implicated in the deaths of 10,000 people each year. It should be noted that several research groups in the UK are currently working on alternative estimates.

Any estimate of this order is certain to create a great deal of concern, particularly amongst those living in areas where pollution levels are believed to be high. However, it is necessary to exercise some care when describing these effects. The exposure–response function used by Schwartz accounts for cases where the effect of pollution is manifested very soon after exposure. The large majority of these cases are likely to be among those who are already seriously ill. Indeed, many may have a life expectancy of the order of only a few days. On the other hand it takes no account of the possibility of an effect of long term exposure to air pollution on longevity.

The work presented in this chapter combines the pollution maps described in Chapter 1 with exposure–response functions identified from the literature for the ExternE Project (European Commission, 1995). Although the selection of exposure–response functions is described here, the reader should refer to the material produced for the ExternE Project for a more complete discussion of this important issue. The set of functions used here was originally recommended for assessment of the impacts associated with small increments in pollution levels arising from the operation of fossil fuel power plants. Their use in the present context, where the pollution increment of interest is much larger, has yet to be thoroughly reviewed. A Geographical Information System, ARCINFO, has been used, enabling a spatial presentation of the output of the analysis.

2.2. The impact pathway

Within this study we are considering the effects of several of the major air pollutants, particulates, ozone, SO_2 and NO_x on public health. For the purpose of modelling, the impact pathway representing a set of complex processes is broken down into a series of stages (see Figure 2.1, taken from Hurley and Krewitt, European Commission, 1995). A similar model has been used for assessment of damage to other receptors in the following chapters.

Although this Figure indicates some early markers of response to inhaled pollutants, there is at present no agreed understanding of the mechanisms whereby small changes in ambient air pollution may lead to increased mortality or severe morbidity (e.g. hospital usage) on the same day or soon after. In general, this seems likely to occur only against a background of severe and pre-existing ill-health. The precise mechanisms are however unknown; and so Figure 2.1 is necessarily incomplete.

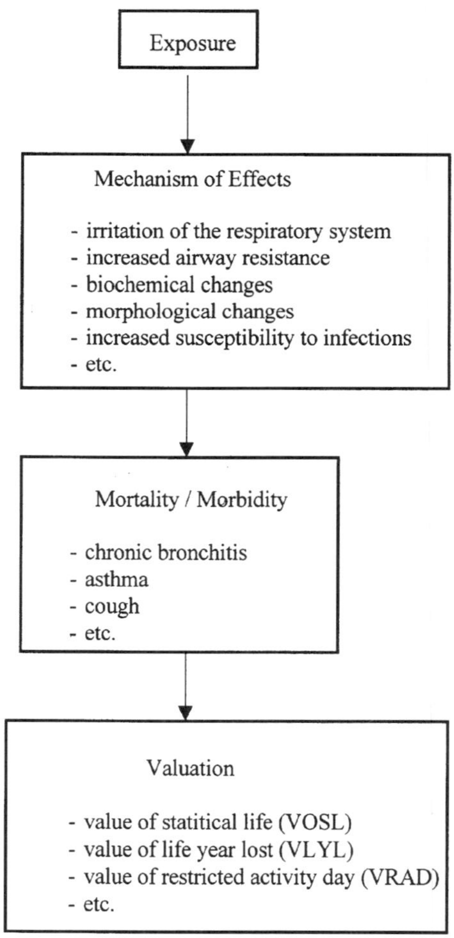

Fig. 2,1. Impact pathway for air pollution impacts on human health.

2.3. Exposure assessment

In general, epidemiological studies of the public health effects of particulates, SO_2, NO_x and ozone do not consider personal exposures of the study group. Rather, they examine relationships between health effects and ambient concentrations of pollutants. Implementation in the present project is therefore similarly in terms of changes in ambient concentrations rather than in personal exposures. This simplifies analysis; but has a disadvantage. There is limited knowledge of how changes in ambient air pollution are related to changes in the various indoor environments where many people typically spend most of their time. A problem of transferability of exposure–response functions arises when the relationship between ambient and personal exposures differs between the epidemiological study populations and the target populations at risk in the present project. Such differences can arise, for example, because of differences in human activity patterns or in housing (especially through the use of air conditioning).

Air pollution levels generally show substantial temporal and spatial variation. Concentrations are integrated over time to give average values (COST, 1992). In studies of acute health effects, the period of interest is usually 24 hours, though sometimes multiple day averages are used. Ozone measures typically refer to maximum 1, 5 or 8-hour concentrations within a day, rather than to the full 24-hour average. In studies of chronic health effects, the averages refer to time-periods of one or several years. The same time scales are used for implementation in the present study. However, estimated daily acute effects are accumulated over a year and expressed as the effect of changes to annual average concentrations.

By focusing on average concentrations within a given period, epidemiological studies give no direct information about the relevance if any to health of short-term peak concentrations within those periods, though clearly the ozone measures are designed to take some account of within-day variations. Insofar as within-day peaks may be relevant to acute health effects, the similarity of 'peak-to-average' ratio is one factor influencing the transferability of exposure–response relationships.

Ambient concentrations are typically measured using fixed-point monitors. Large scale mortality or hospital usage studies may include several million people residing in a region covered by several monitoring sites. Usually, a regional average concentration is used, following examination of the similarity of concentrations across sites. No attempt is made to account for time spent outside the study region; the effect of such a refinement would be small.

Similarly with implementation: The vast geographical area to be considered is sub-divided into smaller regions using a regular grid system. Each grid cell is considered as a micro-environment with a homogeneous pollutant concentration. The associated population at risk is the population resident in that grid cell. As in the epidemiological studies, daily movements of people between grid cells are ignored. Effects are estimated separately by grid cell and then accumulated over grid cells. Population data were taken from the UK Census of 1981, aggregated to the 5 × 5 km grid from a 1 × 1 km grid. At the time of the analysis presented here, data from the 1991 census was not available to the authors of this report in the required format. However, inspection of UK population distribution (CSO, 1993) shows that there has been little change since 1981. Subsequent analysis using the 1991 data shows little difference to the results presented here. Population distribution in the UK is shown in Figure 2.2.

Exposure response functions for children are applied to the population of age 16 and below (22.5% of total UK population). Estimates for the prevalence of asthma among the UK population vary widely. We have used a value of 5% in the impact assessment, based on estimates from the National Asthma Campaign. The baseline death rate (from all-causes) is taken from data held by the Central Statistical Office (CSO, 1993). The central estimate of 11.2 deaths per thousand head of population per year is in a range of 10 (for Northern Ireland) to 12.8 (for Scotland) (data from Philip's Geographical Digest, 1994). Sensitivity analysis on death rate has not been performed because of the small variation in these figures.

Pollution levels were taken from the maps depicted in Figures 1.1, 1.2 and 1.3. Incremental mortality is derived by linking together the baseline deaths, the population weighted pollution increment and the appropriate exposure–response function. Incremental morbidity was calculated by linking the appropriate population at risk (e.g. children, adults, asthmatics), the population weighted pollution increment and the relevant dose–response function.

A fuller discussion of the exposure–response relationships used in the assessment is provided in the Synthesis Report, Section 3. As noted previously, all the country teams attempted to use a common set of functions for as many impacts as possible to ensure some degree of consistency. Similarly, the discussion on the economic valuation of human mortality and the chosen morbidity endpoints need not be repeated again here. The values chosen are detailed and assessed in the Synthesis Report, sections 3.3–3.4.

Results and discussion

Results are presented in Tables 2.1 to 2.6, showing firstly effects of PM_{10}, O_3 and SO_2 on mortality, and then effects on morbidity. Results are presents in terms of the number of cases estimated and the corresponding valuations. Note that the effects of SO_2 and PM_{10} on mortality should not at the present time be regarded as additive. Also note that the exposure metric for PM_{10} and SO_2 is annual mean concentration ($\mu g/m^{-3}$), whereas for ozone it is the annual mean of the daily 1 hour maximum level (ppb).

Our results are certain to attract a strong response from many quarters. For

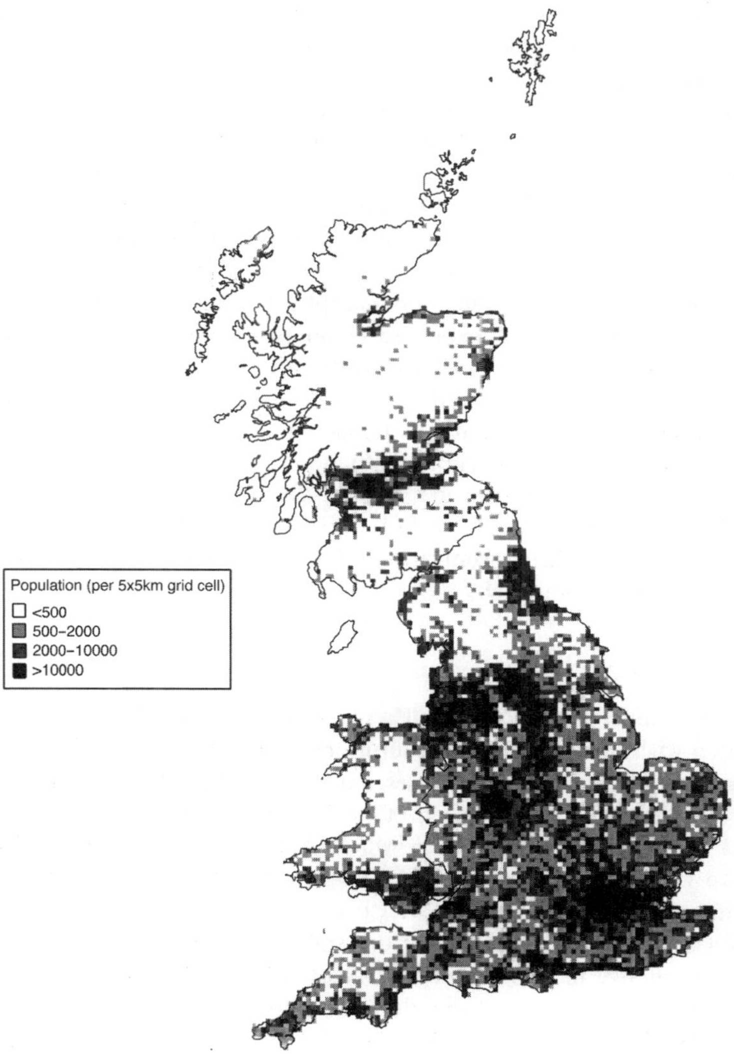

Fig. 2.2. Population distribution in the UK. This map has been plotted at a resolution of 5 × 5 km.

this reason we stress that this section should be read in conjunction with those preceding above, and preferably also in association with the ExternE Project methodology report (European Commission, 1995). This area of work is the subject of much research at the present time and whilst we believe that the methodology adopted is a good reflection of current knowledge others will undoubtedly have different views. It is necessary to repeat that the methodology was originally developed for a different case. This was to assess the effects of relatively small increases in pollutant levels arising from operation of individual power plants, within which context the methodology has been discussed at length. However, there has been much less discussion about its use to quantify the effects of larger changes in pollution levels. Our methodology must therefore be regarded as experimental, and it is to be hoped that discussion will focus as much on the approach taken as upon our results.

The tables of results contain high and low estimates of damage, in addition to best estimates for each case. The range so described shows 95% confidence limits based on the statistical variability of the data from which functions were derived. It does not account for the uncertainty associated with some of the assumptions that we have made, for example, whether the no-threshold model really is applicable for each exposure–response function. This could have significantly reduced damages. In the face of this type of uncertainty it is clear that we have not attempted a full analysis. A methodology for a more comprehensive analysis of uncertainty is being assessed for the ExternE Project (C. Hope, personal communication).

The mean population exposure for each of the three pollutants considered is 21 $\mu g/m^{-3}$ for PM_{10}, 6 ppb for SO_2 and 33 ppb for O_3. For PM_{10} and SO_2 these are annual mean concentrations. For O_3 it is the annual mean of the daily 1 hour maximum level.

2.4.1. *Size of impacts and associated costs*

Mortality (Tables 2.1 and 2.2): The largest effects on mortality are those associated with impacts from long term or chronic exposure to PM_{10}. In total we have estimated 35,000 such cases annually in the UK. After adjusting for background we have a range of 9,800 to 27,000, the wide range here reflecting the fact that levels of PM_{10} can be very variable depending on location. The extent to which background should be taken into account is open to debate, depending on whether 'natural' PM_{10} is as damaging as anthropogenic PM_{10}. This is a question of interpretation of exposure–response data in the context of a major change in pollution levels. Acute effects are added to this chronic effect, although this implies some double counting.

Estimates associated with acute effects of air pollution are roughly similar for PM_{10} and SO_2, with totals of 13,000 and 9,700 deaths respectively. These figures should not be regarded as additive at the present time. Their similarity may not be entirely coincidental, given the strong correlation that typically exists between levels of the two pollutants in the urban environment. The effect of taking out the background level could be very significant for PM_{10} (depending on the assumption made as to how large it is) though not for SO_2, for which we can be confident that background is little more than zero. The results shown in Table 2.1 suggest that, in the UK at least, the effects of ozone on mortality are less important than those of either SO_2 or PM_{10}, with a central estimate of 3,000 cases per year in total, declining to 1,200 or less once the 'natural' background is accounted for.

For the monetary valuation of mortality a figure of ECU 2.6 million for the Value of Statistical Life (VOSL) was used in the initial calculations which are presented below. These can be assumed to represent 1990 prices. In the Synthesis Report all the estimates were updated to 1995 prices. The various endpoint valuations for morbidity impacts also used 1990 prices. Again, the final values used and updated estimates can be found in the Synthesis Report.

The only comparable analyses of air pollution effects on health in the UK were carried out by Schwartz (see Brown, 1994) and Pearce (personal communication). Both used the same methodology and concluded that particulate air pollution can be linked to 10,000 deaths from acute effects of PM_{10} in England and Wales each year. This can be compared directly with our total value of 13,000 deaths from acute exposure to PM_{10}. Given the uncertainties involved in this type of analysis the difference between our results and those of Schwartz and Pearce is of little importance. It must, however, be said that our assessment has been performed in much greater detail – both Schwartz and Pearce assumed a single representative level for urban PM_{10}, and applied this to the UK urban population (68% of the UK total) only. Also, neither sought to account for chronic effects of PM_{10}. From the results given in Table 2.1 we believe that these may well be more severe than acute effects.

For validation we have compared these results against UK mortality statistics. PM_{10} has been associated with the deaths of people with diseases of the respiratory system, and those with diseases of the circulatory system.

Deaths in 1992 = 630,000, of which:

Diseases of circulatory system	= 291,000
Diseases of respiratory system	= 70,000
Total potentially related to PM_{10}	= 361,000

Table 2.2 expresses the results from Table 2.1 as a percentage of each of these figures. This tells us the maximum proportion in each category that could be associated with the different pollutants. The total percentage given for each group in Table 2.2 has been calculated by adding together chronic and acute effects of PM_{10} and acute effects of O_3; effects of SO_2 have not been added as it is thought that they are likely to double count effects of PM_{10}. Results for acute effects of PM_{10} are preferred to an acute effect of SO_2 as the exposure–response function for PM_{10} is based on a larger body of literature. The figures demonstrate that the deaths of 14% of those who die from diseases of the respiratory and circulatory systems, and 8% of all deaths, could be associated with air pollution.

Morbidity (Tables 2.3 to 2.6): Effects of PM_{10} on acute morbidity are shown in Table 2.3. Chronic effects of PM_{10} exposure are shown in Table 2.4. There is consistency with the expectation that there should be substantially more minor symptoms than major ones (defined as any symptom requiring hospital

Table 2.1. Estimated acute and chronic effects on mortality from PM_{10}, SO_2, and ozone in the UK

Endpoint	Back-ground	No. of cases Low	Mid	High	Damages (ECU million) Low	Mid	High
PM_{10} acute effects	0 μg m^{-3}	7,900	13,000	18,000	20,540	33,800	46,800
	5 μg/m^{-3}	6,000	9,800	14,000	15,600	25,480	36,400
	10 μg/m^{-3}	4,100	6,700	9,300	10,660	17,420	24,180
	15 μg/m^{-3}	2,200	3,600	5,000	5,720	9,360	13,000
PM_{10} chronic effects	0 μg/m^{-3}	29,000	35,000	41,000	75,400	91,000	106,600
	5 μg/m^{-3}	22,000	27,000	31,000	57,200	70,200	80,600
	10 μg/m^{-3}	15,000	18,000	21,000	39,000	46,800	54,600
	15 μg/m^{-3}	8,000	9,800	12,000	20,800	25,480	31,200
O_3 acute effects	0 ppb	2,000	3,000	4,000	5,200	7,800	10,400
	20 ppb	790	1,200	1,600	2,054	3,120	4,160
	25 ppb	490	740	980	1,274	1,924	2,548
	30 ppb	190	290	380	494	754	988
SO_2 acute effects	0 μg/m^{-3}	6,600	9,700	13,000	17,610	25,220	33,800
	2.7 μg/m^{-3}	5,600	8,200	11,000	14,560	21,320	28,600

Chronic effects of PM_{10} are given from the function of Pope *et al.* (1995), with figures for acute mortality, calculated from Schwartz (1993a), subtracted. Total damages are shown in rows where back-ground pollution is set to zero. Results for SO_2 and PM_{10} should not be considered to be additive.

Table 2.2. Percentage of deaths from different causes that could be associated with M_{10}, SO_2 and O_3, based on the results shown in Table 2.1

	Deaths caused by diseases of the: Circulatory system (A)	Respiratory system (B)	A + B	Deaths from all causes
1. PM_{10} – chronic	12%	50%	9.7%	5.6%
2. PM_{10} – acute	4.4%	18%	3.6%	2.0%
3. SO_2 –acute	3.3%	14%	2.7%	1.5%
4. O_3 – acute	1.0%	4.3%	0.8%	0.5%
Total (1 + 2 + 4)	15%	72%	14%	8.1%

In each case the mid estimate for total damages (i.e. with background levels set to zero) has been used.

or emergency room treatment). For a partial validation of these results we compared symptoms affecting asthmatics against the number of asthmatics in the UK (estimated above as 5% of the UK population). From our results the average asthmatic in the UK will require 0.0026 emergency room visits in any year as a consequence of exposure to PM_{10}. In addition there are an estimated 2.8 additional ‘shortness of breath days’ for each asthmatic in response to PM_{10}

Table 2.3. Estimated acute effects on morbidity from PM_{10} in the UK

Endpoint	Back-ground	No. of cases (000s) Low	Mid	High	Damages (ECU million) Low	Mid	High
Respiratory hospital admissions	0 $\mu g/m^{-3}$	1.4	2.1	2.8	9.0	14	18
	5 $\mu g/m^{-3}$	1.0	1.6	2.1	6.9	10	14
	10 $\mu g/m^{-3}$	0.7	1.1	1.4	4.7	7.1	9.5
	15 $\mu g/m^{-3}$	0.4	0.6	0.8	2.5	3.8	5.1
Hospital admissions for COPD	0 $\mu g/m^{-3}$	1.8	2.5	3.2	12	17	21
	5 $\mu g/m^{-3}$	1.4	1.9	2.5	8.9	13	16
	10 $\mu g/m^{-3}$	0.9	1.3	1.7	6.1	8.6	11
	15 $\mu g/m^{-3}$	0.5	0.7	0.9	3.3	4.6	6.0
ERVs for COPD	0 $\mu g/m^{-3}$	6.4	8.0	9.5	1.2	1.5	1.8
	5 $\mu g/m^{-3}$	4.9	6.1	7.2	0.9	1.1	1.3
	10 $\mu g/m^{-3}$	3.3	4.1	4.9	0.6	0.8	0.9
	15 $\mu g/m^{-3}$	1.8	2.2	2.7	0.3	0.4	0.5
ERV for asthma	0 $\mu g/m^{-3}$	4.4	7.1	9.5	0.8	1.3	1.8
	5 $\mu g/m^{-3}$	3.4	5.4	7.2	0.6	1.0	1.3
	10 $\mu g/m^{-3}$	2.3	3.7	4.9	0.4	0.7	0.9
	15 $\mu g/m^{-3}$	1.2	2.0	2.7	0.2	0.4	0.5
Hospital visits for childhood croup	0 $\mu g/m^{-3}$	24	32	42	4.5	6.0	7.9
	5 $\mu g/m^{-3}$	18	25	32	3.4	4.6	6.0
	10 $\mu g/m^{-3}$	13	17	22	2.3	3.1	4.1
	15 $\mu g/m^{-3}$	6.7	9	12	1.3	1.7	2.2
RADs	0 $\mu g/m^{-3}$	35,000	55,000	87,000	2,200	3,400	5,400
	5 $\mu g/m^{-3}$	27,000	42,000	66,000	1,700	2,600	4,100
	10 $\mu g/m^{-3}$	18,000	29,000	45,000	1,100	1,800	2,800
	15 $\mu g/m^{-3}$	10,000	15,000	24,000	610	960	1,500
Asthma attacks	0 $\mu g/m^{-3}$	3,900	7,800	11,000	120	240	360
	5 $\mu g/m^{-3}$	2,900	5,900	8,800	92	180	280
	10 $\mu g/m^{-3}$	2,000	4,000	6,000	63	130	190
	15 $\mu g/m^{-3}$	1,100	2,200	3,200	34	68	100
Symptom days	0 $\mu g/m^{-3}$	250,000	520,000	760,000	1,600	3,200	4,800
	5 $\mu g/m^{-3}$	190,000	390,000	580,000	1,200	2,500	3,600
	10 $\mu g/m^{-3}$	130,000	270,000	400,000	800	1,700	2,500
	15 $\mu g/m^{-3}$	68,000	140,000	210,000	430	900	1,300

exposure. All figures are based on the best estimate with no account taken of background levels.

Of the figures shown in Tables 2.3 and 2.4, the hardest to justify seem to be those for symptom days, based on the work of Krupnick *et al.* Limitations of the exposure–response function were noted in the ExternE report (European Commission, 1995). The best estimate, unadjusted for background, translates into roughly 10 symptom days per head of UK population. However, it must be remembered that the symptoms represented by this function are very minor. If the exposure–response function provides a reasonably accurate reflection of the impact, it may well be necessary to reassess the figure used for valuation.

Table 2.4. Estimated chronic effects on morbidity from PM_{10} in the UK

Endpoint	Back-ground	No. of cases (000s)			Damages (ECU million)		
		Low	Mid	High	Low	Mid	High
Chronic bronchitis in adults	0 $\mu g/m^{-3}$	390	600	810	53	83	110
	5 $\mu g/m^{-3}$	290	460	610	41	63	85
	10 $\mu g/m^{-3}$	200	310	420	28	43	58
	15 $\mu g/m^{-3}$	110	170	220	15	23	31
Respiratory illness in adults	0 $\mu g/m^{-3}$	520	820	1,100	72	110	150
	5 $\mu g/m^{-3}$	390	620	840	54	86	120
	10 $\mu g/m^{-3}$	270	420	580	37	58	79
	15 $\mu g/m^{-3}$	140	230	310	20	31	43
Chronic bronchitis in children	0 $\mu g/m^{-3}$	210	400	590	29	55	82
	5 $\mu g/m^{-3}$	160	310	450	22	42	62
	10 $\mu g/m^{-3}$	110	210	310	15	29	43
	15 $\mu g/m^{-3}$	59	110	170	8.2	15	23
Chronic cough in children	0 $\mu g/m^{-3}$	260	520	780	35	71	110
	5 $\mu g/m^{-3}$	200	390	590	27	54	82
	10 $\mu g/m^{-3}$	130	270	410	18	37	56
	15 $\mu g/m^{-3}$	72	140	220	9.9	20	30

Overall, the valuation of acute impacts of PM_{10} on morbidity is dominated by 'symptom days' and by restricted activity days. Each is substantially greater than the sum of all other effects described in the table.

Chronic effects of PM_{10} exposure on morbidity are shown in Table 2.4. In general these are not so well characterised as acute effects, largely because of difficulties with determination of long-term pollution exposure. Estimated total damages for the best estimates with no account taken of background levels are around 300 million ECU – about an order of magnitude less than the damages for acute PM_{10} effects on symptom days and restricted activity days, though higher than for the more severe acute effects.

Estimated effects of acute O_3 exposure are given in Table 2.5. Overall these show similar patterns to the effects of PM_{10}. Again, considering only the best estimates with no account taken of background levels we predict that exposure to O_3 leads to an additional 0.0036 hospital admissions, 0.018 emergency room visits for asthma, and 19 asthma attacks, per UK asthmatic per year. As stated above, when discussing the results of the PM_{10} analysis, these figures can (and should) be checked against more detailed health statistics, which could be part of a continuation of the present study.

Again, the less serious impacts such as 'symptom days' dominate the valuation, though to a slightly lesser extent than they were seen to do for PM_{10}.

We have only estimated one effect of exposure to SO_2 on morbidity, that of acute hospital admissions for chronic bronchitis or emphysema (Table 2.6). Total damages are 170 million ECU per year, higher than for any of the chronic effects estimated for PM_{10} taken individually.

Table 2.5. Estimated acute effects of ozone on mortality

Endpoint	Back-ground	No. of cases (000s)			Damages (ECU million)		
		Low	Mid	High	Low	Mid	High
Hospital adm. for respiratory infections	0 ppb	11	15	18	74	99	120
	20 ppb	4.5	6	7	30	39	47
	25 ppb	2.7	3.7	4.4	18	24	29
	30 ppb	1.1	1.4	1.7	7.2	9.6	11
Hospital admissions for COPD	0 ppb	6.9	11	15	45	72	99
	20 ppb	2.7	4.3	6	18	29	39
	25 ppb	1.7	2.7	3.7	11	18	24
	30 ppb	0.7	1.1	1.4	4.4	6.9	9.6
Hospital admissions for asthma	0 ppb	5.0	10	15	33	67	100
	20 ppb	2.0	4	6	13	27	40
	25 ppb	1.2	2.5	3.8	8.2	16	25
	30 ppb	0.5	1.0	1.5	3.2	6.4	9.7
ERV for asthma	0 ppb	30	47	62	5.7	8.7	12
	20 ppb	12	19	25	2.3	3.4	4.6
	25 ppb	7.5	11.5	15	1.4	2.1	2.8
	30 ppb	2.9	4.5	6	0.5	0.8	1.[illegible]
MRADs	0 ppb	0	27,000	92,000	0	1,700	5,800
	20 ppb	0	11,000	37,000	0	690	2,300
	25 ppb	0	6,800	23,000	0	430	1,400
	30 ppb	0	2,700	8,900	0	170	560
Asthma attacks	0 ppb	32,000	52,000	71,000	1,000	1,600	2,200
	20 ppb	13,000	20,000	28,000	400	640	880
	25 ppb	8,000	13,000	17,000	250	400	550
	30 ppb	3,100	5,000	6,800	97	160	210
Symptom days	0 ppb	48,000	94,000	140,000	300	590	880
	20 ppb	19,000	37,000	55,000	120	230	350
	25 ppb	12,000	23,000	34,000	74	150	220
	30 ppb	4,600	9,000	13,000	29	57	85

Table 2.6. Estimated acute effects on morbidity of SO_2

Endpoint	Back-ground	No. of cases (000s)			Damages (ECU million)		
		Low	Mid	High	Low	Mid	High
Hospital adm. for chronic bronchitis	0 $\mu g/m^{-3}$	1	26	52	6.6	170	340
	2.7 $\mu g/m^{-3}$	0.8	22	44	5.5	150	290

2.4.2. *Distribution of damage*

The use of a Geographical Information System for the analysis allows identification of the areas in which damages are concentrated. Figures 2.3, 2.4 and 2.5 show, for PM_{10}, SO_2 and O_3 respectively, the area in which 75% of total damage (i.e. not accounting for any effect of natural background levels) will

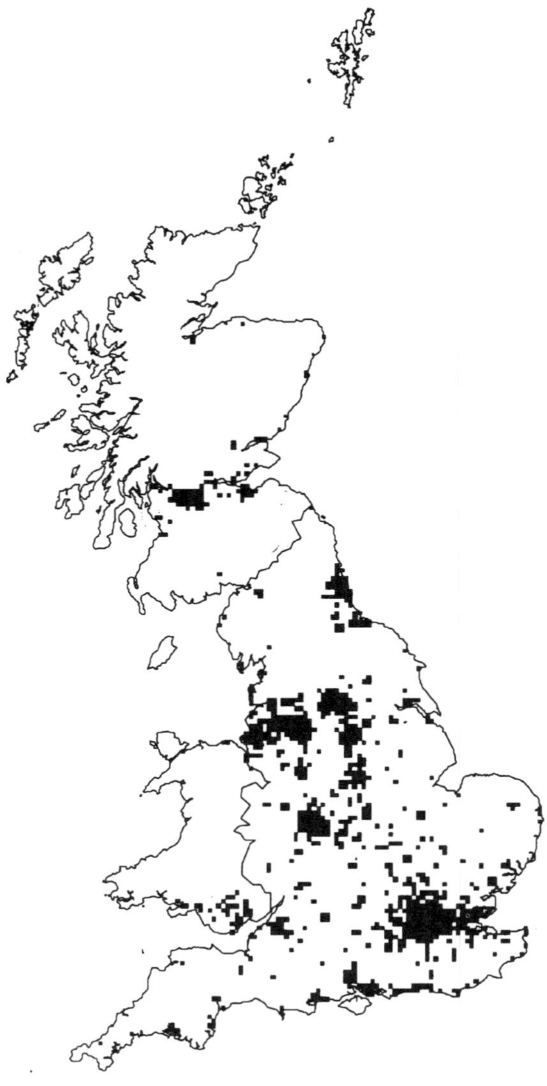

Fig. 2.3. Distribution of health effects of PM_{10} in the UK (75% of damage is contained in the cells shaded black). Map has been plotted at a resolution of 5 × 5 km.

occur, according to our estimations. In all cases damages are concentrated in the areas of greatest population. The distribution is most restricted for SO_2 and least restricted for ozone. This is not surprising given the depression of ozone levels in large urban areas caused by emissions of nitrogen monoxide (NO), which reacts readily with ozone to form NO_2.

Whilst the maps show the distribution of total estimated damage, we have not produced maps for damage that can be attributed solely to anthropogenic

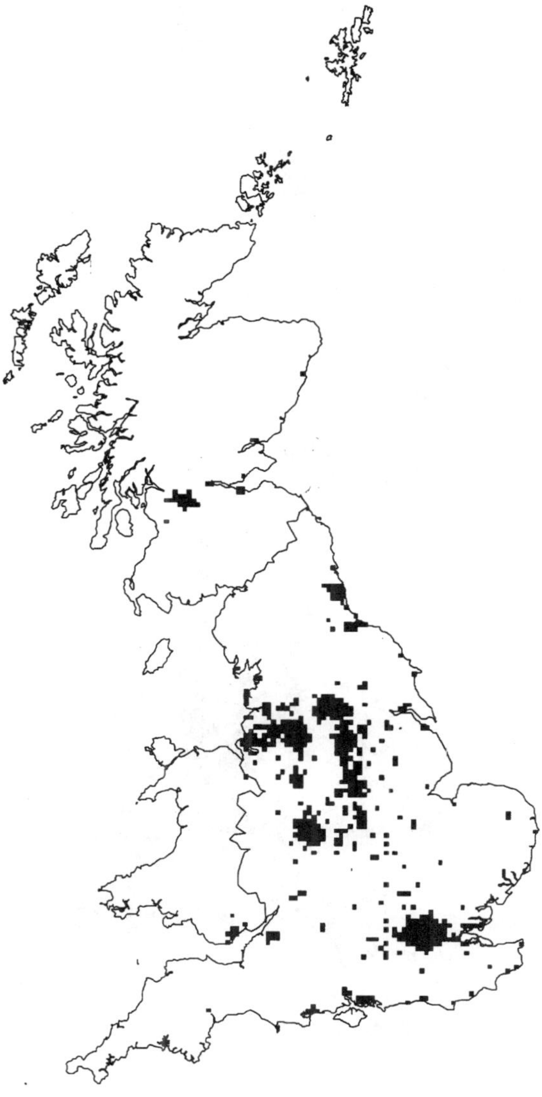

Fig. 2.4. Distribution of health effects of SO_2 in the UK (75% of damage is contained in the cells shaded black). Map has been plotted at a resolution of 5 × 5 km.

emissions. This is likely to make little difference to the maps for SO_2 and PM_{10}. However, it will make a very significant difference to the O_3 map. Depending on precisely which level is assumed for the natural background level, it is possible for O_3 concentration, and hence the net effect of O_3 on health to be lower in urban areas than outside.

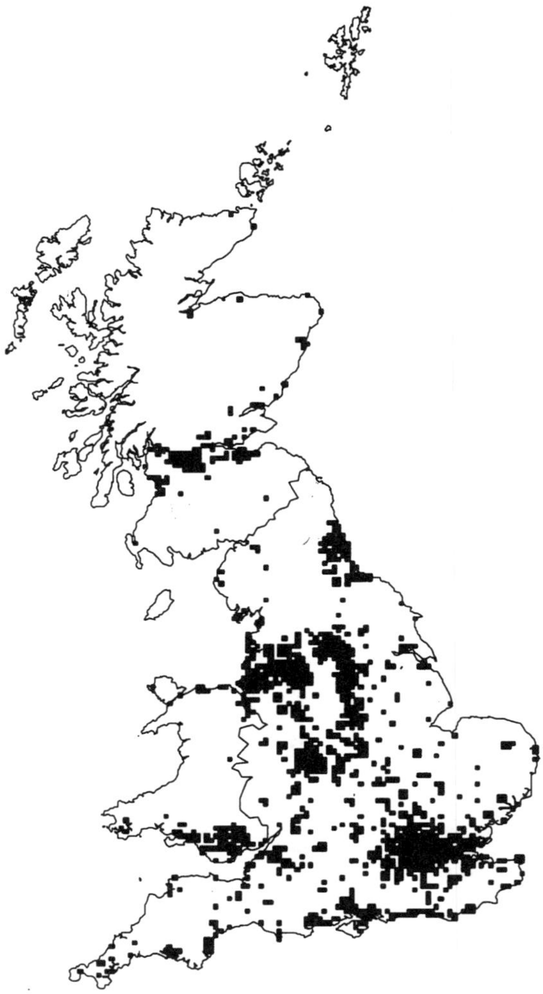

Fig. 2.5. Distribution of health effects of O_3 in the UK (75% of damage is contained in the cells shaded black). Map has been plotted at a resolution of 5 × 5 km.

2.5. Conclusions

According to our estimates, the number of people each year whose deaths are associated with air pollution (though it should not be considered as the only, or even principal, cause of death) is of the order of 8% of all deaths in the UK. This estimate, like the others that we have made here, is sensitive to some important assumptions, the most important of which seems likely to be our belief that it is not appropriate to consider a threshold for air pollution effects on health.

The valuation of acute effects on mortality is complicated by the typically short (though currently unquantified) life expectancy of those affected. Valuation of chronic effects is also complicated because of the small amount of data that is available for the assessment. Chronic effects are thus more uncertain than acute effects, though available information suggests that they are the more serious of the two. For morbidity the most important effects in economic terms seem to be relatively minor symptoms. Although each case is given only a small value, they are so numerous that associated damages are much greater than for the more serious effects.

For the pollutants that we have considered the largest damages are associated with PM_{10}, followed by SO_2 and O_3 for mortality, though the separation of effects for SO_2 and PM_{10} may be artificial (indeed, they are not considered to be additive here). For morbidity the order is PM_{10}, O_3 and then SO_2. Evidence for direct effects of NO_x is very limited, and hence we have not included it in our assessment, other than indirectly through its role in the formation of nitrate aerosol.

Natural levels of SO_2 are so close to zero that we found very little effect of removing background concentrations from total damage estimates. The situation was different for PM_{10} and even more so for O_3 because both of these pollutants have relatively high natural levels. However, there is some debate about what type of PM_{10} is responsible for effects on health. If damages are caused mainly by the very fine fractions, or sulphate aerosols, both of which arise principally from human activities, then it would be wrong to subtract 'natural' PM_{10} as this would already have been done, implicitly, in the derivation of exposure–response functions. In some urban areas annual mean O_3 levels are below background because of O_3 consumption by nitrogen monoxide, a pollutant emitted from road vehicles.

Damage is concentrated in urban areas for all pollutants because of the high concentration of people and high levels of SO_2 and PM_{10}. A similar pattern is observed for O_3, even though concentrations in some urban areas are lower than the supposed background.

We have demonstrated that the ExternE methodology can be successfully applied to quantification of effects of pollution on health at the national level. It is hoped that the approach used and the results will now be debated in detail by those working in this field and by other interested parties. It is to be stressed that, whilst we believe that our results reflect current knowledge, significant uncertainties remain in this analysis. From the policy perspective we believe that discussion of our methodology is at least as important as discussion of the results that we have generated. It is hoped that the issues raised in this Chapter will provide a focus for future research in this field.

CHAPTER 3

NOISE

3.1. Introduction

Noise is the result of a range of activities which produce pressure oscillations in the air – sound waves. Virtually any moving object generates sound. Background noise levels in the environment are dominated by wind noise, which is a strong function of wind speed. Transport systems, industrial and construction activities, and a variety of domestic sources are the most significant anthropogenic sources. The noise level experienced depends on the amplitude of the source, local geography and the distance from the observer. However, it is not possible to understand noise in purely physical terms. Noise is only unwanted sound, and the attitudes towards different sound sources depend on a range of social and psychological variables.

A range of health problems may be related to noise. At high levels, noise can damage the human ear. However, these noise levels are only likely to be experienced by workers in noisy occupations, under which conditions adequate ear protection should be worn. Auditory damage is therefore neglected in this assessment. Chronic effects of noise at lower levels are also of potential concern. Human sensitivity is very variable, but it is clear that there are potential psychological effects of noise, both due to annoyance and sleep disturbance. As with any psychological effects, there is the potential for secondary health problems, notably cardiovascular effects, although studies have found no strong correlation between residential noise levels and heart disease (European Commission, 1991).

It is difficult to distinguish precisely between health effects due to mild disturbance on the one hand, and the less exact concept of loss of amenity on the other. It is likely that loss of amenity covers much of the minor health effects which may be induced, and therefore this analysis concentrates on the amenity impacts of noise.

Noise is one of the most important amenity effects faced by the population of the UK. It is the source of more communication from the general public to the regulatory authorities than any other environmental problem – over 130,000 complaints to Local Authority Environmental Health Officers (EHO) each year (UKDOE, 1994).

Noise is a short range problem. At distances in excess of a few hundred metres from most sources, incremental noise levels due to most sources are very low, due both to dispersion of the energy and absorption in the air and solid objects. Even the aggregate noise impact over a very large population is negligible at distances in excess of few kilometres (European Commission, 1995a). For this reason, control over noise pollution is appropriately handled at a local level. Comprehensive data sets are not available. In addition, the spatial variability of noise levels is very great, so that GIS based systems for noise mapping are not available.

Local control over most noise sources in the UK has resulted in noise being controlled largely through the land use planning system. This system has a non-economistic culture, so that monetarisation of noise damages is still largely a theoretical and academic exercise rather than a practical input to decision making.

3.2. Methodology

The general methodology for including a pollutant in a green accounting exercise needs to be adapted to evaluate noise impacts. For noise, it is assumed that the only significant receptors are people and that they are sensitive to noise only at their homes. Mappings of houses obviously exist: in the UK to the level of the individual dwelling. However, noise mappings are more problematic, as noise levels vary considerably over even short distances. Some surveys have used this approach (e.g. Parkin *et al.*, 1968). However, there are no comprehensive, spatially disaggregated databases on noise levels. In addition, it is not realistically possible to produce noise source maps, as data on noise sources, spatially disaggregated to an adequate level, are also unavailable.

However, this does not prevent some assessment being made, as survey data on noise levels from a statistically representative sample of UK houses is available (Sargent and Fothergill, 1993). This allows spatial disaggregation to be avoided within the assessment, so that the methodology used is:

- identify the affected receptors,
- identify the response of receptors to the pollutant (the dose–response function),
- identify survey data on household external noise levels,
- use national housing data to calculate the numbers of houses in various noise categories,
- use the noise dose–response function to calculate the aggregate national noise,
- apply monetary value to the impact.

The methodology for noise is further simplified, because the only important impact of noise (at least at the levels experienced by the general public) is a loss of amenity, which may be measured directly in financial terms through hedonic pricing studies. The dose–response function therefore provides a monetary value directly.

3.3. The amenity impact of noise

The basis of the noise level needs to be specified more carefully. A variety of measures are used to allow for the fact that noise levels vary over time. For impulsive and intermittent noise sources (e.g. aircraft), indices like the noise nuisance index (NNI) have been developed, which combine peak noise levels with event frequency. These can be used as an input to monetary valuation exercises (European Commission, 1995b). However, this approach is not applicable to more continuous noise sources, such as road traffic, domestic and industrial sources, which are responsible for most noise pollution.

Tonality (the presence or otherwise of distinguishable single frequencies in the sound) is an issue because tonal noise is generally found to be more annoying than 'broad band' (non-tonal) noise. However, most of the important noise sources are not tonal, and therefore this issue can be neglected at the level of national aggregation.

Frequency, even of broad band noise, is a relevant consideration because the human ear has different sensitivities to different frequencies within the audible range. This is directly allowed for in the unit of noise measurement the audio-weighted decibel, or dB(A), which weights sound measurements at different frequencies according to human sensitivity.

Even for most continuous noise sources, the noise level varies on a range of time scales. No single value can capture all possible variations, and a range of options is used to assess different characteristics. Commonly used measures include:

- L_{A90} – the level exceeded for 90% of the time – a reasonable measure of background noise levels,
- L_{A10} – the noise level exceeded for 10% of the time – a measure of peak levels, and
- L_{Aeq} – the noise level corresponding to the average energy in the noise.

For this study the L_{Aeq}, that is the energy weighted average noise level, is used as the measure of noise. This conforms to the recommendations of the International Organisation for Standardisation (ISO, 1987). Most importantly, the L_{Aeq} measure is consistent with the hedonic price studies used for valuation.

The time of day over which noise is measured is another issue. Various periods are used in reported data, including day time (16 hour or 18 hour) and night time (6 hour or 8 hour), as well as the full 24 hour period. These lead to measures variously identified as $L_{Aeq,16}$, $L_{Aeq,18}$, $L_{Aeq,6}$, $L_{Aeq,8}$ and $L_{Aeq,24}$. In some cases a composite day–night adjusted measure, L_{DN} is derived by increasing the night time noise level by 10 dB(A). This allows effectively for the higher annoyance caused by night time noise in residential areas. Whilst the L_{DN} level might be a better measure of annoyance, the $L_{Aeq,24}$ is more commonly used in statistical compilations. The data reported in the research used here (Sargent and Fothergill, 1993) includes 8 hour, 16 hour and 24 hour analyses. By inspection it is clear that the 8 hour levels are approximately 10 dB(A) less than 16 hour levels (on average), and therefore no major error is introduced by using the $L_{Aeq,24}$ level as a measure of loss of amenity.

3.4. Monetary valuation of noise

Monetary valuation of noise is most commonly undertaken using house price data and the hedonic pricing method. It is implicitly assumed that a capitalised measure of loss of amenity is reflected in house prices. A large number of hedonic price studies have been undertaken in recent years in Europe. These have been subjected to extensive review and meta-analysis in the course of a major European Commission study on environmental externalities (European Commission, 1995b). The review is not repeated here. It is concluded that a linear noise sensitivity depreciation index (NSDI) of 0.9% per dB(A) (L_{Aeq}) is the best valuation of environmental noise.

Most of the noise valuation studies undertaken have been at relatively high levels of environmental noise, where there is practical interest in the answers and significant results are expected. Surveys have typically addressed traffic noise levels in excess of 55 dB(A). There remains a problem in extrapolating from these surveys to the monetary valuation of noise at lower levels. No completely satisfactory solution to this problem has been identified (European Commission, 1995a). However, to calculate the aggregate damage it is necessary to identify some noise baseline as a threshold at which to begin implementation of the NSDI identified above.

The baseline could be either:

- the level of noise corresponding to zero economic activity (i.e. non-anthropogenic sources);
- the level of noise corresponding to zero amenity loss, i.e. a threshold for the dose–response function.

In principle, the higher of these is the baseline above which damages should be assessed. The former is difficult to assess but is likely to be very low (< 40 dB(A) L_{Aeq}). The latter is often assumed to be 55 dB(A) (day time L_{Aeq}), because this is the WHO 'no significant community annoyance' standard. It is not axiomatic that this level corresponds to a zero amenity loss – amenity is a highly subjective quality, which is resistant to physical quantification. Certainly studies of impacts of new noise sources in rural areas (e.g. wind turbines) indicates some concerns about noise at levels well below 55 dB(A). However, it seems reasonable to assume that these impacts will cause damages which are small in comparison to those at higher noise levels.

For this study a threshold of 55 dB(A) ($L_{\mathrm{Aeq,24}}$) is used. Because of the uncertainty in the appropriate choice, and as this is likely to be an important parameter, the sensitivity to choices of 50 dB(A) and 60 dB(A) is also investigated and reported.

3.5. Household noise data for the UK

To implement the methodology, household noise data for the UK is required. This is taken from a major study of representative households across the UK (Sargent and Fothergill, 1993). The noise levels are similar to those reported in other, less comprehensive surveys and shows little change from earlier surveys in the UK (e.g. Harland and Abbott, 1977). The major sources of annoyance are road traffic and aircraft (UKDOE, 1994). Noise levels were measured over the whole day at a distance 1 m. from the front facade of the building and 1.2 metres above the ground. The output of this study provides the fraction of households facing noise levels in excess of certain values. The data for $L_{\mathrm{Aeq,24}}$ levels is summarised in Table 3.1.

Most of the population faces noise levels in excess of 50 dB(A), with a significant fraction above 60 dB(A). From the other noise measures reported, it is clear that 7% of the population faces noise levels above 68 dB(A) ($L_{\mathrm{A10,18}}$), which is the qualifying level for noise insulation from new roads under current UK regulations.

The banding of noise levels reported is rather too coarse than is desirable for valuation. However, it is found that the data fit a logistic curve of the form:

$$\text{Percentage exceedence, } P = \frac{100}{1 + \exp\left[\dfrac{L - L_0}{a}\right]}$$

where:

L is the noise level,
L_0 is the median noise level, and
a is the width of the distribution.

Table 3.1. UK noise levels survey data

Noise level (dB(A) L_{Aeq})	% of houses where level is exceeded
30	100.0
40	99.8
50	80.8
60	20.0
70	0.8

The values for best fit are $L_0 = 56$ and $a = 2.78$. These are used to estimate the percentage exceedance at each unit dB(A), and thence the fraction of households in each decibel band. The results are shown in Figure 3.1.

3.6. Results

The household noise distribution data shown in Figure 3.1 has been used as the input to the valuation calculation. For each dB(A) level, the damages are calculated, assuming a stock of 22 million UK houses, an annoyance threshold of 55 dB(A), a NSDI of 0.9% per dB(A), an average house price of ECU 100,000 and an annuity factor of 8%. The damages as a function of noise level are shown in Figure 3.2.

The aggregate damages are ECU 4 billion per year. For the reasons given above, the sensitivity of these to the noise threshold assumed has been investigated. The results are given in Table 3.2.

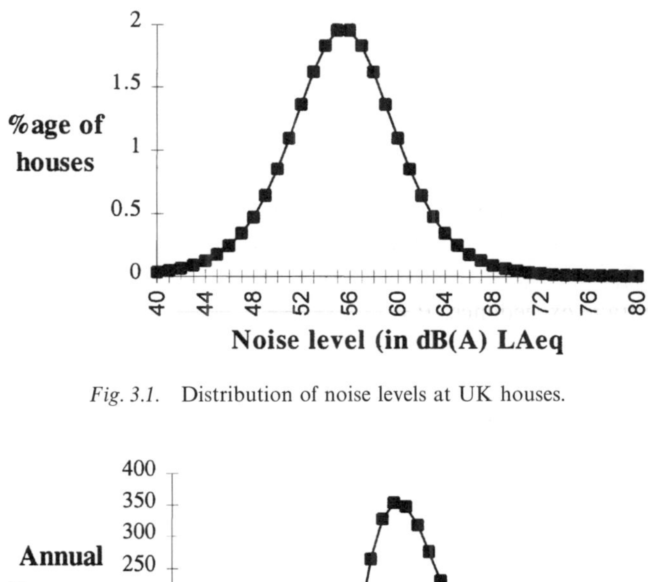

Fig. 3.1. Distribution of noise levels at UK houses.

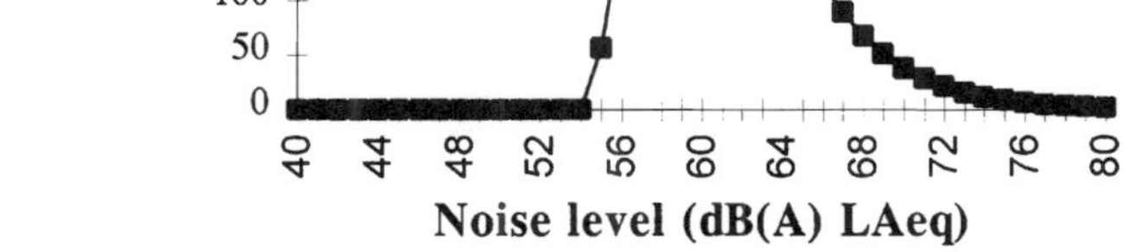

Fig. 3.2. UK noise damages as a function of noise level (assuming a threshold of 55 dB(A)).

Table 3.2. Aggregate UK noise damages as a function of threshold

Noise threshold, dB(A)	Aggregate damages, billion ECU
50	10.4
55	4.0
60	1.0

3.7. Conclusions

Noise damages for the UK have been assessed used a representative sample of household noise data taken from official sources. The valuation uses a meta-analysis from an earlier European Commission project. The best estimate of damages is ECU 4.0 billion per year.

The results are sensitive to the assumption made about the threshold at which human begin to be sensitive to noise, and therefore the level at which noise amenity damages begin. The best estimate is based upon a threshold of 55 dB(A) – the WHO standard for community annoyance. Using a range of 50–60 dB(A) for the threshold, a damage range of ECU 1–10.4 billion has been found.

CHAPTER 4

CROPS

4.1. Introduction

The burning of fossil fuels results in emissions of the gaseous pollutants NO, NO_2 and SO_2. Reactions within the atmosphere can lead to the formation of secondary compounds, primarily ozone, SO_4, HNO_3 and SO_4. This section considers the direct and indirect effects of these pollutants on agricultural production. The methodologies used are based on those developed during the European Commission supported ExternE project (European Commission, 1995). Three aspects are covered:

- direct impacts of gaseous pollutants,
- fertiliser effect of total deposited NO_x,
- indirect effects of deposited acidity.

Quantification of the direct impacts of the gaseous pollutants is based on published dose–response relationships. This approach requires the following input datasets:

- spatially disaggregated data on the stock at risk, (distribution and yield of the relevant crop),
- dose–response relationship for the crop–pollutant combination,
- spatially disaggregated data on atmospheric concentrations of the relevant gaseous pollutants,
- the background concentration of the relevant pollution under non-polluted conditions.

It should be noted that indirect impacts of gaseous air pollutants have not been considered. The most important of these include interactions with climate and pests. They have not been considered here because of a lack of dose–response data. Effects of climate seem likely to be most pronounced for ozone, through an antagonistic interaction with drought. This contrasts with the interaction between SO_2, NO_x and O_3 and pests, however, which are typically synergistic.

Quantification of the fertiliser effect of deposited oxidised nitrogen requires:

- spatially disaggregated data on the 'stock at risk'/receptor distributions,
- spatially disaggregated data on NO_x deposition,
- data on deposition of NO_x under pristine conditions.

The approach developed for the quantification of the indirect effects of deposited acidity requires:

- spatially disaggregated data on the stock at risk,
- spatially disaggregated data on deposited acidity,
- data on the deposition of acidity under pristine conditions.

4.2. Stock at risk and current production

The study focused on intensive agriculture. Intensive agriculture can be broadly divided into two sections, arable and pasture. Table 4.1 shows the breakdown of arable agriculture in the United Kingdom for 1991. The area covered represents 26% of the land that the Ministry of Agriculture, Fisheries and Food (MAFF) record as agricultural holdings. Other land includes short term grass (9%), permanent pasture (31%) and sole rough grazing (27%). Cereals are the dominant crop totalling over 70% of the arable land area.

In order to spatially disaggregate the production of cereals, two other datasets were used, the Land Cover Map of Great Britain (LCM) (Fuller *et al.*, 1993) and Countryside Survey 1990 (CS 90) field survey data (Barr *et al.*, 1993). The LCM offers information on the land cover of every 1 km square in Great Britain in terms of 25 different land cover categories; the occurrences of the cover categories were identified by combining Landsat TM satellite scenes taken at different times of year; the data from which the LCM is derived is centred on 1990. One of the land cover categories is tilled land (target class 18 of the LCM land cover categories); this provided one measure of the spatial distribution of areas covered by arable agriculture.

Although tilled land can be identified using this technique, there may be some confusion with short term grass, especially if it is being ploughed or re-seeded at, or immediately before the time at which the image was produced. To overcome these problems and to provide the data on individual crop

Table 4.1. Area and production of crops in the United Kingdom, 1991

Crop	Area ('000 ha)	% agriculture	% arable	Production ('000 tonnes)
Wheat	1981	11	40	14,363
Barley	1393	8	28	7,627
Oats	104	1	2	523
Potatoes	177	1	4	6267
Oilseed rape	440	3	9	1278
Sugar beet	196	1	4	7673
Peas and beans	193	1	4	647
Total		26	91	

distributions, the proportion of tilled land in each 1 km square was divided into different crop types using CS90 data (Howard *et al.*, 1995). CS90 was a national field survey which recorded land cover across Great Britain in each of a sample of 1 km squares. The sample of 1 km squares surveyed was stratified using the ITE Land Classification (Bunce *et al.*, in press). The average distribution of the various arable crops per 1 km square was determined for each land class, using the field survey data from the relevant land class. These averages, for the relevant land class, were then used to allocate the tilled-land data for each 1 km square, from the LCM, between crops. The resultant figures were totalled by county and compared to MAFF June Census data for 1990. An aggregation of the data into 10 km × 10 km squares was used to match pollutant concentration datasets. The agricultural production of the crops was estimated by weighting the area estimates with production level coefficients (Nix, 1990); these data were held in a GIS on a 10 km grid.

Impacts on grass production were qualified using a dose–response relationship for Italian rye-grass (*Lolium perenne*). This required the estimation of the area of agricultural grasslands under rye-grass. The spatial distribution of areas under managed agricultural grass were again estimated by combining LCM and CS90 data. The LCM provided the distribution of mown and grazed turf (target class 6) and meadow/verge/semi-natural grass (target class 7). The proportion of each 1 km square in this these two land cover classes was estimated. The LCM category pasture/meadow/amenity grass was used to provide the spatial distribution of managed grasslands, as the proportion in each 1 km square in this land cover category. The proportion of these grasslands covered by Italian rye-grass was estimated using data from the CS 90 field survey. The field survey divided managed grasslands up into nine categories (Table 4.2). The field survey data were used to determine the average occurrence per 1 km square of each these nine categories in each ITE Land Class; these data were then used to weight the managed grass land cover information from

Table 4.2. Proportion of grassland categories considered to be under Italian rye-grass

Grassland category	Proportion Italian rye-grass
Recreational (mown) grass	0.00
Short term-grass	0.95
Pure rye-grass	0.75
Well managed grass	0.50
Weedy swards	0.25
Non-agriculturally improved grass	0.00
Calcareous grass	0.00
Upland grass	0.00
Maritime vegetation	0.00

the LCM, and estimate the area under in each 1 km square. As with the arable crops, the resultant data was aggregated on a 10×10 km basis.

The productivity of the grasslands was assessed in a different way to that of the arable crops. Agricultural grassland can be managed for different purposes (namely grazing, fodder production, seed production or fallow); the processes involved differ for different uses. For simplicity, all rye-grass grasslands were weighted with a yield figure for grazed grass (Nix, 1991) which is expressed as tonnes per hectare. The output derived was an estimate of *Lolium* production for each 10×10 km square.

4.3. Direct effects of sulphur dioxide

Although experimental studies have indicated significant direct impacts of SO_2 on the yield of a range of crops, dose–response relationships have only been derived from work on a small number of crops. The impacts on the yield of wheat, barley, oats, field beans, and intensive grass are quantified below. Available data suggested that oil seed rape is insensitive to SO_2 and no suitable dose–response relationship could be found for potatoes.

4.3.1. *Pollution data*

Mean annual concentrations of SO_2 from 1992 were used in the calculations (see Chapter 2). Average values are given for 20 km $\times$ 20 km squares. The background level of SO_2, under pristine conditions, was taken as a value of 1 ppb. Data was provided by D. Fowler and R. Smith (personal communication).

4.3.2. *Dose–response relationships*

Wheat, barley and oats

The change in yield of wheat and barley due to SO_2 were calculated using one of two dose–response equations, dependent upon the SO_2 concentration (European Commission, 1995):

$$y = -0.69(SO_2) + 9.35$$

for concentrations greater than 13.6 ppb, and

$$y = 0.74(SO_2) - 0.055(SO_2)^2$$

for concentrations up to and including 13.6 ppb.

Where:

y is the percentage change in yield and
SO_2 is the atmospheric concentration of SO_2 in ppb.

Potatoes
An equation for potatoes has been published (Pell *et al.*, 1988), but the response is described as the change in the number of Grade One tubers. The response combines both SO_2 and O_3. No statistics could be found to describe the production of tubers of different grades. The model was not applied in the current study; and thus no attempt was made to quantify impacts on potato production.

Field beans and peas.
A linear regression for the relationship between yield reduction due to SO_2 and beans was derived by amalgamating the cultivars tested by Weigel *et al.* (1990):

$$y = 0.785 + 0.203(SO_2)$$

Where:

y is the yield of beans and SO_2 in the concentration of SO_2 in ppb.

Intensive grassland
The following functions for the response in *Lolium* yield to SO_2 were described by Roberts (1984) and European Commission (1995) for concentrations greater than 15.3 ppb:

$$y = 2.75 + 0.18(SO_2)$$

and for concentrations up to and including 15.3 ppb:

$$y = 0.20(SO_2) - 0.013(SO_2)^2$$

Where:

y is the percentage yield loss and SO_2 concentrations are expressed in ppb.

This function has not been used here because of a lack of data on the stock at risk, beyond a knowledge of the area of *Lolium* grown. However, this should be considered in a continuation of the present study.

4.3.3. *Calculation of impacts*

The dose–response relationships were combined with the data on current production of the relevant crop and on SO_2 concentrations within a GIS to calculate the production per 10 km square under pristine conditions. These data were compared with the calculated current production to determine the change in production resulting from pollution. The results suggest reductions in yield of all the crops studied as a result of current pollution levels of SO_2 in central England (around Lincolnshire), but increased production across most of the rest of Britain. This pattern reflects the pattern of pollution concentrations (see Chapter 1). The area of the most intensive arable production in the UK also coincides broadly with the area with the highest concentrations of SO_2, concentrations large enough to produce a reduction in yield. The small increases in yield in the areas of the UK with small concentrations of SO_2 probably result from a combination of suppression of diseases and a fertiliser effect. However, the large number of small gains are outweighed nationally by large losses in a few squares.

The estimated national totals for production and changes in production resulting from pollution are summarised in Table 4.3. A reduction in production is shown for each of the crops as a result of current levels of pollution, compared with the assumed pristine conditions. The largest gross figure, and largest estimated cost is for wheat, reflecting the large area of this crop which is grown. However, beans, oats and barley show larger percentage losses of production than wheat, reflecting their more restricted distribution which is mainly within the area of Great Britain which has the highest rural SO_2 levels. Altogether, the crops included in this assessment total 46% by value of UK arable production.

4.4. Indirect effects of nitrogen oxides

Available data suggests that adverse impacts on crop growth and yield are highly unlikely at current ambient concentrations of nitrogen oxides. No attempt was therefore made to quantify direct impacts of NO_x on yield. However, deposited nitrogen can act as a fertiliser which may enhance crop production. Dose–response functions are available for the impact of N additions on the yield of various crops and these could have been used to calculate responses to atmospheric inputs. However, spatially disaggregated data would also be required on conventional fertiliser inputs as the atmospheric inputs would have to be added to these conventional fertiliser additions to assess the response. We were unable to obtain a suitable national database on fertiliser

Table 4.3. Agricultural production in Great Britain showing estimates of loss due to SO_2 pollution

	'000					
	Current production	Potential production	Loss of production	(%)	Value per tonne	Loss (million ECU)
Wheat	15,600	15,730	130	0.84	96	12.5
Barley	7,290	7,290	90	1.27	54	4.9
Oats	541	549	8	1.45	56	0.4
Beans	347	357	10	2.84	56	0.6
Total						18.4

use and, therefore explored the application of a simpler approach to calculate the possible beneficial effect as of NO_x deposition. The deposition of NO_x to land under intensive agricultural production, arable or grassland, was assumed to be beneficial and act as a fertiliser. The quantity of N deposited, as NO_x, to the relevant area of land was calculated and valued as fertiliser. Ammonia deposition was not considered as emissions from agriculture are beyond the remit of the present study.

4.4.1. *Stock at risk/receptor area*

The area under arable or intensive grassland was defined using the Land Cover Map. The number of hectares within each 1 km square classified in target land cover classes 6, 7 and 18 was derived from the LCM dataset. These data were aggregated to give the total area in the selected land cover classes, from the LCM data, in each 20 × 20 grid square; the data were also combined to give the total area under in these target land cover classes (Table 4.4).

4.4.2. *Pollution/deposition data*

The N deposition as NO_x, in kg/ha/yr was provided on a 20 km grid cell base by D. Fowler and R. Smith from ITE Bush. Deposition of NO_x under pristine conditions was considered to be negligible. These data were combined with the stock at risk and receptor area data to quantify the total deposition of NO_x–N to the areas in the relevant land cover classes. The total estimated deposition was 7.1 kg/ha/yr (Table 4.4).

4.4.3. *Valuation of N inputs*

The value of N added was calculated using a figure of ECU 0.43 per kilo N (Nix, 1990). The total NO_x deposition to agricultural land was multiplied by this figure to derive the total value of the deposited N as a fertiliser. It is interesting that the estimated value, ECU 23 million, is of the same order of magnitude as the estimated value of the loss in wheat production as a result of current levels of pollutant SO_2.

Table 4.4. The quantity and value of nitrogen (N) falling as NO_x on the area of intensive grassland and arable land in Great Britain

	Deposition		
Area, ha	kg/ha/year	Tonnes	Cost (ECU mn.)
8,090,000	7.1	57,600	23

4.5. Direct impacts of ozone

The same methodologies as outlined above for the quantification of the direct impacts of SO_2 on arable crops were used to assess the direct impacts of ozone. The same stock at risk databases were used and these databases were linked with spatially disaggregated atmospheric ozone data and dose–response relationships relating yield to ozone concentration. The analysis presented here accounts for effects on 69% (by value) of UK arable crop production.

4.5.1. *Pollution data*

Two datasets were used of the current patterns of ozone concentrations. The first was the mean concentration at 5 km resolution (see Chapter 1). The second form of distribution was AOT_{40} (accumulation over threshold of 40 ppb); the relationship is specific to different receptors since the response is time dependent (cereals would usually respond to exceedence between April–July, forests more likely April to September). The only AOT_{40} data available for an agricultural crop was for wheat and was provided by D. Fowler and R. Smith (personal communication).

4.5.2. *Dose/response relationships*

This mean concentration data was applied in combination with the NCLAN (National Crop Loss Assessment) dose–response relationships to produce estimates of change for wheat, barley, potatoes, oilseed rape, peas and beans (Heck *et al.*, 1988). Sugar beet was omitted as the threshold for effect was higher than the values in the pollution data. For the NCLAN functions the 'natural' O_3 level assumed in this study is of no consequence as effects are only recorded at levels greater than this value. Further details are given in Table 4.5.

4.5.3. *Calculation of impacts*

The data on current production, the ozone concentration data and the dose response relationships were combined to calculate the production under the

Table 4.5. Rate of loss and thresholds for various crops due to ozone concentrations

Crop	Threshold (ppb)	Rate of loss
Wheat	30	0.15
Barley	40	0.34
Potatoes	30	0.40
Sugar beet	50	0.34
Oilseed rape	30	0.34
Peas and beans	30	0.60

assumed pristine conditions; this was regarded as the potential production. These data were compared with the current production data to determine the loss in production due to current pollution levels of ozone, both absolute and percentage. The values of the crops per tonne were then used to calculate the value of the loss in production (Table 4.6). The estimated loss in wheat production is the same as that estimated to result from current concentrations of SO_2. No impact is seen on barley production and the estimated impact on the production beans is of a similar magnitude to that estimated for SO_2.

4.6. Indirect impacts of deposited acidity on intensive agriculture

The indirect impacts of deposited acidity are soil mediated. Deposited acidity increases the rate of soil acidification which can, in turn impact on crop yields. Agricultural lime, ground limestone or magnesium limestone, is routinely added to agricultural soils when necessary to maintain soil pH conditions for optimal crop production. These additions neutralise acidity that is both generated within the agricultural system and that which is deposited from the atmosphere. A small amount of quantitative data exists on the relationship between pH and crop yield. This could in principle be linked to process models predicting changes in soil pH in response to changes in atmospheric inputs. However, this approach has yet to be tested. A much simpler methodology has, therefore been used here to produce an estimated valuation of the impacts of deposited acidity. This approach values the lime additions required to neutralise the deposited acidity and is based on the methodology used in the ExternE project (European Commission, 1995).

4.6.1. *Pollution data*

Total acid deposition of pollutant origin has been calculated from the sum of the total non-marine sulphur and nitrogen deposition expressed as the equivalent weight of hydrogen ions per hectare (keq H^+ kg^{-1} ha).

Table 4.6. Crop production ('000 tonnes) under present and pristine conditions and calculation of value of loss

	Current	Pristine	Loss	% loss	Price/tonne (ECU)	Damage (ECU mn.)
Wheat	15,600	15,720	120	1	96	11.7
Barley	7,200	7,200	0	0	54	0
Potatoes	5,500	5,680	180	3	115	20.8
Sugar beet				0	45	0
Oilseed rape	1,300	1,313	13	1	350	4.6
Peas	363	375	12	3	56	0.7
Beans	350	360	10	3	56	0.6
Total						38.4

4.6.2. *Stock at risk*

The stock at risk was determined using the Land Cover Map (see above) and a soils database. The area of the UK under intensive agriculture (arable plus intensive grass), was defined by amalgamating the tilled-land category (target class 18) with the managed grass classes (target classes 6 and 7) from the LCM. The response of soils to deposited acidity varies with the chemistry and mineralogy of the soils. Soils containing free carbonates will buffer deposited acidity without additions of extra agricultural lime. The database of land dominated by intensive agriculture was overlain by a database defining areas of the UK dominated by calcareous soils to provide the final definition of stock at risk; areas of the UK where the dominant land use is intensive agriculture on acid sensitive soils. The soils mask defined 1 km squares dominated by the following soil types: rendzinas, sand pararendzinas, rendzina-like alluvial soils, calcareous pelosols, brown calcareous earths, brown calcareous sands and brown calcareous alluvial soils. The total area of agricultural land on acid sensitive soils was estimated at 66,000 km^2. The data from the 1 km grid square database was aggregated to a 20 km grid square basis, to match the pollution data set.

The use of a soil mask based on calcareous soils will result in an overestimate of the amount of lime required as some high pH, clay-rich non-calcareous soils will also buffer all or most of the deposited acidity. However, buffering capacity varies considerably within this broad grouping of soils and the appropriate soil mask is, as a result difficult to define. The overall effect on the final result is likely to be small, though this refinement could be made in the future.

4.6.3. *Calculation of impacts*

The stock at risk database was overlain with the spatially disaggregated total deposited acidity to determine the acidity deposited on the stock at risk. The total deposition was estimated at 1,800 tonnes H^+. Neutralisation of this acidity using calcium carbonate requires 50 times as much lime weight for weight. Thus, 91,000 tonnes of calcium carbonate would be required. At a cost of 16 ECU per tonne, the total value of the lime would be 1.5 million ECU. Additional labour charges for application of this lime do not need to be considered because it would be applied with the lime used to counteract harvest-induced acidification: it seems unlikely that the incremental requirement for lime at the level of the individual farm will increase by an amount that demands a significant increase in labour.

4.7. Conclusions

Estimates of the costs associated with direct effects of SO_2, O_3 and deposited nitrogen on crop yield, and of deposited acidity on soil liming requirements have been made for the UK. Indirect effects, for example through interactions with pests and climate have not been considered because of a lack of suitable dose–response data. The dose–response function used for assessment of SO_2 effects on certain crops allows for increased crop growth at low SO_2 levels, which apply over most of the country.

The total direct effect of pollutants on crop production will depend significantly on the distribution of the crops that have not yet been integrated into our analysis. Future work should seek to identify dose–response functions for these crops and also begin to consider the indirect effects of air pollution on crops mediated through climate and pests.

CHAPTER 5

FORESTS

5.1. Introduction

The problem of forest decline, and its possible association with atmospheric pollution has been the subject of much research in recent years. Although at least 18 major declines were reported in Europe and North America between 1900 and the late 1970s, the fact that damage was identified simultaneously in so many parts of the northern hemisphere makes the present problem particularly noteworthy. Declines have long term consequences on yield or even on the existence of a forest or species.

In Europe unusual symptoms were first noticed on silver fir in Germany during the 1970s, followed by Norway spruce, Scots pine, beech and other species. The widespread nature of the problem, in terms of area and species affected, suggested that the causal factor operated at a regional level. The only agents that seemed to meet this condition were pollution and climate stress. Pollution was further implicated in the observed declines by the fact that population density of Scots pine in Germany and the UK had earlier been shown to be correlated with SO_2 levels (Knabe, 1970; Farrar *et al.*, 1977). SO_2 at annual mean concentrations of between 15 and 40 ppb is also known to be directly responsible for the death of spruce within the 'Black Triangle' of Czechoslovakia, Poland and eastern Germany (Moldan and Schnoor, 1991).

NO_x at current levels is not believed to be capable of causing damage on its own, but may increase damage by interaction with other pollutants or other stresses. Nitrogen deposition is believed by some experts to be responsible for increased forest growth in many areas in the last few decades, though it is questionable whether this is sustainable: some work suggests that excess nitrogen deposition may lead to nutrient imbalances, particularly for base cations. Concern has also focused on O_3, levels of which have risen in recent years, largely as a consequence of the expansion of motor traffic. In areas subject to high insolation this pollutant is quite capable of damaging plants, as experience from the US clearly shows. O_3 levels in the UK frequently exceed those recommended by the World Health Organisation and the UN ECE (Bower *et al.*, 1991).

Although damage to leaves may be caused directly by exposure to pollutants, less direct effects, mediated through the soil, are probably more important. The work of Tamm and Hallbacken (1988) in Sweden has proved that soil acidification is a serious problem even in areas of Europe that are far removed from industrial activity. Soil acidification is known to disrupt nutrient cycling within forests by increasing leaching of essential nutrients such as calcium and magnesium. Increased aluminium mobilisation may exacerbate this problem by interfering antagonistically in the uptake of nutrient base cations. Very high concentrations of aluminium in the soil solution may damage roots.

Valuation of damage to forests should ideally account for the fact that trees and forests perform a wide range of useful functions in addition to timber production. These include use for recreation, protection from avalanches in mountainous areas, CO_2 uptake and storage, water management and wildlife habitat. Indeed, some analysts suggest that the economic damages resulting from reduced non-timber benefits may substantially outweigh the effects of lost timber production (see Nilsson *et al.*, 1992).

Within this Chapter we have applied two available models to estimate the effects of air pollution on timber production for each of the four countries involved in this study. In addition to this further data is given on expenditure in Germany on measures to limit air pollution damage to forestry.

5.2. Modelling forest response

5.2.1. *Available models*

Models of tree response to pollutants have been developed at a number of levels, ranging from the assessment of sub-cellular effects on biochemistry and micro-structure to estimation of damage at the continental scale. Forest growth models are made particularly complex by the fact that trees are long lived and need to be treated in a sustainable manner. To ensure an adequate supply of timber in future years it is thus important that harvests are properly planned. Even under ideal conditions harvesting levels cannot be suddenly increased beyond a point at which the amount of standing timber starts to fall, without either reducing the amount of timber cut in future years or rapidly expanding the growing stock. If acidic deposition has a serious effect on tree growth (which seems likely) it is probable that impacts associated with soil acidification will persist for many years after soils have recovered, whilst the quantity of standing timber recovers to a long term sustainable level.

A number of modelling exercises were reviewed for the ExternE Project (European Commission, 1995):

- models reviewed by the National Acid Precipitation Assessment Programme in the USA (NAPAP, 1990b);
- the IIASA Forest Study Model (Nilsson *et al.*, 1992);
- the forest module of the RAINS model (Makela and Schopp, 1990);

- Sverdrup *et al.* (1993);
- Kuylenstierna and Chadwick (1994).

NAPAP (1990) reported that there were no models able to assess damages caused by the pollutants of interest to this study, at least for application within the USA.

The most ambitious attempt within Europe so far is the IIASA Forest Study Model (Nilsson *et al.*, 1992). Although application of the damage module described is open to serious criticism, it can at least be used to provide some guide to the range of possible effects. A further difficulty with the use of the results of this model is that the emissions scenario used is no longer realistic in the light of a tightening of the Sulphur Protocol within the UN ECE Convention on Long Range Trans-boundary Air Pollution.

The forest module of RAINS (Makela and Schopp, 1990) has been validated to some degree, but is applicable only within a restricted area in Eastern Europe and hence is not relevant in this study. The method presented by Sverdrup *et al.* (1993) needs further evaluation. It is worth noting that all three of these European models tend to predict serious forest damage.

The approach identified by Kuylenstierna and Chadwick is based on the hypothesis that forest damage arises from acidification of the soil giving rise to aluminium concentrations that are high enough to harm the fine roots (Ulrich, 1985; 1990). This in turn affects the ability of trees to take up nutrients and water. The current model is acknowledged by its authors to be preliminary; for example, it does not properly account for the potential for long term damages from acidic deposition, nor does it adequately reflect spatial variation in receptor sensitivity. Also, whilst the dose–response functions that are used are statistically significant, they fail to explain a large amount of the observed variation. However, the model has the advantage of being both simple and transparent. On these grounds, and also to emphasise that re-analysis of the wealth of data that exists on forest condition can provide new insight and should be treated as a research priority, the method was used in preference to the other models developed for the ExternE Project.

5.3. Methodology

For the present study we have used the Kuylenstierna and Chadwick model (hereafter denoted as the KCM), which proceeds through the following stages:

- identification of areas of broadleaved and coniferous forest
- identification of forest areas subjected to critical loads exceedence
- application of a dose–response function to assess effect of critical loads exceedance on condition of the forest canopy
- application of a further dose–response function to assess the effect of (3) on growth
- valuation of lost timber yield

Data on critical loads exceedance of forests in all 4 countries under consideration in the present study have been collected. Resolution of this data is not high, as it is based on the 150 km EMEP grid system. However, given the uncertainty present in the underlying dose–response functions it seems unnecessary to go to a higher resolution. More information on critical loads mapping is provided in the next chapter.

For the purpose of this analysis we have assumed a uniform yield class of 10 (m^3 wood production per hectare of forest per year) and a price of 50 ECU m^3 (from Sinclair and Whiteman, 1992). This is a gross simplification, but in keeping with the low level of accuracy inherent in the method used. Further details of the approach are provided in the methodology report for the ExternE Project (European Commission, 1995).

We have also used the published results of the IIASA Forest Study. Neither method is regarded as ideal, though hopefully the demonstration of their use will stimulate further debate, and hopefully a reassessment of the data that is available. Results are compared to timber production in each of the countries involved in GARP. In neither case has account been taken of non-timber benefits, which could be substantial.

5.4. Results

The results of the use of the KCM are shown in Table 5.1, whilst those of the IIASA Forest Study decline scenario are given in Table 5.2. Table 5.3 compares

Table 5.1. Estimated timber loss in the UK, Germany, Italy and the Netherlands in response to acid deposition, using the KCM

	Best estimate, MECU	Low, MECU	High, MECU
UK	1.5	0.7	2.4
Germany	49	22	77
Italy	2.3	1.0	3.6
The Netherlands	0.04	0.02	0.06

Table 5.2. Estimates of timber loss in response to air pollution using the IIASA forest study (Nilsson *et al.*, 1992)

	Timber loss, Mm^3	Value, MECU
UK	3.6	180
Germany	12	600
Italy	3.3	165
The Netherlands	0.2	10

Table 5.3. Comparison of results with European timber production (total raw wood) and final forestry output statistics for 1984

	Production, Mm^3	Output, MECU	% loss (KCM)	% loss (IIASA)
UK	3.9	200	0.8	90
Germany	29	1,410	3.5	43
Italy	9.2	409	0.6	40
The Netherlands	0.9	n.a.	0.1	22

these results with data on European timber production in the 1980s, with figures taken from Eurostat (1989). There is a difference of 2 orders of magnitude between the results for the two models. To some extent the comparison with current productivity is misleading, as both models assume that the use of forested land is optimised solely for timber production, which is not the case. However, this reflects another limitation of the models; that they do not adequately reflect human behaviour.

The figure for lost production represents the annual fall in harvest resulting from the effects of air pollution on forests being managed in such a way that timber production is maximised throughout the period of interest (100 years), and that a viable standing crop is left at the end of this period. A value of 50 ECU/m^3 has again been assumed.

The results of the two models are very different. These differences are most pronounced for the UK, where the range given covers 1 to 90% of annual timber production! The 90% value, from the IIASA model, is a clear overestimate on the following grounds, leaving aside any doubts about the validity of the decline module of the model itself:

- the emissions scenario assumed by the model does not adequately reflect the decreasing trend in sulphur emissions,
- the model assumes that all European forests are managed optimally for timber production, which is not the case.

Neither model of course takes any account of increased forest growth caused by the deposition of nitrogen, though as stated above there are some doubts about the sustainability of this effect. Of the four countries Germany is predicted to suffer the largest absolute loss of material. The KCM prediction is that damage in each of the other three countries is less than 5 MECU whereas the IIASA model predicts losses in excess of 100 MECU in all countries except the Netherlands.

It is necessary to ask whether these results are of any use more generally within the context of the present study and elsewhere, given the uncertainties involved and the broad ranges that we have derived. At the upper end of the scale the results for timber loss seem small (at least for the UK) compared to other damages described in this report, suggesting that future effort should be directed elsewhere. However, it must be remembered that no account has been taken of non-timber uses of forests. WWF (1992) reports that in both the UK and The Netherlands more than 80% of the forest area is of high or medium importance to recreation.

5.5. Conclusions

Air pollution is responsible, or partially responsible for many of the forest declines observed in recent years in Europe. The pollutants of greatest concern to European forestry are SO_2, NO_x, NH_3, O_3 and acidity. The impact pathway for effects of these pollutants on forests is complex. A large number of variables, above and below ground will be affected. Knowledge of a number of effects is limited. Recent observations of increased tree growth in Europe appear to conflict with concern about forest condition. However, many authorities on this subject believe that the current situation is not sustainable. It appears to result from increased N deposition and increased mobilisation of base cations within the soil. If acid deposition exceeds critical loads (which is the case over much of Europe), ecosystems will become depleted of nutrients. Elevated levels of N will then create nutritional imbalances within plants. This is likely to make them more susceptible to damage.

There are no models available that predict forest response to pollution, that are both widely applicable and well-accepted by the scientific community. The most ambitious attempt to model the effects of acidic deposition on forests was made by Nilsson *et al.* (1991) in the IIASA model. In this study we have used the IIASA model results to provide upper estimates of timber losses, and the estimates by Kuylenstierna and Chadwick (1994) for a lower bound. It is difficult to see how these results could be usefully applied in the development of policy. However, it is hoped that they can be useful in stimulating debate in improvement of the methodology for assessment of forest damage.

CHAPTER 6

ECOSYSTEMS

6.1. Introduction – natural terrestrial systems

The structure and function of ecosystems can be influenced directly and indirectly by the S and N compounds emitted during the burning of fossil fuels and the secondary products arising from atmospheric interactions. Direct effects on vegetation communities or individual plant species can arise from the impacts of the atmospheric concentrations of the gaseous pollutants SO_2, NO_x, NH_3, and O_3; the changes in vegetation can have consequent effects on the associated invertebrate and vertebrate faunas and on soils. Indirect effects result from the deposited load of the pollutant S and N compounds and are soil mediated in the case of terrestrial vegetation and soil and water mediated in the case of aquatic fauna and flora.

6.2. Impacts of the gaseous pollutants

There are few reliable dose–response relationships available for the impact of gaseous pollutants on natural plant species to form the basis of a similar methodology to that used to quantify impacts on crops. However, critical levels have been set for forests and natural vegetation for each of the main gaseous pollutants of interest. These critical levels tend to be general values for natural vegetation rather than for individual species or plant communities. They are being refined and revised as knowledge increases. The impact of current, ambient levels of the various pollutants can be assessed by overlaying spatially disaggregated data of pollutant concentrations on spatially disaggregated information regarding the occurrence of relevant receptors to determine the proportion of the area of the stock at risk for which the critical level is exceeded. Damage is assumed to occur in areas where the critical load for a given receptor is exceeded. This methodology does not, however, provide any quantification of the magnitude of the impact on, for example, growth or flowering. It should also be noted that there is little or no data currently available for valuation of air pollution impacts on ecosystems (European Commission, 1995). In a recent study (Ecotec, 1994) a contingent valuation study was performed to value a reduced threat of acid deposition damage to upland vegetation in the UK. Results seem specific to the exact case in question, and the degree to which they can be extrapolated more generally is open to question. However, the study did demonstrate that valuation work on pollution impacts to natural ecosystems is possible.

6.2.1. *Data requirements*

The approach requires the following input datasets/models:

- spatially disaggregated pollution concentration data for the relevant pollutant,
- spatially disaggregated data on the distribution of the relevant receptor (stock at risk),
- critical levels for the relevant pollutant-receptor combination.

Pollutant data

In the present exercise, pollutant data was provided on the basis of 1 km, 5 km or 20 km grids. For SO_2, NO_x and ammonia, the mean annual concentration for each 1 or 5 km square for 1992 was used and for ozone the total ppb hours above 40 ppb during the growing season for each 20 km grid cell for 1989.

Stock at risk

This was defined using the ITE Land Cover map, based on satellite imagery (Fuller *et al.*, 1994). The database provides information on land cover on the basis of 25 m pixels. These data have been aggregated to provide receptor maps on grids of 1 km, 5 km or 20 km squares. The scale of the aggregation was determined by the scale of the pollutant data available. The area of semi-natural vegetation in each 1 km square was determined; the procedure was repeated to give data for semi-natural grassland, herbaceous semi-natural vegetation, deciduous forest and, by combining these datasets a total cover of semi-natural vegetation.

Critical levels

The critical levels were taken from the UN-ECE Manual on 'Methodologies and Criteria for Mapping Critical Levels/Loads and Geographical Areas Where They are Exceeded' (Federal Environment Agency, 1993) or, in the case of ozone from Fuhrer and Ackerman (1994).

The current critical levels for SO_2 are as follows:

Cyanobacterial lichens:	10 $\mu g\ SO_2\ m^{-3}$ annual mean
Forest ecosystems:	20 $\mu g\ SO_2\ m^{-3}$ annual mean and half-year (March–October) mean
Natural vegetation:	20 $\mu g\ SO_2\ m^{-3}$ annual mean and half-year (March–October) mean

A critical level of 30 $\mu g\ m^{-3}$ of the NO_x gases has been currently set for forests and semi-natural vegetation.

A series of critical levels have been set for ammonia for all vegetation types and depending on the duration of the exposure at a given concentration (Table 6.1).

A critical level of ozone for natural vegetation has been set at 5.3×10^4 ppb hours above a mean concentration of 40 ppb during daylight hours of the growing season.

6.2.2. *Quantification of impacts*

The methodology used in the calculation of the impacts is illustrated using the impact of NO_x on semi-natural vegetation.

Sulphur dioxide

The critical level for forest ecosystems and natural vegetation are currently only exceeded in a very small areas of the Midlands in England. In total, the critical level is exceeded for some 1% of the area of the semi-natural vegetation receptors in the UK, the total area of which was estimated to be 60,184 km^2.

Nitrogen oxides

The critical level for semi-natural vegetation is exceeded over 26% of the land surface of Great Britain but for only 3% of the area of semi-natural vegetation. In this case heathland was also considered separately as a subset of the total semi-natural vegetation; the critical level was exceeded over 9% of the area of heathland, with the area of heathland estimated at 17,635 km^2.

Ammonia

In the UK the critical levels shown in Table 6.1 are only exceeded in small areas adjacent to large intensive animal units.

Ozone

The critical load for natural vegetation is exceeded over 42% of the UK and for some 64% of the semi-natural vegetation stock at risk.

Table 6.1. Critical levels for ammonia

Critical level ($\mu g\ m^{-3}$)	Duration
3300	1 hour
270	1 day
23	1 month
8	1 year

6.3. Pollutant loads

Many approaches are being explored for defining the magnitude and geographical extent of the impacts of acidic deposition, and its components, on natural and semi-natural ecosystems e.g:

- critical load based approaches,
- models of the relationship between environmental factors and soil nutrient status on the one hand and plant species distribution/occurrence,
- mechanistic growth models,
- linked hydrochemical and water chemistry–aquatic biota models.

Mechanistic growth models have, for example, been applied in the Netherlands to assess the impacts of enhanced nitrogen deposition on Calluna heathland (Heil and Bobbink, 1993) but to-date the models have not been tested outside the Netherlands. The use of the nutrient status-plant species occurrence approach has also been explored in the Netherlands (Latour and Reiling, 1993) but similar relationships have not been developed in the UK nor have the Dutch models been tested outside the Netherlands.

Linked hydrochemical and water chemistry–fish status models have been applied to determine the effect of changing atmospheric inputs of pollutants on fishery status (Ormerod *et al.*, 1990) and, in principle, the same approach could be used to assess impacts on other aquatic biota for which the relevant water chemistry–biota status are available, for example invertebrate population status and populations of some water feeding birds. This approach has yet to be fully tested. Critical loads based approaches enable the determination of the area of the stock at risk for which the critical load is exceeded; damage is assumed to be taking place in these areas. A critical load based approach has been explored here.

6.3.1. *Critical load approaches*

Various methodologies have been developed over the last ten years to calculate and map the critical loads of acidity, sulphur and nitrogen for a range of receptors. These approaches can be grouped into the following types (Hettlingh *et al.*, 1991):

- Empirical approaches – where a critical load is assigned on the basis of empirical evidence or data, field observations or additional experiments for example,
- Simple mass balance models – where the sources and sinks of the relevant pollutant at equilibrium are balanced to calculate the pollutant inputs which will not produce imbalance or other damage with respect to the selected receptor,

- Dynamic models – process or mechanistic models which incorporate the main processes or mechanisms controlling the fate and impact of the relevant pollutant in the receptor system.

The approach requires the following data/model inputs:

- spatially disaggregated data on current deposition of the relevant pollutant compounds,
- an estimate of the natural background deposition of the relevant pollutant,
- spatially disaggregated data on the stock at risk,
- critical loads of the relevant pollutant for the stock at risk.

6.3.2. *Use of a critical loads based approach to assess the impact of increased deposition of oxidised nitrogen*

An empirically-based approach was used to determine the impact of current levels of deposition of oxidised nitrogen compounds on a number of N-sensitive types of vegetation.

Pollutant data
For illustrative purposes, this work has concentrated on the effects of oxidised forms of nitrogen, derived from the burning of fossil fuels, primarily from vehicles, power plants and industry. In contrast, the dominant source of reduced nitrogen, ammonia, is agriculture. The pollutant dataset used gave mean annual deposition of oxidised nitrogen compounds on a 10 km grid cell basis for the years 1989–1992.

Stock at risk
Empirical studies have been used to identify N sensitive plant communities and empirically derived critical loads assigned to these communities (Hornung *et al.*, in press). Three of these sensitive communities have been considered in the current study: Calluna heathland, Calcareous species-rich grassland and montane sub-alpine grassland. The plant species typical of these communities were first identified, using published records but chiefly the data held in the National Vegetation Classification (Rodwell *et al.*, 1991). These data were used to identify a number of indicator species for each of the vegetation types. For each of the vegetation types, the number of species, from the indicator list, occurring in each 10 km grid cell in Great Britain was then identified and plotted using the databases of the Biological Records Centre (BRC).

The stock at risk was then defined as those 10 km grid cells in which more than a specified number of the indicator species occurred; the cut off number used for the calcareous grassland was 12 and for the montane-sub-alpine grassland 4. The number of species used for the cut off value was decided after discussions with research workers with extensive knowledge of the relevant vegetation communities.

Critical loads

The critical loads for the relevant vegetation types are shown in Table 6.2. The stock at risk maps were converted to critical load maps by assigning the mid-point value of the relevant critical load from Table 6.2.

Exceedence of critical load

For the purposes of this report it is assumed that there would have been no significant exceedence of critical loads for nitrogen in the 'pristine', pre-industrial revolution environment.

The critical load for the montane grassland is exceeded over only 0.7% of the area of this vegetation type by deposition of oxidised compounds of N. In contrast, the critical load is exceeded over 70% of the area if total nitrogen (oxidised + reduced) deposition is considered. The difference in results reveals two important points. The first is that non-energy sector emissions are of great importance: in order to protect ecosystems against nitrogen, action is also needed in other sectors. The second is that the assumptions made regarding the appropriate baseline for the assessment have a very significant outcome on the results, particularly where there is a well established interaction at the effects level between pollutants.

The critical load is not exceeded for the calcareous species rich grassland by deposition of oxidised N compounds. For comparison, the critical load is exceeded over 19% of the area of this vegetation type if the total nitrogen deposition is considered.

The critical load for heathland is not exceeded by the deposition of oxidised N.

6.4. Other approaches

A recent study by Tickle *et al.* (1995) for the World Wide Fund for Nature provides an assessment of the effects of acidic deposition to ecosystems throughout Europe. The first part of the report provides listings of species known to be affected directly or indirectly by air pollution. These listings were derived by literature review, and should be regarded as indicative only; they are certainly

Table 6.2. Critical loads for nitrogen deposition for selected vegetation types

Vegetation type	Critical load, (kg N ha^{-1} yr^{-1})
Nitrogen limited Calcareous species rich grassland	14–40
Montane subalpine grassland	10–15
Upland Calluna moorland	10–20
Lowland heathland	15–22

Table 6.3. Species of national conservation status identified as being sensitive to acidic air pollution

	UK	Germany	Italy	Netherlands
Mosses	5	–	–	–
Liverworts	1	–	–	–
Charophytes	1	–	–	–
Lichens	4	–	–	–
Higher plants	2	7	2	4

not exhaustive. Of most concern is the identification of 26 species of international conservation status that are believed to be at greater risk through air pollution effects. The number of species identified as being sensitive to acidic air pollution which are also of national conservation status in the GARP study countries (see Table 6.3).

Whilst stressing that this data is not exhaustive, it demonstrates that air pollution is a threat to some species that are regarded as being at risk. It must be said that changes in land use and management and other factors are in many cases a more serious stress for these species than air pollution. Lichens are, however, a notable exception because of their often extreme sensitivity to SO_2. Indeed, following a reduction in SO_2 emissions in Europe in the last 20 years it has been noted that lichen abundance and diversity has increased in many areas.

Tickle *et al.* also provided an analysis of the number of protected sites (taken from the 1990 United Nations List of National Parks and Protected Areas: IUCN, 1990) where air pollution is likely to be a significant threat. The analysis was based around the use of the RAINS model (Alcamo *et al.*, 1990), and the RAINS 'Current Reduction Plans' scenario, which unfortunately does not reflect the 1994 Sulphur Protocol for the UN ECE Convention on Long Range Transboundary Air Pollution. However, the analysis demonstrated that many sites are under threat. Of the countries being assessed within GARP, risk is greatest in the Netherlands and least in Italy.

6.5. Conclusions

Use of the critical loads and critical levels approach has allowed maps to be produced identifying the areas under greatest threat from air pollution. Although we have concentrated on oxidised nitrogen species the methodology is applicable to any pollutant for which a critical load or level has been defined. Emissions from energy use are not the only source of air pollution to affect natural ecosystems. Other sectors, agriculture in particular, also need to be considered.

A number of species of significant conservation interest face increased threat as a result of exposure to air pollution. Economic valuation of the effects of air pollution on natural ecosystems is not possible at the present time. Further review of the results of this work and the methods used to present information needs to be conducted by economists experienced in the use of contingent valuation, to see whether an appropriate valuation study could be designed.

It should be remembered that effects of air pollution are only one of several threats to natural ecosystems. Changes in land use and management are likely to be more important. However, the present study has not considered possible effects resulting from global climate change, which seem likely to have major impacts on the distribution of species particularly in the northern temperate region where many species are at the extreme of their geographical distribution, and in coastal areas.

Introduction

Acidic deposition resulting from the burning of fossil fuels can impact indirectly on freshwater fisheries. Water chemistry is one factor which influences the status of fish stocks and fisheries and this can be affected by acidic deposition onto the catchment of the stream of lake and, in the case of larger lakes, by direct deposition to the surface of the water body. As the acidic inputs do not affect the fish directly, impacts on fishery status cannot be quantified using a dose–response relationship between deposition and fishery status. However, relationships do exist between water chemistry and fish survival, or fishery status, and models have been developed to determine changes in water chemistry in response to variations in atmospheric deposition. The impacts on changes in deposition can therefore be quantified by linking these two types of model. A number of methodologies have been explored and the present study has applied the approach used in ExternE Project (European Commission, 1995). The approach requires the following data and models:

- data on atmospheric deposition of major anions and cations from pollutant and non-pollutant sources,
- hydrochemical model to predict changes in surface water chemistry in response to variations in atmospheric inputs,
- a damage function relating fishery status and water chemistry,
- stock at risk data for (a) fisheries and (b) streams (chemistry) in the area of interest.

The method used in the ExternE project links the MAGIC hydrochemical model (Cosby *et al.*, 1985) and a damage function for fish numbers per unit length of stream derived by Ormerod *et al.* (1990) from a national survey of surface water chemistry and fish populations in Wales.

Data requirements

6.7.1. *Atmospheric inputs*

Background sulphur deposition was set at 40 meq m^{-2} yr^{-1}.

6.7.2. *Stock at risk*

The sensitivity of surface waters to acidification by acidic atmospheric inputs is influenced by a number of factors, but the main controls are the chemistry and mineralogy of the soils and bedrock which underlay the stream catchment. Thus, the streams of catchments underlain by calcareous soils and rocks, which have a large buffering capacity, are insensitive while streams of catchments underlain by shallow, acid soils and massive granitic rocks are highly sensitive. The stock at risk can be broadly defined as the surface waters in those areas underlain by soils and bedrock with low buffering capacities.

A methodology has been developed for predicting the sensitivity of surface waters in Great Britain to acidification which is based on data on the buffering capacity of soils and rocks plus information on land use (Hornung *et al.*, 1995). Datasets on soils, solid geology and land use, each based on 1 km grid cells, were combined in a GIS and each 1 km grid square in Great Britain was then allocated to one of five sensitivity classes. For the present study, the grid cells allocated to the three most sensitive classes were defined as the *stock at risk area.* The area in the two classes totalled 57,070 km^2; equivalent to 22% of the total land area of Great Britain.

The damage function used in the current study requires information on stream length and gives fishery data in terms of numbers of fish per 100 m length of stream. Data on stream length for the two most sensitive classes of land was calculated from a sample of 1 km squares within the stock at risk area. The streams in each of the sample squares were digitised and the total length within each square determined. The mean stream length in the sample squares was multiplied by the total number of 1 km squares in the stock at risk area to give the *stock at risk stream length,* which was 88,857 km.

Detailed investigations in Wales and Scotland have shown that, even in the predominantly sensitive areas, there are less sensitive streams and catchments because of small scale variations in bedrock chemistry and mineralogy. Data from the stream survey within Wales was used to allocate the streams within the stock at risk area within Wales into four sensitivity classes on the basis of stream hardness (Table 6.4). The total calculated stream length in the stock at risk area was allocated to the same four classes and in the same proportion as found in Wales. This is clearly a gross simplification but was used in the absence of a national dataset on which the allocation of the *stock at risk stream length* could be made to hardness classes.

Table 6.4. Distribution of streams in the stock at risk area in Wales (hardness classes and the trout density of streams in each class)

Stream class	Hardness ($mg\,l^{-1}$ $CaCO_3$)	Trout density (per 100 m)	% of streams in class
1	<5	<5	17
2	5–8	5–25	33
3	8–10	26–138	26
4	>10	>138	22

6.7.3. *Application of the linked hydrochemical model and the fish status model*

The MAGIC model has been run for sample streams for each of the sensitivity/hardness classes within the *stock at risk area*. The model was used to generate water chemistry for the sample streams in pre-industrial conditions using the sulphur deposition value of 40 meq m^2 yr^{-1}, for the pre-industrial environment, as the main driving variable.

The water chemistry generated for the sample stream by the application of the MAGIC model were used as input data to the fish density model of Ormerod *et al.* (1990). Stream flow in pre-industrialised times was assumed to be the same as at present. The output from the fish model gave fish densities per 100 m of stream in pristine conditions.

6.8. Quantification of impacts

The application of the combined hydrochemical and fish density models for the pre-industrial environment suggests that all the classes of streams had between 80 and 200 fish per 100 m in this environment. The less sensitive streams, in sensitivity/hardness class 4 showed no significant change in individual stream chemistry or fish density compared with the present day. However, the model outputs suggest that the streams in classes 1, 2 and 3 had similar fish densities to the less sensitive streams in the unpolluted environment.

The current ranges of fish density in each class of stream, as in Table 6.4, and the densities in pre-industrial conditions, 80–200 fish per 100 m stream length, have been multiplied by the total stream length in each stream class within the stock at risk area (Table 6.5). This gives current and past fish stocks in the stock at risk area (Table 6.6). Fish densities in stream class 4 have been assumed to be between 80 and 200 both currently and in the past. These data suggest that the trout stocks within the stock at risk area have been reduced by some 60% since industrialisation.

6.9. Conclusions

The linkage of data on atmospheric deposition, hydrochemical models, damage functions and data on stock at risk has been demonstrated to estimate changes

Table 6.5. Lengths of streams in each sensitivity/hardness class within the stock at risk area of Great Britain

Stream class	Length of streams (m. 10^2)
1	151059
2	293231
3	23103077
4	195488

Table 6.6. Estimated current and pre-industrial fish stocks in the stock at risk area of Great Britain by stream class

Stream class	Current population ($\times 10^6$)	Past population ($\times 10^6$)
1	0–0.76	1.21–3.02
2	1.47–7.33	23.46–56.30
3	6.00–8.78	18.53–46.21
4	15.64–39.04	15.64–39.04
Total	23.11–55.91	58.84–144.57

in fish populations in freshwater systems in Great Britain. Overall, a 60% reduction in fish numbers is predicted by the models, when comparing present day conditions with those predicted to have existed prior to industrialisation. In the long term it is anticipated that conditions will improve, following pan-European agreement to further reduce emissions of sulphur, though it should be noted that the critical load for acidity will still be exceeded in some areas of the UK. The situation with respect to nitrogen deposition and its influence on acidity is not so clear.

Valuation of damage to fisheries has not been performed. There is some literature on this subject which may be of use (e.g. Ecotec, 1994), though this needs further evaluation before application to the present case.

The methodology presented contains several sources of uncertainty, some of which will be answered by ongoing research. It is to be hoped that the demonstration of a coherent methodology for the full impact pathway will provide an impetus for other areas of uncertainty to be addressed, and also for the application of the methodology to be investigated for other countries where freshwater acidification is a problem.

Whilst we have sought to quantify the total effect of pollutant deposition above natural background levels, the change in impact between different years may also be of interest in the context of national accounts. This would establish whether or not there had been improvement from year to year. The same methodology could be applied to such a case.

CHAPTER 7

MATERIALS

7.1. Introduction

The effects of atmospheric pollutants on buildings have for many years provided some of the clearest examples of damage related to the combustion of fossil fuels. In particular, damage to the stonework on cathedrals and other buildings of cultural and aesthetic merit has caused much concern.

Valuation of material damage is complicated by several factors. One of the most important concerns variation in behaviour between individuals. Some people will allow damage to proceed well beyond the point at which action would ideally be taken, allowing secondary damage mechanisms to take effect, perhaps affecting the structural integrity of their property. Acceleration of damage by air pollution may in these cases increase damage costs. Other people will take action earlier than what could be defined as the economic 'ideal' from the perspective of repair costs. This could be because they believe that a given action should be performed at a particular frequency, or, for example, simply because they want to paint their house a different colour. In such cases it is logical to assume that there is no economic effect of pollution damage. Quantification of the proportion of cases that may fall into either category is not currently possible. However, it is clear from the English House Condition Survey that a substantial number of dwellings (the most common type of building) require significant amounts of remedial work.

Another problem concerns valuation of damage to buildings of aesthetic or cultural merit such as ancient cathedrals. Consideration should be given to amenity and existence values for such buildings, as replacement costs do not adequately reflect the full cost of (for example) the loss of fine carvings that are several hundred years old. Costs associated with damages of this type are extremely site-specific, not only in terms of the merit of the item in question, but also in the way in which it can be treated. Unfortunately, therefore, no assessment has been made of damages to buildings in this sector.

A further problem is that there is no available inventory to describe the stock at risk. Given that these buildings make up only a fraction of the total UK building stock there is some justification for the belief that associated damages will be small in comparison to those calculated for utilitarian buildings. However, it is worth noting that the National Audit Office in the UK estimated that 21% of 'listed' buildings (those of special merit, which are legally protected) were unoccupied or in a state of neglect in 1992.

7.2. Scope of the analysis

7.2.1. *Impact pathways for materials damage*

Impact pathways have been used to describe the effects of emissions of acidic pollutants and precursors of photo-oxidants on stone, metals and paints. Impacts for most materials fall into three categories:

- discoloration,
- material loss,
- structural failure.

There are no valuation studies or material inventories from which estimates of the costs of discoloration could be estimated. However, such effects seem likely to be small. Structural failure resulting from pollutant exposure seems unlikely unless either the design of a building is fundamentally flawed, or the owner of a property has not carried out routine maintenance. In the latter case air pollution may increase damages (see above).

7.2.2. *Definition of the impact under assessment*

In this Chapter the main damage assessment, for effects of SO_2 and deposited acidity, is undertaken for 'utilitarian' buildings (houses, shops, factories, offices and schools), and galvanised structures in other sectors, including agriculture, transport (e.g. street furniture and railway gantries) and energy (e.g. electricity pylons) using a repair cost method, which is similar to the impact pathway approach used in previous chapters. Utilitarian structures of other materials are not considered because no suitable inventory exists. Historic buildings have not been assessed for the consequential damage from impacts as there is insufficient data concerning material inventory, repair and existence values.

The materials for which damage caused by acidic deposition is considered are calcareous stone, mortar, paint, concrete, aluminium and galvanised steel. Although not exhaustive, this list includes the most sensitive of the materials commonly used by the construction industry. All steel is assumed to be painted and this stock is therefore transferred into the paint inventory.

Effects of O_3 on rubber and paint in the UK have been taken from a report by Lee *et al.* (1995). A detailed assessment of damages was not possible in this case. Possible effects of O_3 on textiles are reviewed but not quantified. A mechanism for synergistic interaction between O_3 and SO_2 for effects on metals is identified, but at this stage no quantification of impacts has been attempted.

The costs associated with the soiling of buildings by particulates are also included, based on a review by Newby *et al.* (1991).

7.3. Assessment of damages caused by acidic deposition

7.3.1. *Dose–response functions*

Dose–response functions, showing the rate of loss of material for the building materials considered to be at risk, were identified by literature review. A full description of the functions and the studies they were taken from is presented elsewhere (European Commission, 1995). In all cases, we have based results on an average of the best available functions. In general, we have concentrated on three studies: the UK National Materials Exposure Programme (NMEP) (Butlin *et al.*, 1992), Lipfert (1989), and the UN-ECE Integrated Collaborative Programme (Kucera, 1994).

The UN-ECE study (Kucera, 1994) is particularly noteworthy because data is being collected at no fewer than 36 test sites throughout Europe (in the Czech Republic, Estonia, Finland, Germany, Italy, the Netherlands, Norway, Portugal, Sweden, Spain, Russia and the UK) and 3 sites in Canada and the USA. These sites clearly cover a broad geographical region, with substantial variation in climate and pollution exposure regime. The work is not yet complete, and hence finalised relationships are not available. Comparison of the relationships derived after 2 and 4 years of exposure (the experiment is to last for 8 years in all) reveals variation in both magnitude of predicted effect and functional form with time. The study has produced functions for steel, weathering steel, aluminium, copper, bronze, nickel and silver (electric contacts), limestone, sandstone and paint coatings. The effects on glass and polymers are also under analysis although no functions have been derived as yet. The study provides functions for both sheltered and unsheltered material.

The dose–response functions we have considered for each material type are presented below. The following key applies to all equations given in this section:

ER = erosion rate (μm/year)
P = precipitation rate (m/year)
SO_2 = sulphur dioxide concentration (μg/m^3)
O_3 = ozone concentration (μg/m^3)
H^+ = acidity (meq/m^2/year)
R_H = average relative humidity, %
f_1 = $1 - \exp[-0.121 \cdot R_H/(100 - R_H)]$
f_2 = fraction of time relative humidity exceeds 85%
f_3 = fraction of time relative humidity exceeds 80%
TOW = fraction of time relative humidity exceeds 80% and temperature exceeds 0°C
ML = mass loss (g/m^2) per year
MI = mass increase (g/m^2) per year
CD = spread of damage from cut, mm/year
Cl^- = chloride deposition rate in mg/m^2/day
D = dust concentration in mg/m^2/day

To convert mass loss for stone and zinc (equations 3, 9, 10 and 11) into an erosion rate in terms of material thickness, we have assumed densities of 2.0 and 7.14 tonnes/m^3 respectively. For the Kucera functions, the original H^+ concentration term (in mg/l) has been replaced by an acidity term using the conversion:

$$P \cdot H^+ \text{ (mg/l)} = 0.001 \cdot H^+ \text{(acidity in meq/m}^2\text{/year)}$$

We stress that all of the functions given below may characterise only the early stages of damage. To this extent they may underestimate later, more severe damage. For instance, in stone these later mechanisms are known as Stage II and Stage III damages. Stage II damage occurs as the soluble calcium carbonate is washed away and the large insoluble particles left behind drop out as their matrix is removed. Stage III occurs when a large crust is formed with eventual exfoliation (the point when large flakes of crust fall away). However, a recent report (Butlin *et al.*, 1994) suggests that Stages II and III may be reached very quickly (within 4 years), though this seems to contradict earlier evidence.

Stone

For calcareous stone, mass loss in exposed material depends both on dry deposition of SO_2 and time of wetness, as well as on the load of acidity from rain. The functions from all three studies have been evaluated. Equation 1 (Lipfert, 1989) is a theoretical universal damage function for stone loss, supported by a meta-analysis of data. Equation 2 is taken from Butlin *et al.* (1992). Equations 3 to 6 are taken from Kucera (1994). Equations 3 and 5 are derived from unsheltered samples of limestone and sandstone, respectively, whilst equations 4 and 6 are taken from samples that are sheltered from the rain. The analysis of sheltered stone is complicated. Mass of samples initially increases through the formation of a crust of gypsum on the stone surface. This layer eventually flakes off, often removing some underlying material with it (Cooke and Gibbs, 1994).

Unsheltered limestone

$$ER = 18.8P + 0.052 \cdot SO_2 + 0.016 \cdot H^+ \quad (1)$$

$$ER = 2.56 + 5.1 \cdot P + 0.32 \cdot SO_2 + 0.083 \cdot H^+ \quad (2)$$

$$ML = 8.6 + 1.49 \cdot TOW \cdot SO_2 + 0.097 \cdot H^+ \quad (3)$$

Sheltered limestone

$$MI = 0.59 + 0.20 \cdot TOW \cdot SO_2 \quad (4)$$

Unsheltered sandstone

$$ML = 7.3 + 1.56 \cdot TOW \cdot SO_2 + 0.12 \cdot H^+ \quad (5)$$

Sheltered sandstone

$$MI = 0.71 + 0.22 \cdot TOW \cdot SO_2 \quad (6)$$

Mortar

The primary mechanism of mortar erosion is through attack on the calcareous cement binder (UK BERG, 1990; Lipfert, 1987). Assuming that the inert silica aggregate is lost when the binder is attacked, the erosion rate is determined by the erosion of cement. There are no functions available derived from work on mortar, and hence it is necessary to approximate, using those derived for limestone and sandstone.

Concrete

It is believed that the reaction between acidic gases in the atmosphere and the alkaline constituents of the cement matrix in concrete is dominated by carbonation by carbon dioxide. Because of excess CO_2, no costs arise from incremental pollutant concentrations. As a result of this we have not assessed damage to concrete any further. More important mechanisms concern the damage to structural steel elements encased within concrete. The high alkalinity of the cement matrix protects the reinforced steel from corrosion. Carbonation is the main process leading to the loss of alkalinity which in turn can lead to the corrosion of steel. There is some evidence that when the concrete layer is too thin or badly prepared, atmospheric pollutants may have a role in accelerating damage. Thus, the role of acidic pollutants cannot be negated although there are no appropriate quantification methods available at present.

Paint

The only pollution related mechanism for which dose–response functions have been derived is surface erosion (NAPAP, 1990). The most extensive review is from the USA (Haynie, 1986). This identifies a 10-fold difference in acid resistance between carbonate and silicate based paints. In the USA, silicate based paints predominate in the house paint market (NAPAP, 1990). However, discussions with a leading European manufacturer confirm that the overwhelming majority of house paints contain calcium carbonate and that this component is particularly important in the cheaper grades of paints (ICI, 1992). The dose–response function for carbonate based paints (equation 7, Haynie, 1986) is therefore appropriate for UK house paints, in which t_c = the critical thickness loss, which is about 20 μm for a typical application:

$$\Delta ER/t_c = 0.01(P)(8.7)(10^{-\mathrm{pH}} - 10^{-5.2}) + 0.006(SO_2)(f_1) \tag{7}$$

In areas where silicate paints predominate, as seems to be the case in mainland Europe, Haynie's function for silicate paints is more appropriate (equation 8).

$$\Delta ER/t_c = 0.01(P)(1.35)(10^{-\mathrm{pH}} - 10^{-5.2}) + 0.00097(SO_2)(f_1) \tag{8}$$

The Haynie function is derived for laboratory prepared samples on inert substrates. Therefore, it does not take into account corrosion of the substrate and substrate/paint interactions that together provide an alternative mechanism for paint decay. It has been observed that damage to paint predominantly occurs as a result of blistering, cracking, etc., where the presence of flaws or scratches allows the pollutants to attack the substrate. The extent to which pollution is implicated in these processes is not known, though there is good reason to suspect some association.

In the UN-ECE study, the only equation for painted surfaces obtained so far is for the spread of defects on scratched steel panels. It is difficult to see how this function could be implemented within the present study.

Zinc

The mechanism for corrosion of galvanised steel proceeds by a series of parabolic reactions where an oxide film is built up on the surface and subsequently destroyed. When placed in series these parabolic reactions equate to a linear function. Acidic pollutants are responsible for breaking down this film. We have again studied the same three sources for zinc functions. Lipfert (1987) proposes a complex function, with kinetics intermediate between linear and parabolic. There is a strong dependence on time of wetness and an additional term is included to represent the removal of a protective film by wet deposition. The function has a complex time function, which complicates its use within our study.

An alternative function has been derived from the UN-ECE study. This function is dependent on the parameters for ozone and the time of wetness; both of which are not well characterised across Europe. Moreover, the synergistic action of ozone with dry deposition of SO_2 is controversial. The sheltered function is shown in equation 11.

Unsheltered samples

$$M\,L = [t^{0.78} + 0.46\log_e(H^+)] \cdot [4.24 + 0.55 \cdot f_2 \cdot SO_2 + 0.029 \cdot Cl^- + 0.029 \cdot H^+] \qquad (9)$$

$$ML = 14.5 + 0.043 \cdot TOW \cdot SO_2 \cdot O_3 + 0.08 \cdot H^+ \qquad (10)$$

Sheltered samples

$$ML = 5.5 + 0.013 \cdot TOW \cdot SO_2 \qquad (11)$$

Aluminium

Aluminium is the most corrosion resistant of the common building materials. The major pollution related damage mechanism of concern is pitting due to SO_2 (Lipfert, 1987). However, no dose–response functions have been derived for this process. Surface corrosion is less of a problem. A thin surface oxide film is formed, which is highly corrosion resistant down to a pH of 2.5. A few dose–response functions been derived and there is general agreement that the kinetics are approximately linear. However, we have elsewhere concluded that effects of acidic deposition on aluminium are unlikely to cause significant damage (European Commission, 1995).

7.3.2. *Compilation of reference environment data*

The bulk of the data describing the stock at risk was derived from ECOTEC (1986), which provides results of an intensive survey of building types in Birmingham, the second largest city on the UK. An improved inventory was obtained towards the end of this study. Unfortunately this could not be included in this project, though it will be used for future work. Due to limited statistical data being available for the UK as a whole, the 21 common building types identified by ECOTEC were reduced down to 5, and a building materials 'identikit' was prepared for:

- dwellings,
- schools,
- shops,
- offices,
- industrial buildings.

The distribution of these different building types within each 100 × 100 km grid cell of the UK was taken from the following sources:

dwellings:	Central Statistical Office (1991a; 1991b);
schools:	Central Statistical Office (1991b) divided between grid cells in proportion to the population in each (OPCS, 1991);
shops, offices and industry:	Central Statistical Office (1991c), Welsh Office (1990). Data for Wales was extrapolated to Scotland based on population data.

The UK materials inventory for each 100 × 100 km grid cell of the British National Grid was calculated by multiplying the number of each type of building in a grid square by the average area of each material type per building (given by the building identikits). A separate inventory was used for galvanised steel. The inventory was extended to cover the wider use of galvanised steel in general infrastructure. Values were derived from industrial zinc usage in the UK galvanising industry and factored to give a 'stock at risk'. This stock was proportioned to National Grid squares according to population. Both 'dipped' products and continuous process products (such as sheet and strip) were included. Sectors included:

Dipped material:

- building and construction,
- street furniture,
- power sector (electricity pylons),
- transport infrastructure,
- agriculture.

Continuous process (sheet and strip):

- building and construction.

For each sector, the proportion of material likely to be exposed to the elements was estimated. Other industrial sectors were analysed but are considered not to be at significant risk from atmospheric pollution. For instance, galvanised sheet and strip on automobiles will be corroded by road salts far more than by atmospheric pollution. Air pollution will clearly have some impact, but it is likely to be very small in comparison. Incremental costs seem likely to be made negligible by the effect of obligatory annual vehicle inspections in most countries on repair cycles.

The proportion of the galvanised stock which is painted was then estimated. The upper and lower limits on exposed material give the range of our analysis.

For the best estimate of stock at risk, the painted stock has been transferred to a separate paint inventory, which is analysed for paint damage and repair. This is very important in some industrial sectors, for example, construction, where the majority of dipped products will be painted to fit the local architectural colour scheme.

Each sector has also been assigned a specific galvanised film thickness. For example power transmission pylons have very thick coatings (200 μm) whilst sheet and strip have very thin ones (25 μm). In all sectors, there is a coating limit of around 300 microns. Above this level, there is a tendency for the material to become brittle. It should be noted that more resistant coatings are appearing in the market and that this will affect future damage calculations.

In addition, the following sources of data were used:

- precipitation averages; UK Meteorological Office (1977);
- average relative humidity; UK Meteorological Office (1970);
- estimated % of time that relative humidity exceeds 80%, 85%, 90%; UK Meteorological Office (1975);
- UK ozone levels; background levels assumed to be 40 $\mu g/m^3$;
- precipitation outside the UK; 0.6 m/year chosen as a representative value (within a factor of 2 of all major centres of population in Europe);
- average background humidity and ozone levels from the UK were applied to the rest of Europe.

7.3.3. *Atmospheric modelling*

The maps of pollution concentration described in Chapter 1 were not used to estimate damage to materials. Instead, concentrations of SO_2 and deposition of acidity were modelled using the Harwell Trajectory Model (European Commission, 1995). The principal reason for this was to demonstrate the use of alternative methods for describing air pollution. The Harwell Trajectory Model is a receptor oriented Lagrangian type box model, incorporating a wind rose weighted dispersion, atmospheric chemistry of sulphur and nitrogen and both dry and wet deposition. The model output is given at a 20 × 20 km resolution; we have averaged to 100 × 100 km grids based on the British

National Grid to fit our material inventory. Only values for cells containing land mass have been used in this process.

7.3.4. *Calculation of repair frequency*

Replacement of material was judged necessary when the following amount of surface erosion had occurred:

Stone and mortar	5 mm
Paint (Haynie, 1986)	20 μm
Galvanised steel (Short, 1994)	
Dipped – Building and construction	100 μm
Street furniture	100 μm
Power sector (electricity pylons)	200 μm
Transport infrastructure	100 μm
Agriculture	200 μm
Sheet and strip – Building and construction	25 μm

These figures were specified (European Commission, 1995) based partly on published information and partly on expert judgement. Further work is required to reduce the uncertainty associated with the analysis.

Some commentators have said that figures of the order of the 5 mm erosion used for mortar and stone are too low. However, it must be remembered that damage is typically not uniform across a surface, and that the timing of maintenance is likely to be determined by the condition of the worst affected areas. In view of this an average loss of 5 mm depth of mortar certainly seems to be justified. In the case of stone the argument seems to be irrelevant, as quantified damages seem insignificant (see below, though remembering that damage to structures of aesthetic merit has been excluded from this analysis).

For bare galvanised steel, when the repair action is painting rather than replacement (see below), we have assumed repair would take place before total removal of the film. As an approximation, we have taken the value of 50% zinc film erosion (i.e. half the above values), for the initiation of repair action.

7.3.5. *Estimation of repair costs*

Estimated repair costs (Table 7.1) are taken from unit cost factors for each of the materials for which assessment was performed. These figures are based on data from ECOTEC (1986) and Lipfert (1987).

We have applied identical repair costs for all repainting, whether on wood, steel or galvanised steel. This is likely to underestimate impacts. For exposed (bare) galvanised steel, the first round of maintenance is likely to comprise of painting. The upper limit for damages has been calculated assuming 50% replacement of material and 50% painting, as it is unlikely complete replacement would occur.

Table 7.1. Unit cost factors for materials assessed for pollution damage (ECU 1990)

Activity	Cost, ECU/m^2
Repainting	11.3
Repointing (brick)	22.8
Replacing zinc/galvanised	45.1
Refacing stone	248

7.3.6. *Implementation of methodology and sensitivity analysis*

The dose–response functions were used to estimate the change in repair frequency caused by anthropogenic emissions of the acidifying gases. In all cases, we have assumed complete exposure of material. The functions described for sheltered material have thus not been used.

Upper and lower estimates for damage to stone were calculated from the three dose–response functions (equations 1, 2 and 3). The best estimate is the mean calculated from the use of the three functions. The upper limit of the range also uses German data for assessment of the stock at risk. This roughly doubles the quantity of stone predicted by the ECOTEC (1986) inventory, which was produced from a survey of the City of Birmingham, where an unrepresentatively small number of buildings are made of stone.

The best estimate for mortar is taken as the mean of results from three dose–response functions (equations 1, 2 and 5). The upper and lower limits show the extremes calculated using the three functions.

For paint, the best estimate was calculated using the function for carbonate paint (equation 7). The silicate function (equation 8) was also used to calculate the lower estimate. The upper limit is a nominal factor of 2 from the best estimate.

Galvanised steel that is painted before or during construction is treated in the same way as other painted surfaces. All estimates for unpainted galvanised steel use an average of the Kucera and Lipfert functions. The best estimate assumes painting after a 50% erosion of the coating film. The lower and upper estimates are based on the possible proportion of stock exposed (i.e. not painted). Repair costs are calculated in a similar way to the best estimate, though the upper estimate allows for 50% replacement of exposed galvanised steel once the entire galvanised layer has corroded.

7.4. Estimation of soiling costs

The estimates obtained are presented in Table 7.2, showing damage to all materials assessed, and Table 7.3 which provides a more detailed breakdown for galvanised steel. It should be noted that a very high proportion of total damages are associated with impacts on galvanised steel. This suggests that

Table 7.2. Annual effects of acidic deposition and SO_2 on various building materials

	Annual Costs (1990 MECU)		
Material/action	Best estimate	Low	High
Stone refacing	3.0	0.9	10.0
Brick repointing	64	17	109
Paint	502	79	1,004
Galvanised (painted)	326	51	652
Galvanised (unpainted)	617	404	1,191
Total	1,512	552	2,966

Table 7.3. Estimated cost for galvanised steel damages in different sectors in the UK

	Annual costs (1990 MECU)		
Damage category	Best estimate	Range low	Range high
Building construction ('dipped')	29	12	90
Street furniture	45	26	92
Power sector	12	12	17
Agricultural sector	21	14	42
Transport sector	2	1.4	3.9
Construction (sheet and strip)	508	339	946
Total	617	404	1,191

further investigation of materials damages should focus on damages to galvanised steel.

Soiling of buildings results primarily from the deposition of particulates on external surfaces. Three major categories of potential damage cost may be identified; damage to the building fabric, cleaning costs, and amenity costs. In addition there could be effects on building asset values, as a capitalised value of these damages.

Damage to building fabric through synergistic effects of particulates and other pollutants is a possibility. Indeed one dose–response function for steel identified includes such a factor. However, the effect is rather weak and is negligible in comparison to the effects of acid pollution. Cleaning costs and amenity costs need to be considered together. Data on the former is, of course, easier to identify. In an ideal market, the marginal cleaning costs should be equal to the marginal amenity benefits to the building owner or occupier. However, markets are not perfect and amenity benefits to the public as a whole lie outside this equation. It is therefore clear that cleaning costs will be lower than total damage costs resulting from the soiling of buildings. In the absence of willingness to pay data, cleaning cost are used here as an indicator of minimum damage costs.

The total UK building cleaning market is estimated to be £80 million (112 MECU at 1990 prices) annually (Newby *et al.*, 1991). Most of this is in urban areas and it is assumed that it is entirely due to anthropogenic emissions. It has been suggested that amenity costs are equal to cleaning costs at the time of cleaning (A. Rabl, personal communication). This would raise damages to £160 million (224 MECU) annually. No data on the costs associated with cleaning windows, vehicles, etc. are available for the UK.

7.5. Assessment of damages caused by ozone

There is very little work available to allow assessment of the effects of O_3 on materials. A recent review (Lee *et al.*, 1995) makes tentative estimates of the potential costs of annual O_3 damage in the UK. The first estimate, for total damage to rubber goods (including the costs of anti-ozonant protection) and surface coatings was a range of £170 million–£345 million (221–450 MECU). This range was derived from extrapolation of studies performed in the USA in the late 1960s (there have been no more since then). As such the authors recognise the result to be very crude indeed.

Independently of this Lee *et al.* estimated the costs of protecting rubber goods to be between £25 million and £190 million to the consumer (33–247 MECU). The lower end of the range is a more robust figure representing the amount spent on anti-ozonant chemicals by the UK rubber industry each year. The upper end of the range includes additional costs that feed through to the consumer, such as those associated with a more complex compounding process, and was calculated using very generalised factors. Accordingly the upper end of the range has a high uncertainty attached to it, compared to the level of confidence associated with the lower end of the range. O_3 damage to tyres was estimated to be in the range £0 to £4 million (0–5.2 MECU).

For paint a range of £0 to £60 million (0 to 78 MECU) was derived for a 10 ppb reduction in O_3 levels using a damage function approach, though only very low confidence can be associated with the dose–response functions available. No account was taken of potential damage to textiles, or the increased costs of manufacturing textiles so that they are not affected by O_3. At this stage no account has been taken of the effects of ozone on zinc expressed in equation 10, because the ozone map was not available at the time when materials damages were calculated. This can, however, easily be done in future work.

Lee *et al.* essentially underlined the need for further experimental work on O_3 damage to materials. Most of the numbers that they produced are highly uncertain, though they do indicate that UK damages from O_3 could be in excess of 100 MECU per year, which is clearly a significant amount. This result was checked against UK sales of rubber goods and paint and does not seem excessive in comparison.

One of the problems with assessment of ozone damage to receptors that are concentrated in urban areas in the context of the present study, is the fact that vehicle emissions of nitrogen monoxide act as a sink for ozone. This can reduce urban levels below the natural background in some areas, with the net effect that current damages are lower than those that would occur in a (albeit hypothetical) 'pristine' environment. The effect of reducing emissions of NO_x, an agreed policy goal throughout Europe and elsewhere, may therefore lead to some increases in O_3. The extent to which the damages identified for ozone can thus be attributed to human activity for the purposes of the present study is thus debatable. Further work, using modelled O_3 data for different scenarios is therefore necessary to better inform the policy debate.

7.6. Conclusions

The analysis presented here demonstrates that air pollution may cause significant damage to materials in the UK. The most serious damage from SO_2 and acidic deposition appears to affect paint and galvanised steel. Perhaps surprisingly, effects on stone seem to be very small, though we were unable to characterise damage to buildings of cultural merit, in which stone is one of the most important elements. The annual costs of cleaning stonework have been derived directly from industry data. Some commentators suggest that the figures should be adjusted upwards in order to incorporate impacts on amenity. Estimates have been made of damage costs associated with ozone effects on materials, though the methods used were, in the light of a paucity of data, necessarily crude. No account was taken of synergism between effects of SO_2 and O_3, which is suggested by some of the latest dose–response functions for metals.

A number of uncertainties remain. The following are identified as research priorities:

- improvement of inventories, in particular; disaggregation of the inventory for paint to describe the type of paint in use; disaggregation of the inventory for galvanised steel to reflect different uses;
- further development of dose–response functions, particularly for paints, mortar and cement;
- assessment of exposure dynamics of surfaces of differing aspect, and identification of the extent to which different materials can be considered to be sheltered;
- better definition of service lifetimes for stone, concrete and galvanised steel;
- integration of better data on repair techniques and costs, particularly for galvanised steel;

- improvement of awareness of human behaviour with respect to buildings maintenance;
- development of dose–response functions and models of atmospheric transport and chemistry for assessment of O_3 effects.

Although this list of uncertainties is extensive, it would be wrong to conclude that our knowledge of air pollution effects on materials is poor. Indeed, we feel that the converse is true; it is because we know a lot about damage to materials such that we can specify the uncertainties in detail. The factors affecting galvanised steel and paint are of most concern since they constitute a high proportion of total materials damage. Potentially important areas to be considered include: damages to historic buildings and monuments with 'non-utilitarian' benefits, damage to paint work through mechanisms other than surface erosion, damage to reinforcing steel in concrete and synergistic effects of different pollutants. Nevertheless, we believe that this analysis has made an important start in quantifying materials damages.

CHAPTER 8

REFERENCES

Alcamo, J., Shaw, R. and Hordijk, L. (1990). The RAINS Model of Acidification. Kluwer, Dordrecht.

Barr, C.J., Bunce, R.G.H., Clarke, R.T., Fuller, R.M., Furse, M.T., Gillespie, M.K., Groom, G.B., Hallam, C.J., Hornung, M., Howard, D.C. and Ness, M.J. (1993). Countryside Survey 1990 Main Report. Countryside 1990 Series, volume 2 Department of the Environment: London

Bates, D.V., Baker-Anderson, M. and Sizto, R. (1990). Asthma Attack Periodicity: A Study Of Hospital Emergency Visits In Vancouver. Environmental Research, Vol. 51, 51–70.

Behra, P., Sigg, L. and Stumm, W. (1989). Dominating Influence Of NH_3 On The Oxidation Of Aqueous SO_2: The Coupling Of NH_3 And SO_2 In Atmospheric Water. Atmospheric Environment Vol. 23, 2691–2707.

Bower, J.S., Stevenson, K.J., Broughton, G.F.J., Lampert, J.E., Sweeney, B.P. and Wilken, J. (1991). Ozone In The UK: A Review Of 1989/90 Data From Monitoring Sites Operated By Warren Spring Laboratory. Report number LR 793 (AP), for the UK Department of the Environment.

Brown, W. (1994). Dying From Too Much Dust. New Scientist, 12 March 1994, pp 12–13.

Bunce, R.G.H., Barr, C.J., Clarke, R.T., Howard, D.C. and Lane, A.M.J. The ITE Merlewood Land Classification. Forthcoming from Journal of Biogeography.

Burnett, R.T., Dales, R.E., Razienne, M.E., Krewski, D., Summers, P.W., Roberts, G.R., Raad-Young, M., Dann, T. and Brook, J. (1994). Effect Of Low Ambient Levels Of Ozone And Sulfates On The Frequency Of Respiratory Admissions To Ontario Hospitals. Environmental Research, Vol. 65, 172–194.

Butlin, R.N. et al. (1992). Preliminary Results from the Analysis of Stone Tablets from the National Materials Exposure Programme (NMEP). Atmospheric Environment, Vol. 26B, 189.

Butlin, R.N., et al. (1994). Effects Of Pollutants On Buildings. DOE Report No. DoE/HMIP/RR/94/030, for Her Majesty's Inspectorate of Pollution, Department of the Environment, London.

Cody, R.P., Weisel, C.P., Birnbaum, G. and Lioy, P.J. (1992). The Effect Of Ozone Associated With Summertime Photochemical Smog On The Frequency Of Asthma Visits To Hospital Emergency Departments. Environmental Research, Vol. 58, 184–194.

Cosby, B.J., Wright, R.F., Hornberger, G.M. and Galloway, J.N. (1985). Modelling The Effects Of Acid Deposition: Assessment Of A Lumped Parameter Model Of Soil Water And Streamwater Chemistry. Water Resource Research, Vol. 21, 51–63.

COST 613/2 Committee on Air Pollution Epidemiology. (1992). Exposure Assessment. Luxembourg: Commission of the European Communities. (COST Committee Report No. 1) (EUR 14345).

Central Statistical Office (1991a). Housing and Construction Statistics. HMSO, London.

Central Statistical Office (1991b). Annual Abstract of Statistics. HMSO, London.

Central Statistical Office (1991c). Commercial and Industrial Floorspace Statistics, England. HMSO, London.

Cooke, R.U. and Gibbs, G.B., (1994). Crumbling Heritage? Studies of Stone Weathering in Polluted Atmospheres. National Power. Swindon.

CSO (1993). Annual Abstract of Statistics, 1993. Central Statistical Office Report No. 129, (Ed. Geoff Denise), HMSO, London.

Derwent, R.G. and Nodop, K. (1986). Long Range Transport And Deposition Of Acidic Nitrogen Species In North West Europe. Nature Vol. 324, 356–358.

Derwent, R.G., Dollard, G.J. and Metcalfe, S.E. (1988). On The Nitrogen Budget For The United Kingdom And North West Europe. Q.J.R. Meteorol. Soc. Vol. 114, 1127–1152.

Dockery, D.W. et al. (1989). Effects Of Inhalable Particles On Respiratory Health On Children. American Review of Respiratory Disease, Vol. 139, 587–594.
Dockery, D.W., Pope, C.A. III, Xu, X., Spengler, J.D., Ware, J.H., Fay, M.E., Ferris, Jr B.G. and Speizer, F.E. (1993). An Association Between Air Pollution and Mortality in Six US Cities. New England Journal of Medicine, Col. 329, 1753–1759.
Dockery, D.W. and Pope, C.A. III (1994). Acute Respiratory Effects of Particulate Air Pollution. Annual Review of Public Health., Vol. 15, 107–132.
ECOTEC (1986). Identification and Assessment of Materials Damage to Buildings and Historic Monuments by Air Pollution. Report to the UK Department of the Environment.
ECOTEC (1994). An Evaluation of the Benefits of Reduced Sulphur Dioxide Emissions: Annex 4; A Contingent Valuation Study of Aquatic Ecosystems. A Report to the Department of the Environment, London.
Eliassen, A. (1984). Aspects of Lagrangian air pollution modelling in Air Pollution Modelling And Its Application III (ed. C. De Wispelaere) Plenum Publishing Corporation.
Eliassen, A. (1980). A Review Of Long-Range Transport Modelling. J. Appl. Met. Vol. 19, 231–240.
EMEP (1992). Calculated Budgets For Airborne Acidifying Components In Europe 1985–1991. EMEP-MCS-W Report 1/92. Norwegian Meteorological Institute, Oslo.
European Commission (1991). The Effects of Noise: Research Review of the 4th Environmental Action Programme. Proceedings of a European Commission Workshop, Brussels, November 1991.
European Commission DGXII (1995a–f). Science, Research and Development Programme (JOULE). Externalities of Energy 'EXTERNE' Project. Report numbers: EUR 16520 (Summary), EUR 16521 (Methodology), EUR 16522 (Coal & Lignite), EUR 16523 (Oil & Gas), EUR 16524 (Nuclear), EUR 16525 (Wind & Hydro).
Eurostat (1989) Agriculture Statistical Yearbook 1989, Eurostat, Luxembourg.
Farrar, J.F., Relton, J. and Rutter, A.J. (1977). Sulphur Dioxide And The Scarcity Of Pinus Sylvestris L. In The Industrial Pennines. Environmental Pollution Vol. 14, 63–68.
Federal Environment Agency (1993). Manual on Methodologies and Criteria for Mapping Critical Levels/Loads and Geographical Areas where they are Exceeded. Berlin: Federal Environment Agency.
Fuhrer, J. and Achermann, B., Eds. (1994) Critical levels for Ozone; a UN-ECE workshop report. FAC Report no. 16, Swiss Federal Research Station for Agricultural Chemistry and Environmental Hygiene, Liebefeld-Bern, 1994.
Fuller, R.M., Brown, N.J., Ullyett, J.M., Sanders, M.E., Groom, G.B., Howard, D.C. and Gillespie, M.K. (1993). Countryside Survey 1990 – Mapping The Land Cover Of Great Britain Using Landsat Imagery: A Demonstrator Project In Remote Sensing. Final report on pattern analysis and GIS Contract report to the Department of the Environment, London
Fuller, R.M., Groom, G.B. and Jones, A.R. (1994). The Land Cover Map Of Great Britain: An Automatic Classification Of Landsat Thematic Mapper Data. Photogrammatric Engineering and Remote Sensing, Vol. 60, 553–562.
Harland, D.G. and Abbott, P.G. (1977). Noise and Road Traffic Outside Homes in England. Transport and Road Research Laboratory Report No. 770.
Haynie, F.H. (1986). Atmospheric Acid Deposition Damage due to Paints. US Environmental Protection Agency Report EPA/600/M-85/019.
Heck, W.W., Taylor, O.C. and Tingey, D.T. (eds.) (1988). Assessment Of Crop Loss From Air Pollutants. Elsevier, New York.
Heil, G.W. and Bobbink, R. (1993) 'CALLUNA' A Stimulation Model For Evaluation Of Impacts Of Atmospheric Nitrogen Deposition On Dry Heathlands. Ecological Modelling, Vol. 68, 161–182.
Hettelingh, J.-P., Downing, R.J. and De Smet, P.A.M. (1991). Mapping Critical Loads for Europe. CCE Technical Report No. 1. Bilthoven (The Netherlands). National Institute for Public Health and Environmental Protection.
Holguin, A.H., et al. (1984). The Effects of Ozone on Asthmatics in the Houston Area in Evaluation of the Scientific Basis for Ozone/Oxidants Standards (Lee, S.D., ed), Houston, Texas. Air Pollution Control Association, 1984, 262–280.
Hornung, M., Sutton, M.A. and Wilson, R.B. (In press). Mapping and modelling of critical loads for nitrogen – a workshop report. Edinburgh: Institute of Terrestrial Ecology.
Hornung, M., Bull, K.R., Cresser, M., Ullyett, J., Hall, J.R., Langan, S.J. and Loveland, P.J. (1995).

The Sensitivity of Surface Waters of Great Britain to acidification predicted from catchment characteristics. Enivironmental Pollution, Vol. 87, 207–214.
Howard, D.C., Fuller, R.M. and Barr, C.J. (1995). Using GIS To Link Ecological Information Recorded From Ground, Air And Space: Examples From The Countryside 1990 Project, Journal of Biogeography, 22.
ICI (1992). ICI Paints Division. Private Communication
ISO (1987). International Organisation for Standardisation. Acoustics – Description and Measurement of Environmental Noise – Part 2: Acquisition of Data Pertinent to Land Use. ISO 1996-2:1987.
IUCN (1990). 1990 United Nations List of National Parks and Protected Areas. IUCN, Gland, Switzerland.
Jesinghaus, J. (1993). EXTASY: Towards An Environmental Index With The Help Of Expert Surveys. Draft discussion paper prepared for the Commission of the European Communities, EUROSTAT, Luxembourg.
Johnson, W.B. (1983). Inter-Regional Exchanges Of Air Pollution: Model Types And Applications. J. Air Pollution Control Assoc. Vol. 33, 563–574.
Kinney, P.L., Ito, K. and Thurston, G.D. (1994). Daily Mortality And Air Pollution In Los Angeles: An Update (Abstract Only). American Review of Respiratory and Critical Care Medicine; Supplement: p. A660.
Knabe, W. (1970). Kiefernwaldbreitung und Schwefeldioxid Immissionen im Ruhrgebeit. Staub, Reinhalt. Luft Vol. 39, 32–35.
Krupnick, A.J., Harrington, W. and Ostro, B. (1990). Ambient Ozone and Acute Health Effects: Evidence from Daily Data, Journal of Economic Management, Vol. 18, 1–18.
Kucera, V. (1994). The UN-ECE International Cooperative Programme on Effects on Materials, Including Historic and Cultural Monuments. Report to the working group on effects within the UN-ECE in Geneva, Swedish Corrosion Institute, Stockholm.
Kuylenstierna, J. and Chadwick, M. (1994). Research Report on Modelling Forest Damage from Air Pollution, submitted to the ExternE Project.
Latour, J.B. and Reiling, R. (1993). A Multiple Stress Model For Vegetation (Move): A Tool For Scenario Studies And Standard Setting. Science of the Total Environment, Vol. 93, 1513–1526.
Lee, D.S., Holland, M.R. and Falla, N. (1995). An Assessment of the Potential Damage to Materials in the UK from Tropospheric Ozone. Forthcoming from Atmospheric Environment.
Lipfert, F.W. (1987). Effects of Acidic Deposition on the Atmospheric Deterioration of Materials. Materials Performance, Vol. 12, National Association of Corrosion Engineers, 1987.
Lipfert, F.W. (1989). Atmospheric Damage to Calcareous Stones: Comparison and Reconciliation of Recent Findings. Atmospheric Environment, Vol. 23, 415.
Maff, Soafd, Dani and Wo (1994). Agriculture in the United Kingdom HMSO, London.
Makela, A. and Schopp, W. (1990). Regional-scale SO_2 forest-impact calculations in The RAINS Model of Acidification – Science and Strategies in Europe (Eds. J. Alcamo, R. Shaw, L. Hordijk). Kluwer Academic Publishers, Dordrecht.
Moldan, B. and Schnoor, J.L. (1992). Czechoslovakia – Examining A Critically Ill Environment. Environ. Sci. Technol., Vol. 26, 14–21.
NAPAP (1990a). Development and use of tree and forest response models. Report 17. In Acidic Deposition: State of Science and Technology. Summary report of the US National Acidic Precipitation Assessment Program by P.M. Irving (Ed.). Washington.
NAPAP (1990b). National Acid Precipitation Assessment Programme. 1990 Integrated Assessment Report.
Newby, P.T., Mansfield, T.A. and Hamilton, R.S., (1991). Sources and Economic Implications of Building Soiling in Urban Areas. The Science of the Total Environment , Vol. 100, 347.
Nilsson, S., Sallnas, O. and Duinker, P. (1992). Forest Potentials and Policy Implications: A Summary of a Study of Eastern and Western European Forests by the International Institute for Applied Systems Analysis. Executive Report 17, IIASA, Laxenburg, Austria.
Nix, J. (1989). Farm management pocketbook, 20th edition, Wye College, London.
OPCS (1991). Population Estimates. OPCS Series PP1. HMSO, 1991.
Ormerod, S.J., Weatherley, N.S. and Gee, A.S. (1990). Modelling the ecological impact of changing acidity in Welsh streams. In: Acid Waters in Wales. Edited by R.W. Edwards, A.S. Gee and J.H. Stoner. Dordrecht: Kluwer Academic Publishers.
Ostro, B.D. (1987). Air Pollution and Morbidity Revisited: a Specification Test, Journal of Environmental and Economic Management, Vol. 14, 7–98.

Ostro, B.D. and Rothschild, S. (1989). Air Pollution And Acute Respiratory Morbidity; An Observational Study Of Multiple Pollutants. Environmental Research, Vol. 48, 238–247.
Ostro, B.D, Lipsett, M.J., Wiener, M.B. and Selner, J.C. (1991). Asthmatic Response to Acid Aerosols. American Journal of Public Health, Vol. 81, 694–702.
Parkin, P.H., Purkis, H.J., Stephenson, R.J. and Schlaffenberg, B. (1968). The London Noise Survey. Building Research Establishment. HMSO.
Pell, E.J., Pearson, N.S. and Vinten-Johnsen, C. (1988). Qualitative And Quantitative Effects Of Ozone And/Or Sulphur Dioxide On Field Grown Potato Plants. Environmental Pollution. Vol. 53, 171–186.
Philip's Geographical Digest (1994). Heinemann Educational, Oxford.
Pönkä, A. and Virtanen, M. (1994). Chronic Bronchitis, Emphysema, And Low Level Air Pollution In Helsinki, 1987–1989. Environmental Research, Vol. 65, 207–217.
Pope, C.A. III, Thun, M.J., Namboodri, M.M., Dockery, D.W., Evans, J.S., Speizer, F.E. and Heath, C.W. (1995). Particulate Air Pollution As A Predictor Of Mortality In A Prospective Study Of US Adults. American Journal of Respiratory Critical Care Medicine, Vol. 151, 669–674.
PORG (1993). Ozone In The United Kingdom, 1993. Third Report of the Photochemical Oxidants Review Group, Department of the Environment, London.
QUARG (1993). Urban Air Quality In The United Kingdom. First Report of the Quality of Urban Air Review Group, Department of the Environment, London.
RGAR (1990). Acid Deposition In The United Kingdom, 1986–1988. Third report of the Review Group on Acid Rain. Department of the Environment, London.
Roberts, T.M. (1984). Long Term Effects Of Sulphur Dioxide On Crops; An Analysis Of Dose Response Relations. Philosophical Transactions of the Royal Society, London, Ser. B., 305, 299–316.
Rodwell, J.S., Pigott, C.D., Ratcliffe, D.A., Malloch, A.J.C., Birks, H.J.B., Proctor, M.C.F., Shimwell, D.W., Huntley, J.P., Radford, E., Wiggington, M.J. and Wilkins, P. (1991). British Plant Communities. Volume 3. Grasslands and Montane Communities. Cambridge: Cambridge University Press. 540 pp.
Sargent, J.W. and Fothergill, L.C. (1993). A Survey of Environmental Noise Levels in the UK. Proceedings of Noise 93 Vol. 7 161–166.
Schwartz, S.E. (1989). Acid Deposition: Unravelling A Regional Phenomenon. Science Vol. 243, 753–763.
Schwartz, J. (1993a). Air Pollution And Daily Mortality In Birmingham, Alabama. American Journal of Epidemiology, Vol. 137, 1136–1147.
Schwartz, J. (1993b). Particulate Air Pollution And Chronic Respiratory Disease. Environmental Research, Vol. 62, 7–13.
Schwartz J. (1994). Air Pollution And Hospital Admissions For The Elderly In Detroit, Michigan. American Review of Respiratory and Critical Care Medicine, Vol. 150, 648–655.
Schwartz, J., Spix, C., Wichmann, H.E. and Malin, E. (1991). Air Pollution And Acute Respiratory Illness In Five German Communities. Environmental Research, Vol. 56(1), 1–14.
Schwartz, J., Slater, D., Larson, T.V., Pierson, W.E. and Koenig, J.Q. (1993). Particulate Air Pollution And Hospital Emergency Room Visits For Asthma In Seattle. American Review of Respiratory Disease, Vol. 147, 826–831.
Short, N. (1994). Personal communication.
Sinclair, J. and Whiteman, A. (1992) . Price-Size Curves For Conifers. The Forestry Authority, Research Information Note 226, Farnham, UK.
Stedman, J.R. (1995). Estimated High Resolution Maps Of The United Kingdom Air Pollution Climate. Report no. AEA/CS/16419035/001/DRAFT, produced under contract for the Department of the Environment, London.
Sunyer, J., Sáez, M., Murillo, C., Castellsague, J., Martinez, F. and Antó, J.M. (1993). Air Pollution And Emergency Room Admissions For Chronic Obstructive Pulmonary Disease: A 5-Year Study. American Journal of Epidemiology, Vol. 137, 701–705.
Sverdrup, H., Warfvinge, P. and Nihlgaard, B. (1993). Assessment Of Ecological And Economic Effects Of Acidification On Forestry In Sweden. Unpublished research report.
Tamm, C.O. and Hallbacken, L. (1988). Changes In Soil Acidity From The 1920s To The 1980s In Two Forest Areas With Different Acid Deposition. Ambio Vol. 17, 56–61.
Thurston, G.D., Kazuhiko, I., Hayer, C.G., Bates, D.V. and Lippmann, M. (1994). Respiratory Hospital Admissions and Summertime Haze Air Pollution in Toronto, Ontario: Consideration of the Role of Acid Aerosols. Environmental Research, Vol. 65, 271–290.

Tickle, A., Ferguson, M. and Drucker, G. (1995). Acid Rain and Nature Conservation in Europe. World Wide Fund for Nature, Gland, Switzerland.

Touloumi, G., Pocock, S.J., Katsouyanni, K. and Trichopoulos, D. (1994). Short Term Effects Of Air Pollution On Daily Mortality In Athens: A Time-Series Analysis. International Journal of Epidemiology, Vol. 23, 957–967.

UKBERG (1990). The Effects of Acid Deposition on Buildings and Building Materials. UK Building Effects Review Group, HMSO, London.

UKDOE (1994). UK Department of the Environment. Digest of Environmental Protection and Water Statistics. HMSO, 1994.

UK Meteorological Office (1970). Averages of Humidity for the British Isles. UK Meteorological Office, HMSO, London.

UK Meteorological Office (1975). Maps of Mean Vapour Pressure and of Frequencies of High Relative Humidity over the United Kingdom. Climatological Memorandum No. 75, UK Meteorological Office.

UK Meteorological Office (1977). Averages Annual Rainfall 1941–1970 (Met O.886 SB), UK Meteorological Office.

UKBERG (1990). The Effects of Acid Deposition on Buildings and Building Materials. UK Building Effects Review Group, HMSO, London.

Ulrich, B. (1985). Interaction of Indirect and Direct Effects of Air Pollutants in Forests, in Air Pollution and Plants, Ed C. Troyanowsky. 149-81. VCH Publishers, Weinheim.

Ulrich, B. (1990). Waldsterben: Forest Decline in West Germany. Environ. Sci. Technol., Vol. 24(4), 436–441.

Weigel, H.J., Adaros, G. and Jäger, H.J. (1990). Yield Responses Of Different Crop Species To Long-Term Fumigation With Sulphur Dioxide In Open-Top Chambers. Environmental Pollution, Vol. 67, 15–28.

Welsh Office (1990). Commercial and Industrial Floorspace Statistics, Wales. HMSO, London.

WWF (1992). Forests in Trouble: A Review of the Status of Temperate Forests Worldwide. World Wide Fund for Nature, Gland, Switzerland.

Index

FONDAZIONE ENI ENRICO MATTEI (FEEM) SERIES ON ECONOMICS, ENERGY AND ENVIRONMENT

This series serves as an outlet for the main results of FEEM's research programmes in the areas of economics, energy and environment.

---

1. C. Carraro and D. Siniscalco (eds.), *The European Carbon Tax: An Economic Assessment.* 1993 ISBN 0-7923-2520-6
2. C. Carraro (ed.), *Trade, Innovation, Environment.* 1994 ISBN 0-7923-3033-1
3. C. Dosi and T. Tomasi (eds.), *Nonpoint Source Pollution Regulation: Issues and Analysis.* 1994 ISBN 0-7923-3121-4
4. C. Carraro, Y. Katsoulacos and A. Xepapadeas (eds.), *Environmental Policy and Market Structure.* 1996 ISBN 0-7923-3656-9
5. C. Carraro and A. Haurie (eds.), *Operations Research and Environmental Management.* 1996 ISBN 0-7923-3767-7
6. I. Musu and D. Siniscalco (eds.), *National Accounts and the Environment.* 1996 ISBN 0-7923-3741-7
7. C. Carraro and D. Siniscalco (eds.), *Environmental Fiscal Reform and Unemployment.* 1996 ISBN 0-7923-3750-6
8. A. Beltratti: *Models of Economic Growth with Environmental Assets.* 1996 ISBN 0-7923-4032-9
9. G. Chichilnisky, G. Heal and A. Vercelli (eds.), *Sustainability: Dynamics and Uncertainty.* 1998 ISBN 0-7923-4698-X
10. R. Roson and K.A. Small (eds.), *Environment and Transport in Economic Modelling.* 1998 ISBN 0-7923-4913-X
11. A. Markandya and M. Pavan (eds.): *Green Accounting in Europe - Four Case Studies.* 1998 ISBN 0-7923-5470-2
12. A. Lanza (ed.): *Resources Accounting in China.* 1999 ISBN 0-7923-5485-0
13. C. Carraro (ed.): *International Environmental Agreements on Climate Change.* 1999 ISBN 0-7923-5515-6
14. C. Carraro and F. Lévêque (eds.): *Voluntary Approaches in Environmental Policy.* 1999 ISBN 0-7923-5516-4

KLUWER ACADEMIC PUBLISHERS – DORDRECHT / BOSTON / LONDON